RESIDENTIAL ROOF FRAMING

Residential Roof Framing

ALONZO WASS

GORDON A. SANDERS

RESTON PUBLISHING COMPANY, INC.
A Prentice-Hall Company, Reston, Virginia

Library of Congress Cataloging in Publication Data

Wass, Alonzo.
Residential roof framing.

Includes index
1. Roofs. 2. House construction. I. Sanders, Gordon A., joint author. II. Title.
TH2391.W37 695 80-16306
ISBN 0-8359-6655-0

10 9 8 7 6 5 4 3 2 1

Printed in the United States of America.

Contents

the first line of the rafter table. This table makes it possible for the carpenter to determine rafter lengths accurately without having to extract the square root of a number.

In preparing to lay out rafters, the unit rise is usually determined by consulting the wall cross section in the building plans for the roof slope triangle, but sometimes either a fractional pitch or a total rise is given.

When the slope triangle is used, the horizontal leg of the slope triangle is always 12″ which represents one unit of run. The vertical leg represents the unit rise and varies.

When only a fractional pitch is listed, the unit rise may be determined by applying the pitch formula:

$$\text{Pitch} = \frac{\text{rise}}{\text{span}}$$

EXAMPLE:

$$P = \frac{1}{3}$$

$$\text{Pitch} = \frac{\text{unit rise}}{\text{unit span}}$$

$$\frac{1}{3} = \frac{\text{unit rise}}{24''}$$

$$\text{Unit rise} = 8''$$

Once the pitch formula is understood, there is no need to work it in the manner shown in the previous example. All that is necessary is to multiply the fractional pitch by the unit of span.

EXAMPLE:

$$P = \frac{1}{3} \qquad \text{Unit span} = 24''$$

$$\text{Unit rise} = P \times 24''$$

$$\text{Unit rise} = \frac{1}{3} \times 24 = \frac{24}{3} = 8''$$

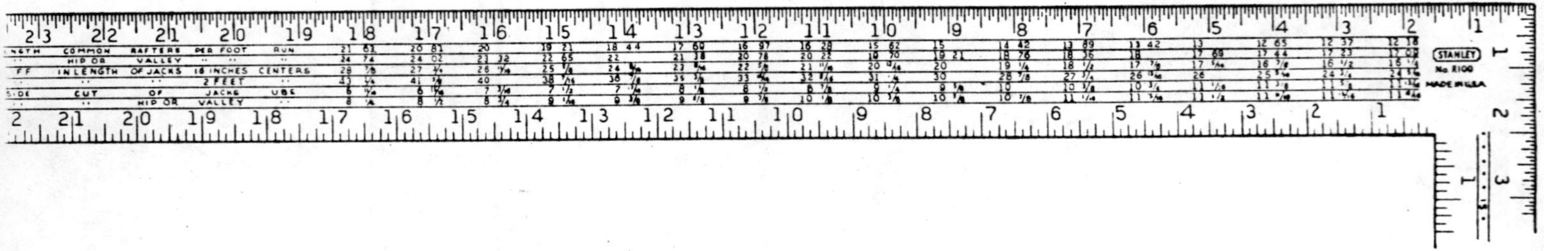

Figure 2-7
Rafter tables for
roof framing.

List of Plates

Preface

Our endeavor has been to bring together in this text two methods of learning roof framing: the simple and direct *graphic* method, and the more technical *framing square* method.

The graphic method is used extensively in many western countries. With it, the layman, the farmer, the summer cottage builder, or the carpenter may erect a roof with little use of mathematics or knowledge of formal descriptive geometry. The framing square method is a natural development of the former and is faster; hence, its greater use in carpentry.

The accomplished builder should have a full knowledge of roof framing. He must understand framing theory and know how the layout rules were developed. He must be able to read blueprints. He should be able to develop descriptive drawings for the solving of problems, and he must calculate for necessary measurements. Finally, he must be able to cut the parts and erect the roof.

The beginning builder, in order to develop the abilities and skills required of an accomplished workman, should have experiences in many areas. His trade school training should be supplemented by his on-the-job experiences. In the school, he should attend theory lessons, mathematics lessons, and shop drawing lessons. He should observe shop demonstrations and have time to put into practice many of the things that he has learned. Mistakes in roof framing should be made during layout practice sessions—not when cutting an expensive roof.

This book is organized into four distinct yet integrated parts for learning and teaching:

1. *Rafter Layout Theory.* These chapters explain construction terms and general methods of rafter layout.
2. *Mathematics.* All the chapters describe in detail how the calculations are done.
3. *Shop Drawings.* Descriptive drawings are developed to aid in the understanding of roofing principles. The hopper construction chapters are included to provide a broader understanding of roof principles.
4. *Shop Assignments.* The shop jobs provide the necessary practical experiences in rafter layout and roof framing.

In recognition of an apparent trend toward metrification, this book offers assistance to persons in the construction industry, or to those about to enter it, who need to increase their knowledge of the metric system. Metric conversion tables and illustrations of metric framing squares are provided. With the data presented, a residential construction worker will be able to use any of the following carpenter's framing squares:

1. The traditional Stanley framing square;
2. The Frederickson metric framing square; and
3. The Stanley metric framing square.

The book may be used in a number of ways. The person attempting to learn roofing without the help of an instructor may work from front to back. In small schools, the student may do the same, or he may, under one instructor's guidance, move through the four areas of theory, mathematics, shop drawing, and shop work simultaneously. In larger institutions, the instruction may be given by four separate instructors, each specializing in one aspect. With careful planning and timing, and each instructor working toward a common goal, an excellent learning situation should be provided.

A.W.
G.A.S.

Acknowledgments

The following sources of technical data are gratefully acknowledged: the Canadian Wood Council; the Metric Commission of Canada; Ottenheimer Publishers, Inc.; Paul H. Frederickson (Canada); Stanley Tool Company, Inc. (USA); and TECO Products and Testing Corporation (USA). The sources are indicated as the data are presented in the text.

For critical reading of the textual material, our thanks are directed to Edna M. Sanders and Rosamond Gladwys Cox. We are also grateful to a senior carpenter apprentice, who, while attending the Northern Alberta Institute of Technology, read several chapters.

Time given to a project means less time for other aspects of living. We thank our respective wives, Joan Rose Wass and Edna M. Sanders, for their encouragement and forbearance.

A. Wass
G. A. Sanders

1

Imperial to Metric

Chapter 1 provides seven conversion tables of imperial to metric measurements, general information about symbols, and a set of questions designed to be helpful when learning the relationships between the two measurement systems. The measurements given are mainly linear—those of thickness, width and length as used in the light construction industry.

The titles of the tables, listed in section 1, which follows, are descriptive and should be used as a ready reference when reading the book.

1. Selected imperial/metric tables[1] for the light construction industry

Table 1.1 Metric Units of International Standards (SI) Used for Wood Products

[1]The authors acknowledge, with appreciation, permission to publish data for the tables of Chapter 1: Tables 1.1 and 1.2—the Canadian Wood Council: Tables 1.3, 1.4 and 1.5—the Metric Commission of Canada; Tables 1.6 and 1.7—Ottenheimer Publishers, Inc.

General statements are made with respect to how metric measurements are written.

2. Metric units commonly used for wood products

Table 1.1 is a data base with respect to units of length, length/time, area, and volume. Provided with each unit are two sets of information: metric units and imperial equivalents; and imperial units and metric equivalents. Units of length have applications to residential roof framing. All units have application to the light construction industry.

3. Metric nomenclature for surfaced softwood lumber products

The next four tables (1.2, 1.3, 1.4, and 1.5) provide measurement data in considerable detail with respect to rough lumber; studs, light framing, joist and plank; boards, shiplap, and center match; and lengths. Care should be taken to become familiar, by reading the tables, with the current meanings of terms such as nominal, actual, nomenclature, dry, and green. The tables suggest greater use of actual measurements. For example, a surfaced 2 × 4 in metric measurements is currently 38 mm × 89 mm.

4. Conversion tables

Table 1.6 is a reference table for conversions of fine measurements from $\frac{1}{64}$ inch/0.3969 mm/0.0397 cm to 1 inch/25.4000 mm/2.5400 cm. Such measurements are useful when establishing tolerances for line lengths of rafters; and when describing tools such as points per unit length for a saw, thickness of a steel blade, and size of gauge.

For conversions from one inch (1 in.) to one hundred feet (100 ft.) use Table 1.7. Measurements are given in feet and inches, centimeters, and meters.

By using the two tables most common measurements can be converted, as in the following set of examples:

Dimension	
Imperial	*Metric Equivalent*
Residence 26′–0″ × 42′–0′	7924.8 mm × 12801.6 mm
Center to center rafter spacing (OC) 16″ OC	406.4 mm
Trusses 24″ OC	609.6 mm
Line length of a common rafter (CR) 14′–9½″	4508.5 mm or 45.085 cm or 4.5085 m Check: 4508.5 ÷ 304.8 = 14.79″

Actual conversions produce some awkward measurements, making necessary some practical changes in industry. Framing-member spacings, for example, become 400 mm and 600 mm rather than 406.4 mm and 609.6 mm. Wood panel dimensions change from 4′–0″ × 8′–0″ to 1200 mm × 2400 mm, a slightly different size of panel which fits either 400 mm or 600 mm spacings of framing members.

The metric conversion tables on pages 10–14 have been excerpted from *Metric Conversion Tables,* Barnes and Noble, and are reproduced here with the permission of the copyright holder, Ottenheimer Publishers, Inc.

5. Metric measure

Symbols for units of measure are important to the builders:

mm for millimeter,
m for meter, and
m^2 for square meter

Metric symbols are used without periods and plurals:

38 mm, 600 mm, 7 200 mm or 7200 mm, 7.200 m and 0.836 127 m^2.

TABLE 1.1*

Metric (SI) Units Commonly Used for Wood Products

Metric Units		*Imperial Equivalents*		*Imperial Units*		*Metric Equivalents*	
Length							
1 millimeter (mm)	=	0.039 370 1	inch	1 inch	=	25.4	mm
1 centimeter (cm)	=	0.393 701	inch		=	2.54	cm
1 meter (m)	=	39.370 1	inches		=	0.025 4	m
	=	3.280 84	feet	1 foot	=	0.304 8	m
	=	1.093 61	yards	1 yard	=	0.914 4	m
	=	0.049 709 7	chain	1 chain (66 ft)	=	20.116 8	m
1 kilometer (km)	=	0.621 371	mile	1 mile	=	1.609 34	km
Length/time							
1 meter per second (m/s)	=	3.280 84	feet per second	1 foot per second	=	0.304 8	m/s
1 kilometer per hour (km/h)	=	0.621 371	mile per hour	1 mile per hour	=	1.609 34	km/h
Area							
1 square millimeter (mm^2)	=	0.001 550	square inch	1 square inch	=	645.16	mm^2
1 square centimeter (cm^2)	=	0.155 000	square inch		=	6.451 6	cm^2

1 square meter (m^2)	=	10.763 9	square feet	1 square foot	= 0.092 903 0	m^2
	=	1.195 99	square yards	1 square yard	= 0.836 127	m^2
	=	0.247 105	mil-acre	1 mil acre	= 4.046 86	m^2
1 hectare (ha)	=	2.471 05	acres	1 acre	= 0.404 686	ha
1 square kilometer (km^2)	=	0.386 102	square miles	1 square mile	= 2.589 99	km^2
Volume						
1 cubic millimeter (mm^3)	=	0.000 061 023 7	cubic inch	1 cubic inch	= 16 387.1	mm^3
1 cubic centimeter (cm^3)	=	0.061 023 7	cubic inch		= 16.387 1	cm^3
1 cubic meter (m^3)	=	35.314 7	cubic feet	1 cubic foot	= 0.028 316 8	m^3
	=	1.307 95	cubic yard	1 cubic yard	= 0.764 555	m^3
	=	0.353 147	cunit	1 cunit	= 2.831 68	m^3
1 milliliter (mL)	=	0.035 195 1	fluid ounce	1 fluid ounce	= 28.413 1	mL
1 liter (L) or 1 cubic decimeter (dm^3)	=	0.219 969	gallon	1 gallon	= 4.546 09	L
1 cubic meter per hectare (m^3/ha)	=	14.291 3	$foot^3$/acre	1 $foot^3$/acre	= 0.069 972 5	m^3/ha

**Metric Manual for Wood Products* (10M78.09, 10M79.03), p. 46, published by the Canadian Wood Council, 170 Laurier Avenue West, Ottawa, Ontario, Canada K1P 5V5, and reproduced here with permission of the originators.

Metric Nomenclature for Surfaced Softwood Lumber Products

TABLE 1.2*

Rough Lumber

Item	*Thickness*			*Width*		
	Nominal inches	*Actual Size inches*	*Metric Nomenclature mm*	*Nominal inches*	*Actual Size inches*	*Metric Nomenclature mm*
Rough Lumber	1	1	25	2	2	51
	$1\frac{1}{4}$	$1\frac{1}{4}$	32			
	$1\frac{1}{2}$	$1\frac{1}{2}$	38	3	3	76
	$1\frac{5}{8}$	$1\frac{5}{8}$	41	4	4	102
	$1\frac{3}{4}$	$1\frac{3}{4}$	44	5	5	127
	$1\frac{7}{8}$	$1\frac{7}{8}$	48	6	6	152
	2	2	51	7	7	178
	$2\frac{1}{2}$	$2\frac{1}{2}$	64	8	8	203
	3	3	76	9	9	229
	$3\frac{1}{2}$	$3\frac{1}{2}$	89	10	10	254
	4	4	102	11	11	279
	$4\frac{1}{2}$	$4\frac{1}{2}$	114	12	12	305
	5	5	127	14	14	356
	6	6	152	16	16	406
	7	7	178	18	18	457
	8	8	203	20	20	508
	9	9	229			
	10	10	254			
	11	11	279			
	12	12	305			
	14	14	356			
	16	16	406			
	18	18	457			
	20	20	508			

**Metric Manual for Wood Products* (10M78.09, 10M79.03), p. 12.

Metric Nomenclature for Surfaced Softwood Lumber Products

TABLE 1.3*

Studs, Light Framing, Joist and Plank

Item	*Thickness*				*Width*			
		Actual				*Actual*		
	Nominal inches	*Dry inches*	*Green inches*	*Metric Nomenclature mm*	*Nominal inches*	*Dry inches*	*Green inches*	*Metric Nomenclature mm*
Studs,	2	$1\frac{1}{2}$	$1\frac{9}{16}$	38	2	$1\frac{1}{2}$	$1\frac{9}{16}$	38
Light	$2\frac{1}{2}$	2	$2\frac{1}{16}$	51	3	$2\frac{1}{2}$	$2\frac{9}{16}$	64
Framing,	3	$2\frac{1}{2}$	$2\frac{9}{16}$	64	4	$3\frac{1}{2}$	$3\frac{9}{16}$	89
Joist	$3\frac{1}{2}$	3	$3\frac{1}{16}$	76	5	$4\frac{1}{2}$	$4\frac{5}{8}$	114
and	4	$3\frac{1}{2}$	$3\frac{9}{16}$	89	6	$5\frac{1}{2}$	$5\frac{5}{8}$	140
Plank	$4\frac{1}{2}$	4	$4\frac{1}{16}$	102	7	$6\frac{1}{2}$	$6\frac{5}{8}$	165
					8	$7\frac{1}{4}$	$7\frac{1}{2}$	184
					10	$9\frac{1}{4}$	$9\frac{1}{2}$	235
					12	$11\frac{1}{4}$	$11\frac{1}{2}$	286
					14	$13\frac{1}{4}$	$13\frac{1}{2}$	337
					16	$15\frac{1}{4}$	$15\frac{1}{2}$	387

Note: Shiplap and center match will use the same nomenclature as the S4S boards above, regardless of the length of the lap or tongue.

**Metric Handbook for Canadian Softwood Lumber* (First Edition, November, 1977) p. 7. Compiled by Metric Sub-Sector Committee 8.21—Softwood Lumber for Metric Commission of Canada.

The millimeter is a basic construction measurement. Unless otherwise indicated, all dimensions of a metric residence may be given in millimeters. The top plate measurement for a roof may, for example, be 12 250 mm by 7400 mm.

Calculation and measurement errors may be made when working with dual terminology, imperial and metric. When a knowledge conversion base has been achieved, then one measurement system only should be used.

The objective of conversion should be immersion in one system. Use only a metric tape measure and only a metric framing (roofing) square.

Metric Nomenclature for Surfaced Softwood Lumber Products

TABLE 1.4*

Boards, Shiplap and Center Match

	Thickness				*Width*			
		Actual				*Actual*		
Item	*Nominal inches*	*Dry inches*	*Green inches*	*Metric Nomenclature mm*	*Nominal inches*	*Dry inches*	*Green inches*	*Metric Nomenclature mm*
Boards,	1	$\frac{11}{16}$	$\frac{3}{4}$	17	2	$1\frac{1}{2}$	$1\frac{9}{16}$	38
Shiplap	1	$\frac{3}{4}$	$\frac{25}{32}$	19	3	$2\frac{1}{2}$	$2\frac{9}{16}$	64
and	$1\frac{1}{4}$	1	$1\frac{1}{32}$	25	4	$3\frac{1}{2}$	$3\frac{9}{16}$	89
Center	$1\frac{1}{2}$	$1\frac{1}{4}$	$1\frac{9}{32}$	32	5	$4\frac{1}{2}$	$4\frac{5}{8}$	114
Match					6	$5\frac{1}{2}$	$5\frac{5}{8}$	140
					7	$6\frac{1}{2}$	$6\frac{5}{8}$	165
					8	$7\frac{1}{4}$	$7\frac{1}{2}$	184
					9	$8\frac{1}{4}$	$8\frac{1}{2}$	210
					10	$9\frac{1}{4}$	$9\frac{1}{2}$	235
					11	$10\frac{1}{4}$	$10\frac{1}{2}$	260
					12	$11\frac{1}{4}$	$11\frac{1}{2}$	286
					14	$13\frac{1}{4}$	$13\frac{1}{2}$	337
					16	$15\frac{1}{4}$	$15\frac{1}{2}$	387

Note: Shiplap and center match will use the same nomenclature as the S4S boards above, regardless of the length of the lap or tongue.

**Metric Handbook for Canadian Softwood Lumber* (First Edition, November, 1977) p. 7. Compiled by Metric Sub-Sector Committee 8.21—Softwood Lumber for Metric Commission of Canada.

Metric Nomenclature for Surfaced Softwood Lumber Products

TABLE 1.5*

Lengths

Nominal Length feet	*Metric Length* meters
3	0.91
4	1.22
5	1.52
6	1.83
7	2.13
8	2.44
9	2.74
10	3.05
11	3.35
12	3.66
13	3.96
14	4.27
15	4.57
16	4.88
17	5.18
18	5.49
19	5.79
20	6.10
21	6.40
22	6.71
23	7.01
24	7.32
26	7.92
28	8.53
30	9.14

(continued on page 10)

Metric Nomenclature for Surfaced Softwood Lumber Products

TABLE 1.5*—CONTINUED

Lengths

Nominal Length feet	*Metric Length meters*
32	9.75
34	10.36
36	10.97
38	11.58
40	12.19
42	12.80
44	13.41
46	14.02
48	14.63
50	15.24

**Metric Handbook for Canadian Softwood Lumber* (First Edition, November, 1977) p.12.

TABLE 1.6*

Fractions of Inch to Millimeters and Centimeters

Conversion factor: 1 inch = 25.4 millimeters = 2.54 centimeters
Inches: in millimeters:mm centimeters:cm

in	mm	cm	in	mm	cm	in	mm	cm	in	mm	cm	in	mm	cm
1/64	.3969	.0397	1/4	6.3500	.6350	31/64	12.3031	1.2303	23/32	18.2563	1.8256	61/64	24.2094	2.4209
1/32	.7938	.0794	17/64	6.7469	.6747	1/2	12.7000	1.2700	47/64	18.6531	1.8653	31/32	24.6063	2.4606
3/64	1.1906	.1191	9/32	7.1438	.7144	33/64	13.0969	1.3097	3/4	19.0500	1.9050	63/64	25.0031	2.5003
1/16	1.5875	.1588	19/64	7.5406	.7541	17/32	13.4938	1.3494	49/64	19.4469	1.9447	1	25.4000	2.5400
5/64	1.9844	.1984	5/16	7.9375	.7938	35/64	13.8906	1.3891	25/32	19.8438	1.9844			
3/32	2.3813	.2381	21/64	8.3344	.8334	9/16	14.2875	1.4288	51/64	20.2406	2.0241			
7/64	2.7781	.2778	11/32	8.7313	.8731	37/64	14.6844	1.4684	13/16	20.6375	2.0638			
1/8	3.1750	.3175	23/64	9.1281	.9128	19/32	15.0813	1.5081	53/64	21.0344	2.1034			
9/64	3.5719	.3572	6/16	9.5250	.9525	39/64	15.4781	1.5478	27/32	21.4313	2.1431			
5/32	3.9688	.3969	25/64	9.9219	.9922	5/8	15.8750	1.5875	55/64	21.8281	2.1828			
11/64	4.3656	.4366	13/32	10.3188	1.0319	41/64	16.2719	1.6272	7/8	22.2250	2.2225			
3/16	4.7625	.4763	27/64	10.7156	1.0716	21/32	16.6688	1.6669	57/64	22.6219	2.2622			
13/64	5.1594	.5159	7/16	11.1125	1.1113	43/64	17.0656	1.7066	29/32	23.0188	2.3019			
7/32	5.5563	.5556	29/64	11.5094	1.1509	11/16	17.4625	1.7463	59/64	23.4156	2.3416			
15/64	5.9531	.5953	15/32	11.9063	1.1906	45/64	17.8594	1.7859	15/16	23.8125	2.3813			

**Metric Conversion Tables,* Barnes & Noble, 1975 (Ottenheimer Publishers, Inc.).

TABLE 1.7*

Feet and Inches to Centimeters and Meters

Conversion factor: 1 inch = 2.54 centimeters = .0254 meters
1 foot = 30.48 centimeters = .3048 meters
feet inches:ft in centimeters:cm meters:m

ft	in	cm	m	ft	in	cm	m	ft	in	cm	m	ft	in	cm	m	ft	in	cm	m
	1	2 54	0254	5	1	154 94	1 5494	10	1	307 34	3 0734	15	1	459 74	4 5974	20	1	612 14	6 1214
	2	5 08	0508	5	2	157 48	1 5748	10	2	309 88	3 0988	15	2	462 28	4 6228	20	2	614 68	6 1468
	3	7 62	0762	5	3	160 02	1 6002	10	3	312 42	3 1242	15	3	464 82	4 6482	20	3	617 22	6 1722
	4	10 16	1016	5	4	162 56	1 6256	10	4	314 96	3 1496	15	4	467 36	4 6736	20	4	619 76	6 1976
	5	12 70	1270	5	5	165 10	1 6510	10	5	317 50	3 1750	15	5	469 90	4 6990	20	5	622 30	6 2230
	6	15 24	1524	5	6	167 64	1 6764	10	6	320 04	3 2004	15	6	472 44	4 7244	20	6	624 84	6 2484
	7	17 78	1778	5	7	170 18	1 7018	10	7	322 58	3 2258	15	7	474 98	4 7498	20	7	627 38	6 2738
	8	20 32	2032	5	8	172 72	1 7272	10	8	325 12	3 2512	15	8	477 52	4 7752	20	8	629 92	6 2992
	9	22 86	2286	5	9	175 26	1 7526	10	9	327 66	3 2766	15	9	480 06	4 8006	20	9	632 46	6 3246
	10	25 40	2540	5	10	177 80	1 7780	10	10	330 20	3 3020	15	10	482 60	4 8260	20	10	635 00	6 3500
	11	27 94	2794	5	11	180 34	1 8034	10	11	332 74	3 3274	15	11	485 14	4 8514	20	11	637 54	6 3754
1	0	30 48	3048	6	0	182 88	1 8288	11	0	335 28	3 3528	16	0	487 68	4 8768	21	0	640 08	6 4008
1	1	33 02	3302	6	1	185 42	1 8542	11	1	337 82	3 3782	16	1	490 22	4 9022	21	1	642 62	6 4262
1	2	35 56	3556	6	2	187 96	1 8796	11	2	340 36	3 4036	16	2	492 76	4 9276	21	2	645 16	6 4516
1	3	38 10	3810	6	3	190 50	1 9050	11	3	342 90	3 4290	16	3	495 30	4 9530	21	3	647 70	6 4770
1	4	40 64	4064	6	4	193 04	1 9304	11	4	345 44	3 4544	16	4	497 84	4 9784	21	4	650 24	6 5024
1	5	43 18	4318	6	5	195 58	1 9558	11	5	347 98	3 4798	16	5	500 38	5 0038	21	5	652 78	6 5278
1	6	45 72	4572	6	6	198 12	1 9812	11	6	350 52	3 5052	16	6	502 92	5 0292	21	6	655 32	6 5532
1	7	48 26	4826	6	7	200 66	2 0066	11	7	353 06	3 5306	16	7	505 46	5 0546	21	7	657 86	6 5786
1	8	50 80	5080	6	8	203 20	2 0320	11	8	355 60	3 5560	16	8	508 00	5 0800	21	8	660 40	6 6040
1	9	53 34	5334	6	9	205 74	2 0574	11	9	358 14	3 5814	16	9	510 54	5 1054	21	9	662 94	6 6294
1	10	55 88	5588	6	10	208 28	2 0828	11	10	360 68	3 6068	16	10	513 08	5 1308	21	10	665 48	6 6548
1	11	58 42	5842	6	11	210 82	2 1082	11	11	363 22	3 6322	16	11	515 62	5 1562	21	11	668 02	6 6802
2	0	60 96	6096	7	0	213 36	2 1336	12	0	365 76	3 6576	17	0	518 16	5 1816	22	0	670 56	6 7056
2	1	63 50	6350	7	1	215 90	2 1590	12	1	368 30	3 6830	17	1	520 70	5 2070	22	1	673 10	6 7310
2	2	66 04	6604	7	2	218 44	2 1844	12	2	370 84	3 7084	17	2	523 24	5 2324	22	2	675 64	6 7564
2	3	68 58	6858	7	3	220 98	2 2098	12	3	373 38	3 7338	17	3	525 78	5 2578	22	3	678 18	6 7818
2	4	71 12	7112	7	4	223 52	2 2352	12	4	375 92	3 7592	17	4	528 32	5 2832	22	4	680 72	6 8072
2	5	73 66	7366	7	5	226 06	2 2606	12	5	378 46	3 7846	17	5	530 86	5 3086	22	5	683 26	6 8326
2	6	76 20	7620	7	6	228 60	2 2860	12	6	381 00	3 8100	17	6	533 40	5 3340	22	6	685 80	6 8580
2	7	78 74	7874	7	7	231 14	2 3114	12	7	383 54	3 8354	17	7	535 94	5 3594	22	7	688 34	6 8834
2	8	81 28	8128	7	8	233 68	2 3368	12	8	386 08	3 8608	17	8	538 48	5 3848	22	8	690 88	6 9088
2	9	83 82	8382	7	9	236 22	2 3622	12	9	388 62	3 8862	17	9	541 02	5 4102	22	9	693 42	6 9342
2	10	86 36	8636	7	10	238 76	2 3876	12	10	391 16	3 9116	17	10	543 56	5 4356	22	10	695 96	6 9596
2	11	88 90	8890	7	11	241 30	2 4130	12	11	393 70	3 9370	17	11	546 10	5 4610	22	11	698 50	6 9850
3	0	91 44	9144	8	0	243 84	2 4384	13	0	396 24	3 9624	18	0	548 64	5 4864	23	0	701 04	7 0104
3	1	93 98	9398	8	1	246 38	2 4638	13	1	398 78	3 9878	18	1	551 18	5 5118	23	1	703 58	7 0358
3	2	96 52	9652	8	2	248 92	2 4892	13	2	401 32	4 0132	18	2	553 72	5 5372	23	2	706 12	7 0612
3	3	99 06	9906	8	3	251 46	2 5146	13	3	403 86	4 0386	18	3	556 26	5 5626	23	3	708 66	7 0866
3	4	101 60	1 0160	8	4	254 00	2 5400	13	4	406 40	4 0640	18	4	558 80	5 5880	23	4	711 20	7 1120
3	5	104 14	1 0414	8	5	256 54	2 5654	13	5	408 94	4 0894	18	5	561 34	5 6134	23	5	713 74	7 1374
3	6	106 68	1 0668	8	6	259 08	2 5908	13	6	411 48	4 1148	18	6	563 88	5 6388	23	6	716 28	7 1628
3	7	109 22	1 0922	8	7	261 62	2 6162	13	7	414 02	4 1402	18	7	566 42	5 6642	23	7	718 82	7 1882
3	8	111 76	1 1176	8	8	264 16	2 6416	13	8	416 56	4 1656	18	8	568 96	5 6896	23	8	721 36	7 2136
3	9	114 30	1 1143	8	9	266 70	2 6670	13	9	419 10	4 1910	18	9	571 50	5 7150	23	9	723 90	7 2390
3	10	116 84	1 1684	8	10	269 24	2 6924	13	10	421 64	4 2164	18	10	574 04	5 7404	23	10	726 44	7 2644
3	11	119 38	1 1938	8	11	271 78	2 7178	13	11	424 18	4 2418	18	11	576 58	5 7658	23	11	728 98	7 2898
4	0	121 92	1 2192	9	0	274 32	2 7432	14	0	426 72	4 2672	19	0	579 12	5 7912	24	0	731 52	7 3152
4	1	124 46	1 2146	9	1	276 86	2 7686	14	1	429 26	4 2926	19	1	581 66	5 8166	24	1	734 06	7 3406
4	2	127 00	1 2700	9	2	279 40	2 7940	14	2	431 80	4 3180	19	2	584 20	5 8420	24	2	736 60	7 3660
4	3	129 54	1 2954	9	3	281 94	2 8194	14	3	434 34	4 3434	19	3	586 74	5 8674	24	3	739 14	7 3914
4	4	132 08	1 3208	9	4	284 48	2 8448	14	4	436 88	4 3688	19	4	589 28	5 8928	24	4	741 68	7 4168
4	5	134 62	1 3462	9	5	287 02	2 8702	14	5	439 42	4 3942	19	5	591 82	5 9182	24	5	744 22	7 4422
4	6	137 16	1 3716	9	6	289 56	2 8956	14	6	441 96	4 4196	19	6	594 36	5 9436	24	6	746 76	7 4676
4	7	139 70	1 3970	9	7	292 10	2 9210	14	7	444 50	4 4450	19	7	596 90	5 9690	24	7	749 30	7 4930
4	8	142 24	1 4224	9	8	294 64	2 9464	14	8	447 04	4 4704	19	8	599 44	5 9944	24	8	751 84	7 5184
4	9	144 78	1 4478	9	9	297 18	2 9718	14	9	449 58	4 4958	19	9	601 98	6 0198	24	9	754 38	7 5438
4	10	147 32	1 4732	9	10	299 72	2 9972	14	10	452 12	4 5212	19	10	604 52	6 0452	24	10	756 92	7 5692
4	11	149 86	1 4986	9	11	302 26	3 0226	14	11	454 66	4 5466	19	11	607 06	6 0706	24	11	759 46	7 5946
5	0	152 40	1 5240	10	0	304 80	3 0480	15	0	457 20	4 5720	20	0	609 60	6 0960	25	0	762 00	7 6200

TABLE 1.7—CONTINUED

ft	in	cm	m
25	1	764.54	7.6454
25	2	767.08	7.6708
25	3	769.62	7.6962
25	4	772.16	7.7216
25	5	774.70	7.7470
25	6	777.24	7.7724
25	7	779.78	7.7978
25	8	782.32	7.8232
25	9	784.86	7.8486
25	10	787.40	7.8740
25	11	789.94	7.8994
26	0	792.48	7.9248
26	1	795.02	7.9502
26	2	797.56	7.9756
26	3	800.10	8.0010
26	4	802.64	8.0264
26	5	805.18	8.0518
26	6	807.72	8.0772
26	7	810.26	8.1026
26	8	812.80	8.1280
26	9	815.34	8.1534
26	10	817.88	8.1788
26	11	820.42	8.2042
27	0	822.96	8.2296
27	1	825.50	8.2550
27	2	828.04	8.2804
27	3	830.58	8.3058
27	4	833.12	8.3312
27	5	835.66	8.3566
27	6	838.20	8.3820
27	7	840.74	8.4074
27	8	843.28	8.4328
27	9	845.82	8.4582
27	10	848.36	8.4836
27	11	850.90	8.5090
28	0	853.44	8.5344
28	1	855.98	8.5598
28	2	858.52	8.5852
28	3	861.06	8.6106
28	4	863.60	8.6360
28	5	866.14	8.6614
28	6	868.68	8.6868
28	7	871.22	8.7122
28	8	873.76	8.7376
28	9	876.30	8.7630
28	10	878.84	8.7884
28	11	881.38	8.8138
29	0	883.92	8.8392
29	1	886.46	8.8646
29	2	889.00	8.8900
29	3	891.54	8.9154
29	4	894.08	8.9408
29	5	896.62	8.9662
29	6	899.16	8.9916
29	7	901.70	9.0170
29	8	904.24	9.0424
29	9	906.78	9.0678
29	10	909.32	9.0932
29	11	911.86	9.1186
30	0	914.40	9.1440

ft	in	cm	m
30	1	916.94	9.1694
30	2	919.48	9.1948
30	3	922.02	9.2202
30	4	924.56	9.2456
30	5	927.10	9.2710
30	6	929.64	9.2964
30	7	932.18	9.3218
30	8	934.72	9.3472
30	9	937.26	9.3726
30	10	939.80	9.3980
30	11	942.34	9.4234
31	0	944.88	9.4488
31	1	947.42	9.4742
31	2	949.96	9.4996
31	3	952.50	9.5250
31	4	955.04	9.5504
31	5	957.58	9.5758
31	6	960.12	9.6012
31	7	962.66	9.6266
31	8	965.20	9.6520
31	9	967.74	9.6774
31	10	970.28	9.7028
31	11	972.82	9.7282
32	0	975.36	9.7536
32	1	977.90	9.7790
32	2	980.44	9.8044
32	3	982.98	9.8298
32	4	985.52	9.8552
32	5	988.06	9.8806
32	6	990.60	9.9060
32	7	993.14	9.9314
32	8	995.68	9.9568
32	9	998.22	9.9822
32	10	1000.76	10.0076
32	11	1003.30	10.0330
33	0	1005.84	10.0584
33	1	1008.38	10.0838
33	2	1010.92	10.1092
33	3	1013.46	10.1346
33	4	1016.00	10.1600
33	5	1018.54	10.1854
33	6	1021.08	10.2108
33	7	1023.62	10.2362
33	8	1026.16	10.2616
33	9	1028.70	10.2870
33	10	1031.24	10.3124
33	11	1033.78	10.3378
34	0	1036.32	10.3632
34	1	1038.86	10.3886
34	2	1041.40	10.4140
34	3	1043.94	10.4394
34	4	1046.48	10.4648
34	5	1049.02	10.4902
34	6	1051.56	10.5156
34	7	1054.10	10.5410
34	8	1056.64	10.5664
34	9	1059.18	10.5918
34	10	1061.72	10.6172
34	11	1064.26	10.6426
35	0	1066.80	10.6680

ft	in	cm	m
35	1	1069.34	10.6934
35	2	1071.88	10.7188
35	3	1074.42	10.7442
35	4	1076.96	10.7696
35	5	1079.50	10.7950
35	6	1082.04	10.8204
35	7	1084.58	10.8458
35	8	1087.12	10.8712
35	9	1089.66	10.8966
35	10	1092.20	10.9220
35	11	1094.74	10.9474
36	0	1097.28	10.9728
36	1	1099.82	10.9982
36	2	1102.36	11.0236
36	3	1104.90	11.0490
36	4	1107.44	11.0744
36	5	1109.98	11.0998
36	6	1112.52	11.1252
36	7	1115.06	11.1506
36	8	1117.60	11.1760
36	9	1120.14	11.2014
36	10	1122.68	11.2268
36	11	1125.22	11.2522
37	0	1127.76	11.2776
37	1	1130.30	11.3030
37	2	1132.84	11.3284
37	3	1135.38	11.3538
37	4	1137.92	11.3792
37	5	1140.46	11.4046
37	6	1143.00	11.4300
37	7	1145.54	11.4554
37	8	1148.08	11.4808
37	9	1150.62	11.5062
37	10	1153.16	11.5316
37	11	1155.70	11.5570
38	0	1158.24	11.5824
38	1	1160.78	11.6078
38	2	1163.32	11.6332
38	3	1165.86	11.6586
38	4	1168.40	11.6840
38	5	1170.94	11.7094
38	6	1173.48	11.7348
38	7	1176.02	11.7602
38	8	1178.56	11.7856
38	9	1181.10	11.8110
38	10	1183.64	11.8364
38	11	1186.18	11.8618
39	0	1188.72	11.8872
39	1	1191.26	11.9126
39	2	1193.80	11.9380
39	3	1196.34	11.9634
39	4	1198.88	11.9888
39	5	1201.42	12.0142
39	6	1203.96	12.0396
39	7	1206.50	12.0650
39	8	1209.04	12.0904
39	9	1211.58	12.1158
39	10	1214.12	12.1412
39	11	1216.66	12.1666
40	0	1219.20	12.1920

ft	in	cm	m
40	1	1221.74	12.2174
40	2	1224.28	12.2428
40	3	1226.82	12.2682
40	4	1229.36	12.2936
40	5	1231.90	12.3190
40	6	1234.44	12.3444
40	7	1236.98	12.3698
40	8	1239.52	12.3952
40	9	1242.06	12.4206
40	10	1244.60	12.4460
40	11	1247.14	12.4714
41	0	1249.68	12.4968
41	1	1252.22	12.5222
41	2	1254.76	12.5476
41	3	1257.30	12.5730
41	4	1259.84	12.5984
41	5	1262.38	12.6238
41	6	1264.92	12.6492
41	7	1267.46	12.6746
41	8	1270.00	12.7000
41	9	1272.54	12.7254
41	10	1275.08	12.7508
41	11	1277.62	12.7762
42	0	1280.16	12.8016
42	1	1282.70	12.8270
42	2	1285.24	12.8524
42	3	1287.78	12.8778
42	4	1290.32	12.9032
42	5	1292.86	12.9286
42	6	1295.40	12.9540
42	7	1297.94	12.9794
42	8	1300.48	13.0048
42	9	1303.02	13.0302
42	10	1305.56	13.0556
42	11	1308.10	13.0810
43	0	1310.64	13.1064
43	1	1313.18	13.1318
43	2	1315.72	13.1572
43	3	1318.26	13.1826
43	4	1320.80	13.2080
43	5	1323.34	13.2334
43	6	1325.88	13.2588
43	7	1328.42	13.2842
43	8	1330.96	13.3096
43	9	1333.50	13.3350
43	10	1336.04	13.3604
43	11	1338.58	13.3858
44	0	1341.12	13.4112
44	1	1343.66	13.4366
44	2	1346.20	13.4620
44	3	1348.74	13.4874
44	4	1351.28	13.5128
44	5	1353.82	13.5382
44	6	1356.36	13.5636
44	7	1358.90	13.5890
44	8	1361.44	13.6144
44	9	1363.98	13.6398
44	10	1366.52	13.6652
44	11	1369.06	13.6906
45	0	1371.60	13.7160

ft	in	cm	m
45	1	1374.14	13.7414
45	2	1376.68	13.7668
45	3	1379.22	13.7922
45	4	1381.76	13.8176
45	5	1384.30	13.8430
45	6	1386.84	13.8684
45	7	1389.38	13.8938
45	8	1391.92	13.9192
45	9	1394.46	13.9446
45	10	1397.00	13.9700
45	11	1399.54	13.9954
46	0	1402.08	14.0208
46	1	1404.62	14.0462
46	2	1407.16	14.0716
46	3	1409.70	14.0970
46	4	1412.24	14.1224
46	5	1414.78	14.1478
46	6	1417.32	14.1732
46	7	1419.86	14.1986
46	8	1422.40	14.2240
46	9	1424.94	14.2494
46	10	1427.48	14.2748
46	11	1430.02	14.3002
47	0	1432.56	14.3256
47	1	1435.10	14.3510
47	2	1437.64	14.3764
47	3	1440.18	14.4018
47	4	1442.72	14.4272
47	5	1445.26	14.4526
47	6	1447.80	14.4780
47	7	1450.34	14.5034
47	8	1452.88	14.5288
47	9	1455.42	14.5542
47	10	1457.96	14.5796
47	11	1460.50	14.6050
48	0	1463.04	14.6304
48	1	1465.58	14.6558
48	2	1468.12	14.6812
48	3	1470.66	14.7066
48	4	1473.20	14.7320
48	5	1475.74	14.7574
48	6	1478.28	14.7828
48	7	1480.82	14.8082
48	8	1483.36	14.8336
48	9	1485.90	14.8590
48	10	1488.44	14.8844
48	11	1490.98	14.9098
49	0	1493.52	14.9352
49	1	1496.06	14.9606
49	2	1498.60	14.9860
49	3	1501.14	15.0114
49	4	1503.68	15.0368
49	5	1506.22	15.0622
49	6	1508.76	15.0876
49	7	1511.30	15.1130
49	8	1513.84	15.1384
49	9	1516.38	15.1638
49	10	1518.92	15.1892
49	11	1521.46	15.2146
50	0	1524.00	15.2400

TABLE 1.7—CONTINUED

ft	in	cm	m	ft	in	cm	m	ft	in	cm	m	ft	in	cm	m	ft	in	cm	m
50	1	1526.54	15.2654	55	1	1678.94	16.7894	60	1	1831.34	18.3134	65	1	1983.74	19.8374	70	1	2136.14	21.3614
50	2	1529.08	15.2908	55	2	1681.48	16.8148	60	2	1833.88	18.3388	65	2	1986.28	19.8628	70	2	2138.68	21.3868
50	3	1531.62	15.3162	55	3	1684.02	16.8402	60	3	1836.42	18.3642	65	3	1988.82	19.8882	70	3	2141.22	21.4122
50	4	1534.16	15.3416	55	4	1686.56	16.8656	60	4	1838.96	18.3896	65	4	1991.36	19.9136	70	4	2143.76	21.4376
50	5	1536.70	15.3670	55	5	1689.10	16.8910	60	5	1841.50	18.4150	65	5	1993.90	19.9390	70	5	2146.30	21.4630
50	6	1539.24	15.3924	55	6	1691.64	16.9164	60	6	1844.04	18.4404	65	6	1996.44	19.9644	70	6	2148.84	21.4884
50	7	1541.78	15.4178	55	7	1694.18	16.9418	60	7	1846.58	18.4658	65	7	1998.98	19.9898	70	7	2151.38	21.5138
50	8	1544.32	15.4432	55	8	1696.72	16.9672	60	8	1849.12	18.4912	65	8	2001.52	20.0152	70	8	2153.92	21.5392
50	9	1546.86	15.4686	55	9	1699.26	16.9926	60	9	1851.66	18.5166	65	9	2004.06	20.0406	70	9	2156.46	21.5646
50	10	1549.40	15.4940	55	10	1701.80	17.0180	60	10	1854.20	18.5420	65	10	2006.60	20.0660	70	10	2159.00	21.5900
50	11	1551.94	15.5194	55	11	1704.34	17.0434	60	11	1856.74	18.5674	65	11	2009.14	20.0914	70	11	2161.54	21.6154
51	0	1554.48	15.5448	56	0	1706.88	17.0688	61	0	1859.28	18.5928	66	0	2011.68	20.1168	71	0	2164.08	21.6408
51	1	1557.02	15.5702	56	1	1709.42	17.0942	61	1	1861.82	18.6182	66	1	2014.22	20.1422	71	1	2166.62	21.6662
51	2	1559.56	15.5956	56	2	1711.96	17.1196	61	2	1864.36	18.6436	66	2	2016.76	20.1676	71	2	2169.16	21.6916
51	3	1562.10	15.6210	56	3	1714.50	17.1450	61	3	1866.90	18.6690	66	3	2019.30	20.1930	71	3	2171.70	21.7170
51	4	1564.64	15.6464	56	4	1717.04	17.1704	61	4	1869.44	18.6944	66	4	2021.84	20.2184	71	4	2174.24	21.7424
51	5	1567.18	15.6718	56	5	1719.58	17.1958	61	5	1871.98	18.7198	66	5	2024.38	20.2438	71	5	2176.78	21.7678
51	6	1569.72	15.6972	56	6	1722.12	17.2212	61	6	1874.52	18.7452	66	6	2026.92	20.2692	71	6	2179.32	21.7932
51	7	1572.26	15.7226	56	7	1724.66	17.2466	61	7	1877.06	18.7706	66	7	2029.46	20.2946	71	7	2181.86	21.8186
51	8	1574.80	15.7480	56	8	1727.20	17.2720	61	8	1879.60	18.7960	66	8	2032.00	20.3200	71	8	2184.40	21.8440
51	9	1577.34	15.7734	56	9	1729.74	17.2974	61	9	1882.14	18.8214	66	9	2034.54	20.3454	71	9	2186.94	21.8694
51	10	1579.88	15.7988	56	10	1732.28	17.3228	61	10	1884.68	18.8468	66	10	2037.08	20.3708	71	10	2189.48	21.8948
51	11	1582.42	15.8242	56	11	1734.82	17.3482	61	11	1887.22	18.8722	66	11	2039.62	20.3962	71	11	2192.02	21.9202
52	0	1584.96	15.8496	57	0	1737.36	17.3736	62	0	1889.76	18.8976	67	0	2042.16	20.4216	72	0	2194.56	21.9456
52	1	1587.50	15.8750	57	1	1739.90	17.3990	62	1	1892.30	18.9230	67	1	2044.70	20.4470	72	1	2197.10	21.9710
52	2	1590.04	15.9004	57	2	1742.44	17.4244	62	2	1894.84	18.9484	67	2	2047.24	20.4724	72	2	2199.64	21.9964
52	3	1592.58	15.9258	57	3	1744.98	17.4498	62	3	1897.38	18.9738	67	3	2049.78	20.4978	72	3	2102.18	21.0218
52	4	1595.12	15.9512	57	4	1747.52	17.4752	62	4	1899.92	18.9992	67	4	2052.32	20.5232	72	4	2104.72	21.0472
52	5	1597.66	15.9766	57	5	1750.06	17.5006	62	5	1902.46	19.0246	67	5	2054.86	20.5486	72	5	2107.26	21.0726
52	6	1600.20	16.0020	57	6	1752.60	17.5260	62	6	1905.00	19.0500	67	6	2057.40	20.5740	72	6	2109.80	21.0980
52	7	1602.74	16.0274	57	7	1755.14	17.5514	62	7	1907.54	19.0754	67	7	2059.94	20.5994	72	7	2112.34	21.1234
52	8	1605.28	16.0528	57	8	1757.68	17.5768	62	8	1910.08	19.1008	67	8	2062.48	20.6248	72	8	2114.88	21.1488
52	9	1607.82	16.0782	57	9	1760.22	17.6022	62	9	1912.62	19.1262	67	9	2065.02	20.6502	72	9	2117.42	21.1742
52	10	1610.36	16.1036	57	10	1762.76	17.6276	62	10	1915.16	19.1516	67	10	2067.56	20.6756	72	10	2119.96	21.1996
52	11	1612.90	16.1290	57	11	1765.30	17.6530	62	11	1917.70	19.1770	67	11	2070.10	20.7010	72	11	2222.50	22.2250
53	0	1615.44	16.1544	58	0	1767.84	17.6784	63	0	1920.24	19.2024	68	0	2072.64	20.7264	73	0	2225.04	22.2504
53	1	1617.98	16.1798	58	1	1770.38	17.7038	63	1	1922.78	19.2278	68	1	2075.18	20.7518	73	1	2227.58	22.2758
53	2	1620.52	16.2052	58	2	1772.92	17.7292	63	2	1925.32	19.2532	68	2	2077.72	20.7772	73	2	2230.12	22.3012
53	3	1623.06	16.2306	58	3	1775.46	17.7546	63	3	1927.86	19.2786	68	3	2080.26	20.8026	73	3	2232.66	22.3266
53	4	1625.60	16.2560	58	4	1778.00	17.7800	63	4	1930.40	19.3040	68	4	2082.80	20.8280	73	4	2235.20	22.3520
53	5	1628.14	16.2814	58	5	1780.54	17.8054	63	5	1932.94	19.3294	68	5	2085.34	20.8534	73	5	2237.74	22.3774
53	6	1630.68	16.3068	58	6	1783.08	17.8308	63	6	1935.48	19.3548	68	6	2087.88	20.8788	73	6	2240.28	22.4028
53	7	1633.22	16.3322	58	7	1785.62	17.8562	63	7	1938.02	19.3802	68	7	2090.42	20.9042	73	7	2242.82	22.4282
53	8	1635.76	16.3576	58	8	1788.16	17.8816	63	8	1940.56	19.4056	68	8	2092.96	20.9296	73	8	2245.36	22.4536
53	9	1638.30	16.3830	58	9	1790.70	17.9070	63	9	1943.10	19.4310	68	9	2095.50	20.9550	73	9	2247.90	22.4790
53	10	1640.84	16.4084	58	10	1793.24	17.9324	63	10	1945.64	19.4564	68	10	2098.04	20.9804	73	10	2250.44	22.5044
53	11	1643.38	16.4338	58	11	1795.78	17.9578	63	11	1948.18	19.4818	68	11	2100.58	21.0058	73	11	2252.98	22.5298
54	0	1645.92	16.4592	59	0	1798.32	17.9832	64	0	1950.72	19.5072	69	0	2103.12	21.0312	74	0	2255.52	22.5552
54	1	1648.46	16.4846	59	1	1800.86	18.0086	64	1	1953.26	19.5326	69	1	2105.66	21.0566	74	1	2258.06	22.5806
54	2	1651.00	16.5100	59	2	1803.40	18.0340	64	2	1955.80	19.5580	69	2	2108.20	21.0820	74	2	2260.60	22.6060
54	3	1653.54	16.5354	59	3	1805.94	18.0594	64	3	1958.34	19.5834	69	3	2110.74	21.1074	74	3	2263.14	22.6314
54	4	1656.08	16.5608	59	4	1808.48	18.0848	64	4	1960.88	19.6088	69	4	2113.28	21.1328	74	4	2265.68	22.6568
54	5	1658.62	16.5862	59	5	1811.02	18.1102	64	5	1963.42	19.6342	69	5	2115.82	21.1582	74	5	2268.22	22.6822
54	6	1661.16	16.6116	59	6	1813.56	18.1356	64	6	1965.96	19.6596	69	6	2118.36	21.1836	74	6	2270.76	22.7076
54	7	1663.70	16.6370	59	7	1816.10	18.1610	64	7	1968.50	19.6850	69	7	2120.90	21.2090	74	7	2273.30	22.7330
54	8	1666.24	16.6624	59	8	1818.64	18.1864	64	8	1971.04	19.7104	69	8	2123.44	21.2344	74	8	2275.84	22.7584
54	9	1668.78	16.6878	59	9	1821.18	18.2118	64	9	1973.58	19.7358	69	9	2125.98	21.2598	74	9	2278.38	22.7838
54	10	1671.32	16.7132	59	10	1823.72	18.2372	64	10	1976.12	19.7612	69	10	2128.52	21.2852	74	10	2280.92	22.8092
54	11	1673.86	16.7386	59	11	1826.26	18.2626	64	11	1978.66	19.7866	69	11	2131.06	21.3106	74	11	2283.46	22.8346
55	0	1676.40	16.7640	60	0	1828.80	18.2880	65	0	1981.20	19.8120	70	0	2133.60	21.3360	75	0	2286.00	22.8600

TABLE 1.7—CONTINUED

ft	in	cm	m	ft	in	cm	m	ft	in	cm	m	ft	in	cm	m	ft	in	cm	m
75	1	2288.54	22.8854	80	1	2440.94	24.4094	85	1	2593.34	25.9334	90	1	2745.74	27.4574	95	1	2898.14	28.9814
75	2	2291.08	22.9108	80	2	2443.48	24.4348	85	2	2595.88	25.9588	90	2	2748.28	27.4828	95	2	2900.68	29.0068
75	3	2293.62	22.9362	80	3	2446.02	24.4602	85	3	2598.42	25.9842	90	3	2750.82	27.5082	95	3	2903.22	29.0322
75	4	2296.16	22.9616	80	4	2448.56	24.4856	85	4	2600.96	26.0096	90	4	2753.36	27.5336	95	4	2905.76	29.0576
75	5	2298.70	22.9870	80	5	2451.10	24.5110	85	5	2603.50	26.0350	90	5	2755.90	27.5590	95	5	2908.30	29.0830
75	6	2301.24	23.0124	80	6	2453.64	24.5364	85	6	2606.04	26.0604	90	6	2758.44	27.5844	95	6	2910.84	29.1084
75	7	2303.78	23.0378	80	7	2456.18	24.5618	85	7	2608.58	26.0858	90	7	2760.98	27.6098	95	7	2913.38	29.1338
75	8	2306.32	23.0632	80	8	2458.72	24.5872	85	8	2611.12	26.1112	90	8	2763.52	27.6352	95	8	2915.92	29.1592
75	9	2308.86	23.0886	80	9	2461.26	24.6126	85	9	2613.66	26.1366	90	9	2766.06	27.6606	95	9	2918.46	29.1846
75	10	2311.40	23.1140	80	10	2463.80	24.6380	85	10	2616.20	26.1620	90	10	2768.60	27.6860	95	10	2921.00	29.2100
75	11	2313.94	23.1394	80	11	2466.34	24.6634	85	11	2618.74	26.1874	90	11	2771.14	27.7114	95	11	2923.54	29.2354
76	0	2316.48	23.1648	81	0	2468.88	24.6888	86	0	2621.28	26.2128	91	0	2773.68	27.7368	96	0	2926.08	29.2608
76	1	2319.02	23.1902	81	1	2471.42	24.7142	86	1	2623.82	26.2382	91	1	2776.22	27.7622	96	1	2928.62	29.2862
76	2	2321.56	23.2156	81	2	2473.96	24.7396	86	2	2626.36	26.2636	91	2	2778.76	27.7876	96	2	2931.16	29.3116
76	3	2324.10	23.2410	81	3	2476.50	24.7650	86	3	2628.90	26.2890	91	3	2781.30	27.8130	96	3	2933.70	29.3370
76	4	2326.64	23.2664	81	4	2479.04	24.7904	86	4	2631.44	26.3144	91	4	2783.84	27.8384	96	4	2936.24	29.3624
76	5	2329.18	23.2918	81	5	2481.58	24.8158	86	5	2633.98	26.3398	91	5	2786.38	27.8638	96	5	2938.78	29.3878
76	6	2331.72	23.3172	81	6	2484.12	24.8412	86	6	2636.52	26.3652	91	6	2788.92	27.8892	96	6	2941.32	29.4132
76	7	2334.26	23.3426	81	7	2486.66	24.8666	86	7	2639.06	26.3906	91	7	2791.46	27.9146	96	7	2943.86	29.4386
76	8	2336.80	23.3680	81	8	2489.20	24.8920	86	8	2641.60	26.4160	91	8	2794.00	27.9400	96	8	2946.40	29.4640
76	9	2339.34	23.3934	81	9	2491.74	24.9174	86	9	2644.14	26.4414	91	9	2796.54	27.9654	96	9	2948.94	29.4894
76	10	2341.88	23.4188	81	10	2494.28	24.9428	86	10	2646.68	26.4668	91	10	2799.08	27.9908	96	10	2951.48	29.5148
76	11	2344.42	23.4442	81	11	2496.82	24.9682	86	11	2649.22	26.4922	91	11	2801.62	28.0162	96	11	2954.02	29.5402
77	0	2346.96	23.4696	82	0	2499.36	24.9936	87	0	2651.76	26.5176	92	0	2804.16	28.0416	97	0	2956.56	29.5656
77	1	2349.50	23.4950	82	1	2501.90	25.0190	87	1	2654.30	26.5430	92	1	2806.70	28.0670	97	1	2959.10	29.5910
77	2	2352.04	23.5204	82	2	2504.44	25.0444	87	2	2656.84	26.5684	92	2	2809.24	28.0924	97	2	2961.64	29.6164
77	3	2354.58	23.5458	82	3	2506.98	25.0698	87	3	2659.38	26.5938	92	3	2811.78	28.1178	97	3	2964.18	29.6418
77	4	2357.12	23.5712	82	4	2509.52	25.0952	87	4	2661.92	26.6192	92	4	2814.32	28.1432	97	4	2966.72	29.6672
77	5	2359.66	23.5966	82	5	2512.06	25.1206	87	5	2664.46	26.6446	92	5	2816.86	28.1686	97	5	2969.26	29.6926
77	6	2362.20	23.6220	82	6	2514.60	25.1460	87	6	2667.00	26.6700	92	6	2819.40	28.1940	97	6	2971.80	29.7180
77	7	2364.74	23.6474	82	7	2517.14	25.1714	87	7	2669.54	26.6954	92	7	2821.94	28.2194	97	7	2974.34	29.7434
77	8	2367.28	23.6728	82	8	2519.68	25.1968	87	8	2672.08	26.7208	92	8	2824.48	28.2448	97	8	2976.88	29.7688
77	9	2369.82	23.6982	82	9	2522.22	25.2222	87	9	2674.62	26.7462	92	9	2827.02	28.2702	97	9	2979.42	29.7942
77	10	2372.36	23.7236	82	10	2524.76	25.2476	87	10	2677.16	26.7716	92	10	2829.56	28.2956	97	10	2981.96	29.8196
77	11	2374.90	23.7490	82	11	2527.30	25.2730	87	11	2679.70	26.7970	92	11	2832.10	28.3210	97	11	2984.50	29.8450
78	0	2377.44	23.7744	83	0	2529.84	25.2984	88	0	2682.24	26.8224	93	0	2834.64	28.3464	98	0	2987.04	29.8704
78	1	2379.98	23.7998	83	1	2532.38	25.3238	88	1	2684.78	26.8478	93	1	2837.18	28.3718	98	1	2989.58	29.8958
78	2	2382.52	23.8252	83	2	2534.92	25.3492	88	2	2687.32	26.8732	93	2	2839.72	28.3972	98	2	2992.12	29.9212
78	3	2385.06	23.8506	83	3	2537.46	25.3746	88	3	2689.86	26.8986	93	3	2842.26	28.4226	98	3	2994.66	29.9466
78	4	2387.60	23.8760	83	4	2540.00	25.4000	88	4	2692.40	26.9240	93	4	2844.80	28.4480	98	4	2997.20	29.9720
78	5	2390.14	23.9014	83	5	2542.54	25.4254	88	5	2694.94	26.9494	93	5	2847.34	28.4734	98	5	2999.74	29.9974
78	6	2392.68	23.9268	83	6	2545.08	25.4508	88	6	2697.48	26.9748	93	6	2849.88	28.4988	98	6	3002.28	30.0228
78	7	2395.22	23.9522	83	7	2547.62	25.4762	88	7	2700.02	27.0002	93	7	2852.42	28.5242	98	7	3004.82	30.0482
78	8	2397.76	23.9776	83	8	2550.16	25.5016	88	8	2702.56	27.0256	93	8	2854.96	28.5496	98	8	3007.36	30.0736
78	9	2400.30	24.0030	83	9	2552.70	25.5270	88	9	2705.10	27.0510	93	9	2857.50	28.5750	98	9	3009.90	30.0990
78	10	2402.84	24.0284	83	10	2555.24	25.5524	88	10	2707.64	27.0764	93	10	2860.04	28.6004	98	10	3012.44	30.1244
78	11	2405.38	24.0538	83	11	2557.78	25.5778	88	11	2710.18	27.1018	93	11	2862.58	28.6258	98	11	3014.98	30.1498
79	0	2407.92	24.0792	84	0	2560.32	25.6032	89	0	2712.72	27.1272	94	0	2865.12	28.6512	99	0	3017.52	30.1752
79	1	2410.46	24.1046	84	1	2562.86	25.6286	89	1	2715.26	27.1526	94	1	2867.66	28.6766	99	1	3020.06	30.2006
79	2	2413.00	24.1300	84	2	2565.40	25.6540	89	2	2717.80	27.1780	94	2	2870.20	28.7020	99	2	3022.60	30.2260
79	3	2415.54	24.1554	84	3	2567.94	25.6794	89	3	2720.34	27.2034	94	3	2872.74	28.7274	99	3	3025.14	30.2514
79	4	2418.08	24.1808	84	4	2570.48	25.7048	89	4	2722.88	27.2288	94	4	2875.28	28.7528	99	4	3027.68	30.2768
79	5	2420.62	24.2062	84	5	2573.02	25.7302	89	5	2725.42	27.2542	94	5	2877.82	28.7782	99	5	3030.22	30.3022
79	6	2423.16	24.2316	84	6	2575.56	25.7556	89	6	2727.96	27.2796	94	6	2880.36	28.8036	99	6	3032.76	30.3276
79	7	2425.70	24.2570	84	7	2578.10	25.7810	89	7	2730.50	27.3050	94	7	2882.90	28.8290	99	7	3035.30	30.3530
79	8	2428.24	24.2824	84	8	2580.64	25.8064	89	8	2733.04	27.3304	94	8	2885.44	28.8544	99	8	3037.84	30.3784
79	9	2430.78	24.3078	84	9	2583.18	25.8318	89	9	2735.58	27.3558	94	9	2887.98	28.8798	99	9	3040.38	30.4038
79	10	2433.32	24.3332	84	10	2585.72	25.8572	89	10	2738.12	27.3812	94	10	2890.52	28.9052	99	10	3042.92	30.4292
79	11	2435.86	24.3586	84	11	2588.26	25.8826	89	11	2740.66	27.4066	94	11	2893.06	28.9306	99	11	3045.46	30.4546
80	0	2438.40	24.3840	85	0	2590.80	25.9080	90	0	2743.20	27.4320	95	0	2895.60	28.9560	100	0	3048.00	30.4800

Metric Conversion Tables, Barnes & Noble, 1975, pp. 3–16 (Ottenheimer Publishers, Inc.).

Review Questions

Make use of tables 1.1 to 1.7 and the general statements to convert measurements from metric to imperial or imperial to metric.

1. Complete the following statements:

4.75 m =	cm =	mm
600 mm =	cm =	m
400 mm =	cm =	m
7900 mm =	cm =	m
12 800 mm =	cm =	m

2. Convert the following imperial measurements to metric measurements:

14′–9$\frac{1}{2}$″ =	mm
16″ =	mm
24″ =	mm
4′–0″ =	mm
8′–0″ =	mm

3. What is the difference in size, measured in millimeters, between the nominal size 2 × 4 and actual metric size 38 mm × 89 mm?

4. Complete the following set of statements pertaining to softwood studs, light framing, joist and plank:

Imperial Nominal Size	*Metric Actual Size*	
2 × 4	mm ×	mm
2 × 6	mm ×	mm
2 × 8	mm ×	mm
2 × 10	mm ×	mm
2 × 12	mm ×	mm

5. What are the dimensions of a traditional framing square? Of a metric framing square?

Mathematical conversions between imperial and metric units have been discussed in Chapter 1. Some specific applications of the metric system to roof framing are introduced in Chapters 6, 7, and 8. Imperial measurements only are used in Chapter 2—ROOFS—which follows.

2

Roofs

Men have always found it necessary to have roofs over their heads. Roofs are made and have been made to shed rain or snow and to give protection from the sun and wind. The city dweller, the Bedouin, and the Eskimo, all build according to their needs and the material at hand.

Where lumber is available, carpenters have learned to cut rafters and to assemble them into strong frameworks. The framing methods have been tested over the years. The assemblies have produced shapes that make our buildings both pleasing to the eye and functional.

1. Types of roofs

a. *Flat Roof.* This is a roof in which heavy ceiling joists are used as rafters. It has a minimum slope for drainage. It must be well supported at the walls. It must be heavily waterproofed. See Figure 2.1.

b. *Shed or Lean-to Roof.* This is a roof which has one sloped surface only. The slope is across the width of the building. The horizontal distance over which the slope passes is the RUN. The lean-to roof is so named because it may lean against another

FIGURE 2.1. Flat roof.

building or wall. It saves the expense of building one wall. It is the simplest roof that a carpenter may have to build. See Figure 2.2.

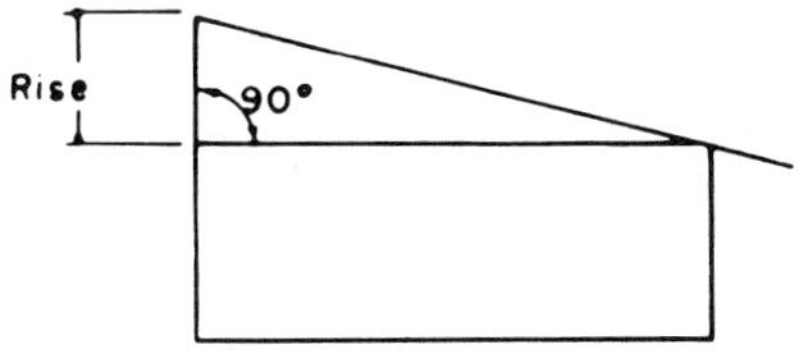

FIGURE 2.2. Lean-to or shed roof.

c. *Gable Roof.* This is a roof which has two sloped surfaces meeting at the RIDGE. The horizontal distance from the foot of one rafter to the foot of the opposite rafter is called the SPAN. Half the span over which each rafter passes is the RUN. This roof is like two lean-to roofs placed together. See Figure 2.3.

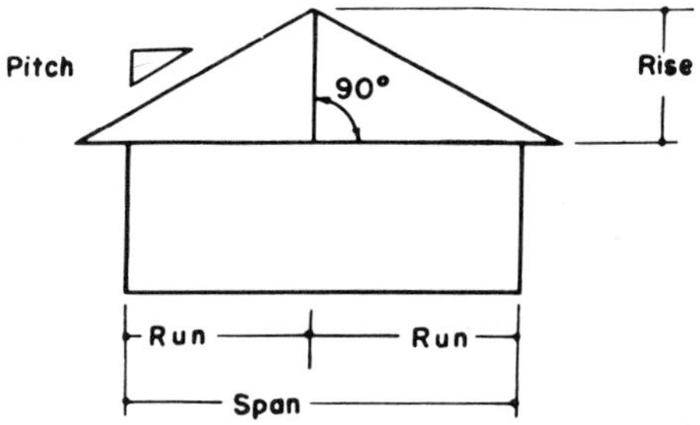

FIGURE 2.3. Gable roof.

d. *Hip Roof.* This is a roof which has a sloped surface from each wall towards the RIDGE. The rafters that fit against the ridge are called COMMON RAFTERS (these will also be referred to as CRs). The main part of the roof whose members are CRs is like two lean-to roofs placed together. See Figure 2.4.

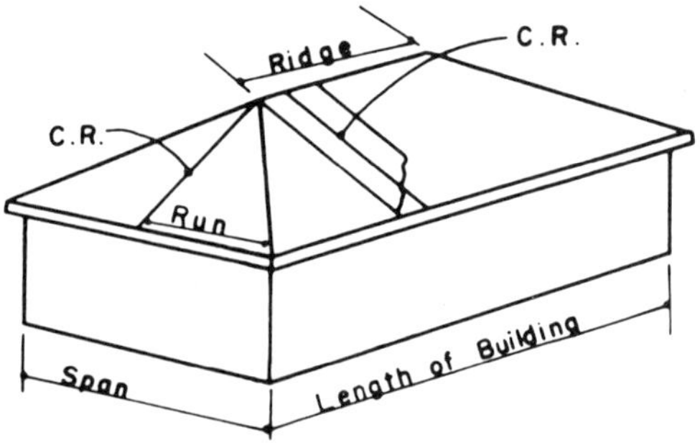

FIGURE 2.4. Hip roof.

e. *Gambrel or Barn Roof.* This is a gable-type roof which has more than one slope on one face. It is used extensively by farmers. The upper portion is like two lean-to roofs placed together. The lower slopes are like lean-to roofs. See Figure 2.5.

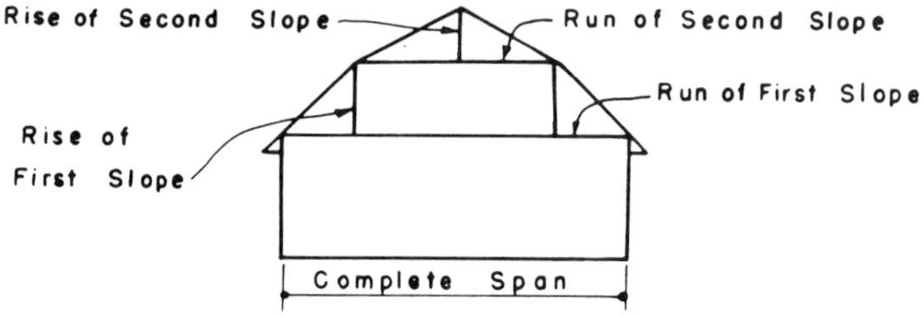

FIGURE 2.5. Gambrel or barn roof.

f. *Sawtooth Roof.* This is a series of roofs which, when viewed from the ends, resembles the angles of saw teeth. It is a factory type of roof. It allows for the placing of glass on one slope to light the floor area below. Each pair of slopes resembles two lean-to roofs of different runs and slopes, but of the same height. See Figure 2.6.

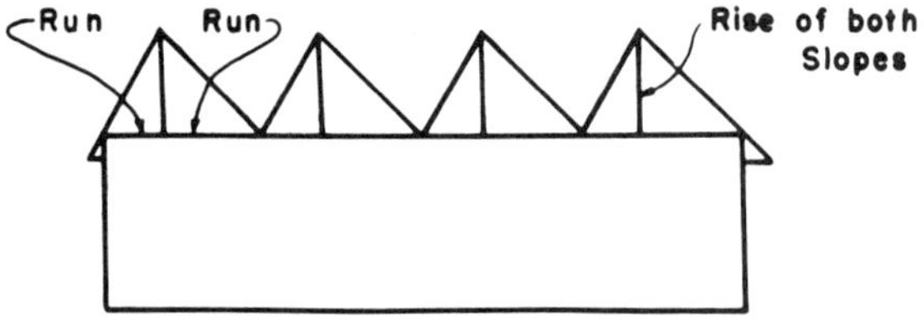

FIGURE 2.6. Sawtooth roof.

2. The right-angle triangle

A plumb line is a vertical line. A run line is a horizontal line. The angle between a plumb line and a run line is 90°.

Roof framing principles are based on right-angle triangulation.

a. If two sides of a right-angle triangle are known, the third may be found by using one of the following statements:

$$\text{Hypotenuse} = \sqrt{\text{base}^2 + \text{altitude}^2}$$

$$\text{Base} = \sqrt{\text{hypotenuse}^2 - \text{altitude}^2}$$

$$\text{Altitude} = \sqrt{\text{hypotenuse}^2 - \text{base}^2}$$

b. It will be seen that the sum of the squares on the base added to the sum of the squares on the height (altitude) equals the sum of the squares on the hypotenuse.

c. In roof framing, the base of the right-angle triangle is called the RUN, the altitude is called the RISE, and the hypotenuse is called the LINE LENGTH (also referred to in this text as LL).

3. Definitions and roof terms as applied to regular roofs

a.	*Span*	The distance between two opposite Common Rafter (CR) Birdsmouths.
b.	*Run*	Half the span.
c.	*Wall plate*	The member on the top of the wall to which the rafters are secured.
d.	*Rise*	The theoretical height of the roof over the tops of the wall plates.
e.	*Pitch*	The slope of the roof. It is the relationship of the rise of the roof to the span. A $\frac{1}{4}$ pitch roof has a rise equal to $\frac{1}{4}$ of its span.
f.	*Ridge*	The uppermost horizontal member of a roof against which the CRs fit.
g.	*Line length*	The theoretical length of any rafter excluding the tail. See Figure 2.7.
h.	*Plumb and level lines*	When a rafter is in position, any vertical line is a plumb line and any horizontal line is a level line.

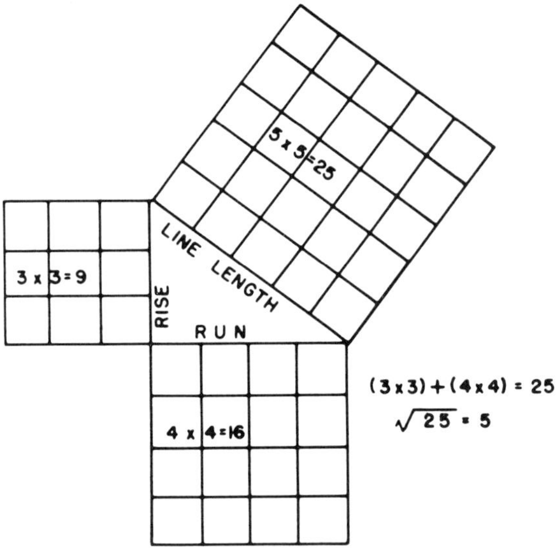

FIGURE 2.7.

i. *Birdsmouth* — The part of a rafter with a plumb and level cut that fits and is secured to the wall plates.

j. *Rafter overhang* — That part of a rafter that extends beyond the wall or building line.

k. *Tail cuts* — The shape of the cut or cuts at the overhang end of a rafter.

4. The span

a. The span is the distance between two opposite rafter seats.

b. It is the total rise divided by the pitch.

Example 1: Rise 6′–0″ divided by the pitch $\frac{1}{4}$.

$$6'\text{–}0'' \div \tfrac{1}{4} = \frac{6'\text{–}0'' \times 4}{1} = 24'\text{–}0'' \quad \text{(Span)}$$

Example 2. Rise 8′–0″ divided by the pitch $\frac{1}{3}$.

$$8'\text{–}0'' \div \tfrac{1}{3} = \frac{8'\text{–}0'' \times 3}{1} = 24'\text{–}0'' \quad \text{(Span)}$$

5. The run

a. The run is half the span.
b. It is the horizontal distance over which any rafter passes.

6. The rise

a. The rise is the theoretical height of the roof over the plates.
b. The rise is the span multiplied by the pitch.

Example 1: Span 24′–0″ multiplied by the pitch $\frac{1}{4}$.

$$\frac{24'\text{–}0'' \times 1}{4} = 6'\text{–}0'' \quad \text{(Rise)}$$

Example 2: Span 24′–0″ multiplied by the pitch $\frac{1}{3}$.

$$\frac{24'\text{–}0'' \times 1}{3} = 8'\text{–}0'' \quad \text{(Rise)}$$

7. The pitch

a. The pitch is the slope of a roof.
b. It is the rise divided by the span, thus:

$$\frac{\text{Rise over}}{\text{Span}} \quad \text{(Remember if you can)}$$

Example 1:

$$\frac{6'\text{–}0''\ \text{(Rise)}}{24'\text{–}0''\ \text{(Span)}} = \tfrac{1}{4}\ \text{Pitch}$$

Example 2:

$$\frac{8'\text{–}0''\ \text{(Rise)}}{24'\text{–}0''\ \text{(Span)}} = \tfrac{1}{3}\ \text{Pitch}$$

Review Questions

1. Name six different types of roofs.

2. Define: span ____________________
run ____________________
wall plate ____________________
rise ____________________
pitch ____________________
ridge ____________________
line length ____________________
plumb line ____________________
level line ____________________
birdsmouth ____________________
rafter overhang ____________________
rafter tail cut ____________________

3. Using two words, define the symbol ℄.

4. A right triangle has a base of 8 units and a rise of 6 units. How many units are there in the hypotenuse?

5. Make a neat, dimensioned, freehand sketch of the section of a gambrel or barn roof.

6. Underline the correct one of the last two words in the following three sentences:

   a. The rise of the adjoining frames of a sawtooth roof are usually the (same) (different).

   b. The run of the adjoining frames of a sawtooth roof are the (same) (different).

   c. The slopes of the adjoining frames of a sawtooth roof are usually the (same) (different).

7. A $\frac{1}{3}$ pitch roof has a span of 35′–0″. What is the rise of the roof?

3

Lean-to or Shed Roof

There are several ways of finding the length and the cuts of the members of any roof. The first and basic method from which all others have been developed is the graphic method.

The graphic method uses descriptive scale drawings to illustrate roof construction theory. This is done by means of right-angle triangles. These triangles show lengths, slopes, and cuts of roof members.

To be able to make anything, a carpenter must be given the intentions of the architect or designer by word of mouth or by drawings and the written word. Plans are the drawings of the intended structure. Specifications are the written details about the intended structure. When read together, the plans and specifications give fully the intent of the architect or designer for the erection of the building.

This chapter deals with the drawing-to-scale of the outline of a lean-to shed roof.

1. Views of a lean-to shed

Figure 3.1 shows the plan and elevations of a lean-to shed, where

(a) is the plan
(b) is the front elevation
(c) is the left elevation
(d) is the right elevation
(e) is the rear elevation

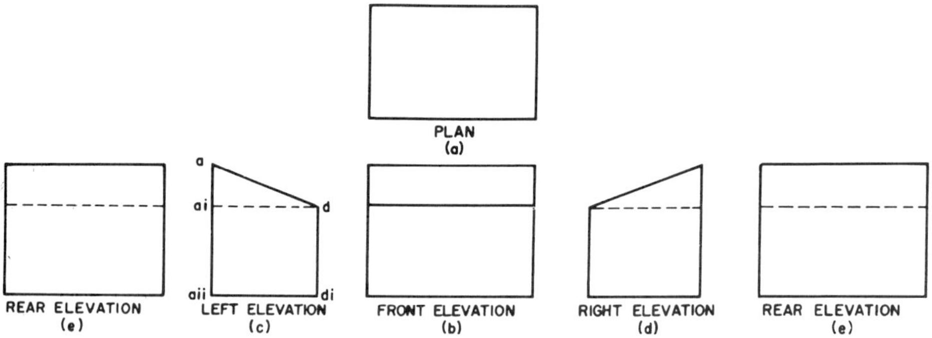

FIGURE 3.1. Views of a lean-to shed.

2. Specifications

Plan view rectangular 12′–0″ × 20′–0″. Wall height 8′–0″. Pitch $\frac{1}{3}$. Overhang 12″. Rafters 16″ on centers (OC). Rafters 2 × 4 common lumber. Rafters at top to be secured to a piece of 1 × 6 nailed to the existing wall. Exposed rafter tail to be plumb and level cut.

3. The development drawing

a. On building paper draw to the scale of $\frac{1}{4}$″ to 1′–0″ a plan and an attached end view of the roof. See Figure 3.2.

b. It will be seen that *ai–b–c–d* is the plan of the building and that:

   ai–d is the run 12′–0″ over which *a–d* the CR passes
   a–ai is the rise 8′–0″ the theoretical rise of the CR over the plate and
   a–d is the LL of the CR

c. The most important feature of the drawing is that *a–ai–d* is a right-angle triangle.

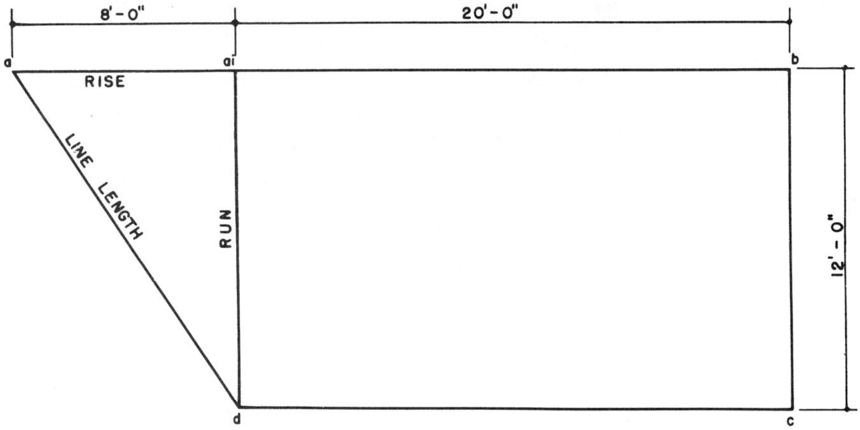

FIGURE 3.2.

d. Upon examination of the drawing with the development it will be seen that:

ai–b measures 5″ and represents 20′–0″, the length of the actual building
ai–d measures 3″ and represents 12′–0″, the actual width of the building
a–ai measures 2″ and represents 8′–0″, the actual rise of the roof
a–d is the LL of the CR.

4. The shop drawing

a. In the Shop Drawing section the scale will be $\frac{1}{12}$, 1″ to 1′–0″. The foregoing development would be drawn on a piece of building paper or on a composition board if drawn in the field. In this case it would be seen that:

20″ represents 20′–0″, the length of the building
12″ represents 12′–0″, the width of the building
8″ represents 8′–0″, the rise of the roof

b. From the information given and the development of the drawing:

What is the LL of the CR?

c. **Measure it!** Too great an emphasis could not be placed on the necessity for accuracy in all drawings.

Using a scale rule (or the framing square), measure very carefully the LL (line length) of the CR. It will be understood that the degree of accuracy in measuring the LL would be greater when measuring from the $\frac{1}{12}$ scale than it would be when measuring from the $\frac{1}{4}''$ represents 1′–0″ scale.

5. The layout problem

Using the graphic method, lay out the common rafter for the lean-to shed as specified in Figure 3.3.

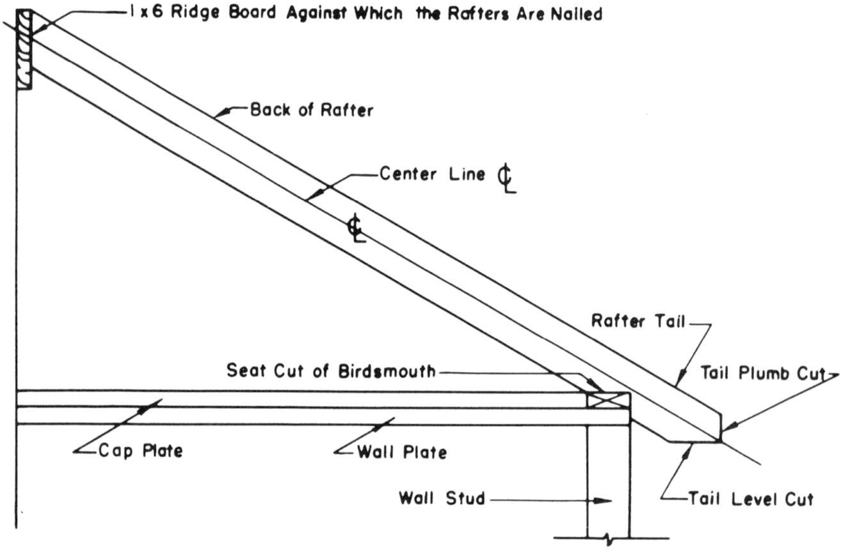

FIGURE 3.3. The CR for a lean-to shed.

Review Questions

1. Using a ruler, make a neat (no scale) sketch of a lean-to roof showing the:

   a. plan

   b. front elevation

c. left elevation
d. rear elevation
e. right elevation.

2. Draw to any convenient scale a right triangle with a base of 8 units and a perpendicular height of 6 units. Measure the hypotenuse and write the number of units of length here. ____________________

3. Draw a rafter for a lean-to shed building and arrow mark the following:

   a. ridge board
   b. back of rafter
   c. ℄ of rafter
   d. seat cut of rafter
   e. wall plate
   f. cap plate
   g. rafter tail
   h. rafter plumb cut of the birdsmouth
   i. rafter tail level cut
   j. rafter tail plumb cut

4. What is the crown side of a rafter? ____________________
   ____

5. By how much should the rafters of a lean-to shed roof be shortened for the ridge board?

4

Layout of the Common Rafter

This chapter deals with the graphic method of laying out a CR. At Plate I the numbered lines give the sequence in which the lines are drawn. It must be thoroughly understood that the same sequence is used for both the graphic and the framing square methods of layout.

1. Pitch board

Cut a piece of $\frac{1}{8}''$ hardened composition board to the exact size of the right-angle triangle developed on your 1″ to 1′–0″ working drawing. Mark on it the run, the rise, and the line length. This object is called a pitch board. See Figure 4.1.

2. Rafter stock

Select a straight clean piece of 2 × 4 of suitable length. Place this piece of material on a pair of sawhorses. Examine the stock for the crown side by holding it with the wide face flat in the palm of the hand and sighting along the narrow face. Place the stock on the sawhorses with the crown side away from you.

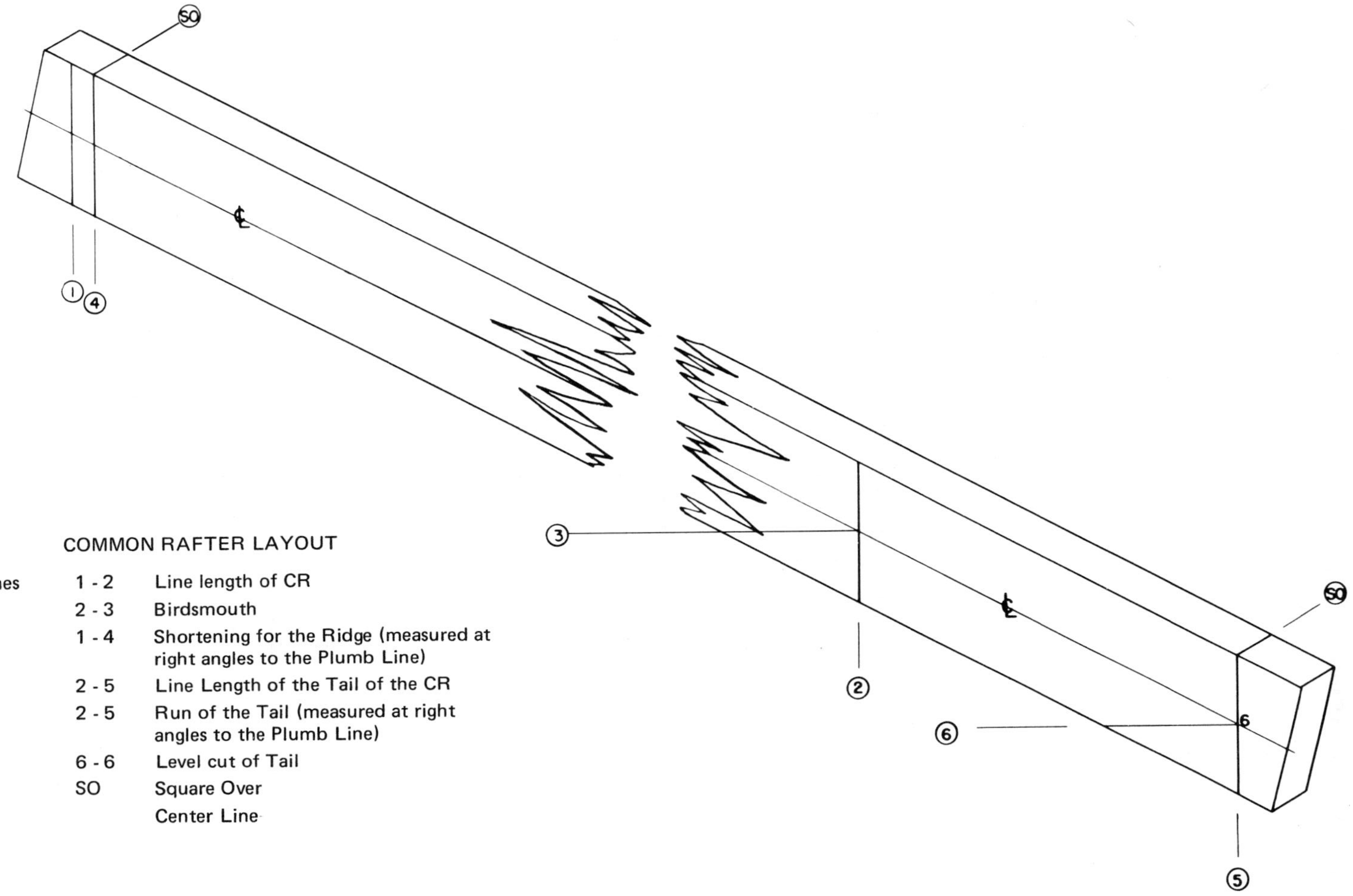

COMMON RAFTER LAYOUT

Lines		
	1 - 2	Line length of CR
	2 - 3	Birdsmouth
	1 - 4	Shortening for the Ridge (measured at right angles to the Plumb Line)
	2 - 5	Line Length of the Tail of the CR
	2 - 5	Run of the Tail (measured at right angles to the Plumb Line)
	6 - 6	Level cut of Tail
	SO	Square Over
		Center Line

PLATE I. Layout of the Common Rafter.

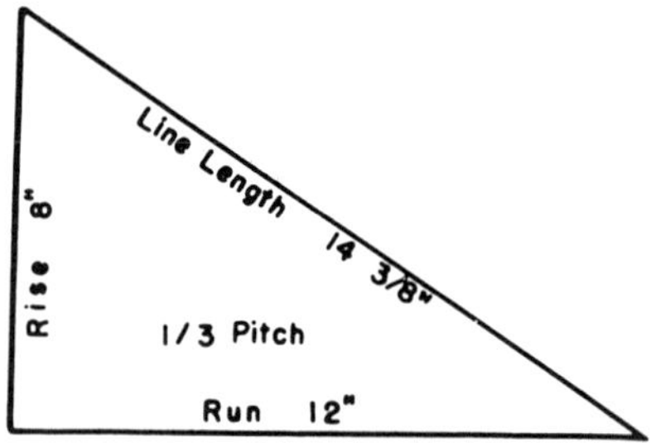

FIGURE 4.1. Pitch board.

3. Center line (afterwards shown in the text as ℄)

a. To find the center line of a piece of 2 × 4, place a tape, square across the $3\frac{5}{8}''$ face of the stock. Traverse the tape until the 4″ mark may be read on the top edge of the stock. Spot the center. Set an adjustable square to the center spot and draw in a center line the full length of the 2 × 4. See Figure 4.2.

b. When the CR is raised into place, the ℄ is the LL of the CR.

c. The horizontal distance over which it passes is the run.

d. The height that it is raised over the plate is the rise.

e. The theoretical center on this CR is at the center of the end section of the 2 × 4. See Figure 4.3.

f. This center will become of ever-increasing importance with all the roofing theory up to and including the Regular Hip Roof in this text.

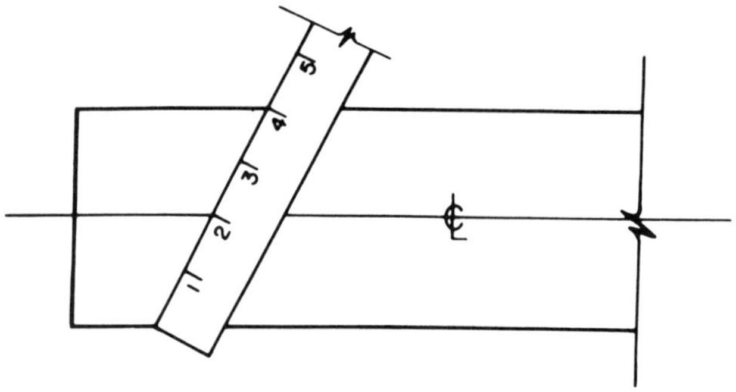

FIGURE 4.2.

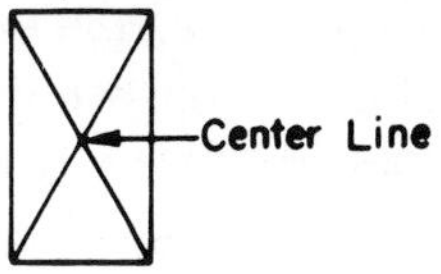

FIGURE 4.3. ℄ of the CR.

4. Plumb cut

a. Place the rafter stock on the sawhorses with the crown side away from you.
b. Lay the hypotenuse of the pitch board, with the plumb cut to the left, on the back of the rafter. See Figure 4.4.
c. Draw in the plumb line (Line 1).

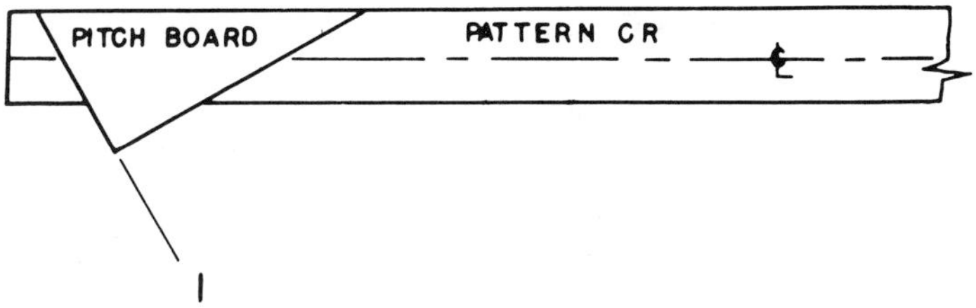

FIGURE 4.4.

5. Line length, plumb cut

a. Measure the LL of the CR *(a–d)* on the Development Drawing. See Figure 3.2, page 25.
b. Check the LL of the CR to a tolerance of $\frac{1}{8}''$ by mathematics. The line length (LL) of the CR is 14′–5″.
c. Check also by measuring very accurately the LL (14′–5″) from your development drawing.
d. Always check the LL of any rafter by as many methods as are known. This will be (when laid out) the pattern rafter from which all other similar rafters are to be cut.
e. Study Plate I, page 29.
f. From the plumb line (Line 1), measure on the top edge of the face of the rafter 14′–5″.

g. Place the hypotenuse of the pitch board to the back of the rafter with the plumb cut of the pitch board to the left. At the 14′–5″ mark draw in another plumb line (Line 2). See Figure 4.5.

h. Observe that the LL is the same length if measured on the back (14′–5″), the center line, or the bottom of the rafter.

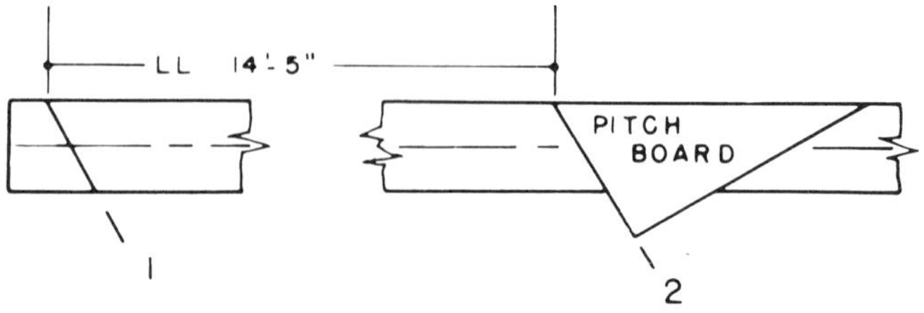

FIGURE 4.5.

6. Line length check

a. As a further check step the pitch board twelve times. Place its hypotenuse along the top edge of the face of the rafter stock.

b. The first application of the pitch board is at Line 1 and the twelfth application is at the terminal point 14′–5″ from Line 1 at Line 2. The measurement is taken from plumb line to plumb line.

c. It should be carefully noted that the stepping off is merely a further check and ought not to be used to obtain the true LL.

d. When stepping off, there is a marked tendency to creep with each application of the pitch board. It is excellent practice always to step off in this manner, but only as an additional check.

7. Line length measurement

a. Let it be assumed that the LL of the rafter is 11′–9$\frac{1}{12}$″; it is a very simple and a common error to read on the tape 9′–11$\frac{1}{12}$″.

b. Stepping off with the pitch board twelve times will most assuredly avert this simple but possible error.

c. The twelfth step should reach within $\frac{3}{8}$″ of the mathematical LL mark.

8. Order of checking line length

Accept LL measurements in the order shown below:

a. mathematical measurement.
b. scaled measurement from the drawing.
c. stepped-off measurement, i.e., the pitch board stepped off twelve times.

In all cases check by as many methods as possible.

9. Golden rules

a. Measure twice and cut once for ordinary work.
b. Measure three times and cut once for roofs.
c. Measure four times and cut once for quality stairs.

10. Seat cut

To draw in the level cut for the birdsmouth:

a. Place the pitch board on the back of the rafter with the plumb cut to the left.
b. Adjust the hypotenuse of the pitch board to the center line of the rafter. The run of the pitch board must intersect the plumb line (Line 2).
c. Draw in the seat cut Line 3. See Figure 4.6.

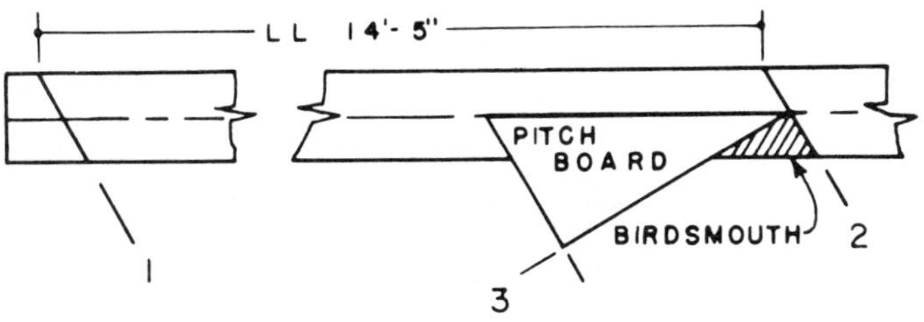

FIGURE 4.6.

11. Birdsmouth

a. The shaded area formed by Lines 2 and 3 meeting at the center line is the birdsmouth. When the rafter is in position the birdsmouth fits the wall plate.

b. Lines 2 and 3 should now be squared under the bottom of the rafter.

12. Shortening

a. Referring to the specifications, Figure 3.2, page 25, you will see that the rafters at the top are to be secured to a piece of 1 × 6 nailed to the side of the building (Figure 3.3).

b. It is possible to erect a lean-to shed against a wooden building without a 1 × 6 nailed to the side, but it is more convenient to have a ridge board.

c. The actual size of a piece of 1 × 6 is $\frac{3}{4}''$ × $5\frac{1}{2}''$. It is evident therefore, that if a $\frac{3}{4}''$ board is nailed to the building, the CR will have to be shortened by the thickness of the board.

d. Figure 4.7 shows an end section of the lean-to shed, and Figure 4.8 shows an enlargement of the CR shortening.

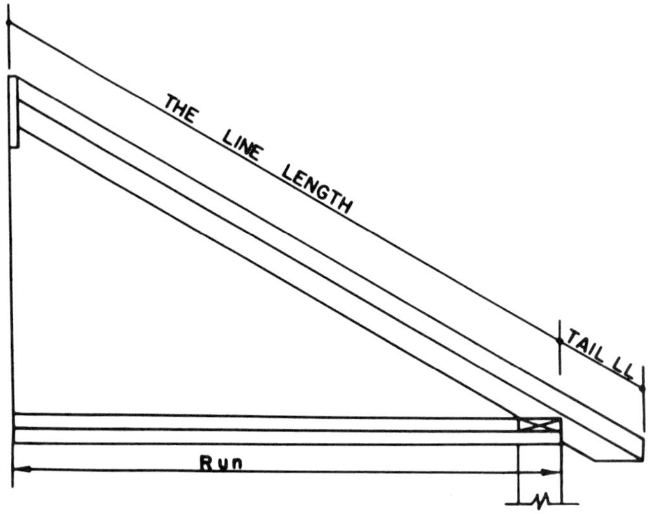

FIGURE 4.7. End section of a lean-to shed.

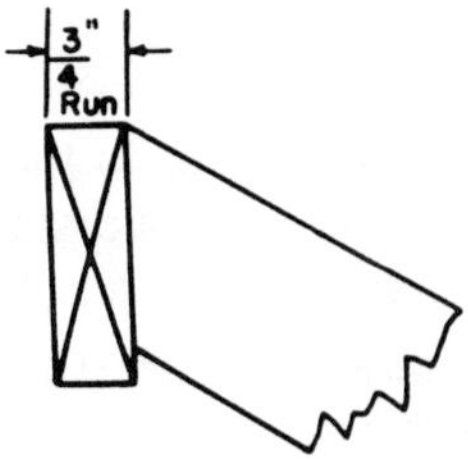

FIGURE 4.8. CR shortening.

13. Measuring the shortening

The shortening is measured at right angles (that is, on the run) to the plumb cut Line 1.

a. Place the pitch board hypotenuse to the back of the rafter at the plumb Line 1.
b. Slide the pitch board back until it assumes a position $\frac{3}{4}''$ back (measured at right angles) to Line 1.
c. Draw in Line 4 (Figure 4.9)

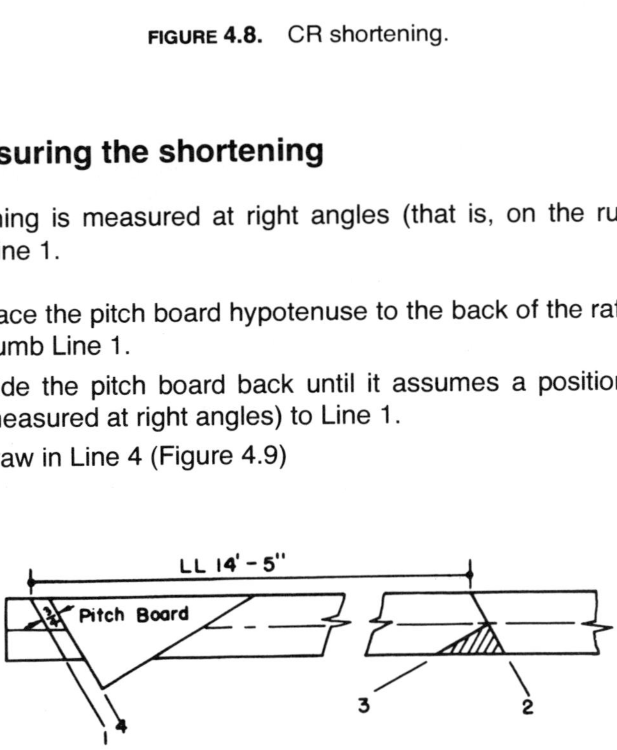

FIGURE 4.9.

14. Ridge board for a lean-to shed roof

a. Any ridge board nailed to the side of a building for a lean-to shed roof will shorten the actual run. The shortening will be equal to the thickness of the lean-to shed roof ridge board and measured on the CR at right angles to the LL plumb cut.
b. To obtain the actual height at which to nail the lean-to roof ridge board, the detail should be drawn full size. The theoretical height is governed by the center line of the CR. See Figure 3.3.

15. Tail length

a. The tail length (LL) of a rafter is the additional length required for it to pass over the horizontal distance from the birdsmouth to the measuring point at the tail end of the rafter.

b. This horizontal distance is called the run of the tail; it is also known as the overhang and is so described in the specifications.

NOTE: *All measurements of run are horizontal or level measurements.* They may be read directly if a full size section is drawn. The thickness of the ridge board is $\frac{3}{4}$″. When measured on the plan and at right angles to Line 1 on the CR layout, this is described as the run of the ridge board.

16. Layout of the tail of a common rafter

Referring to the specifications, you will recall that the overhang is 12″.

a. Place the pitch board on the tail of the rafter with the plumb cut to the left.

b. The hypotenuse of the pitch board must be to the back of the rafter, as in Figure 4.10.

c. The plumb line of the pitch board must be aligned to the plumb line of the birdsmouth Line 2.

d. Measure and mark 12″ along the run of the pitch board, that is, along the base of the triangle.

e. Place the pitch board with the hypotenuse on the bottom edge of the rafter, plumb cut to the right. Slip the pitch board until the plumb cut intersects the 12″ mark on the top edge of the rafter as marked on the layout in Figure 4.11. Draw in Line 5. See Figure 4.11.

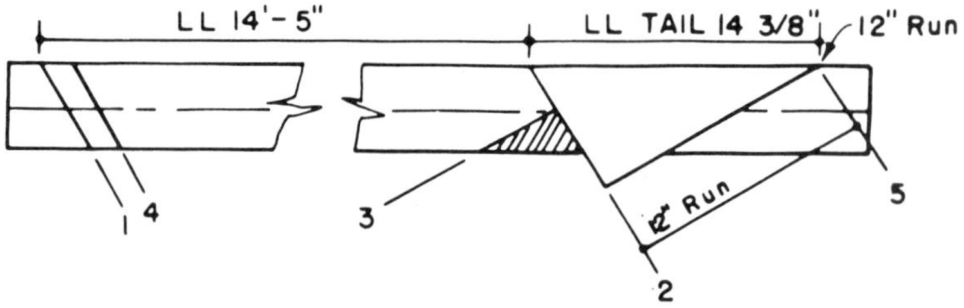

FIGURE 4.10.

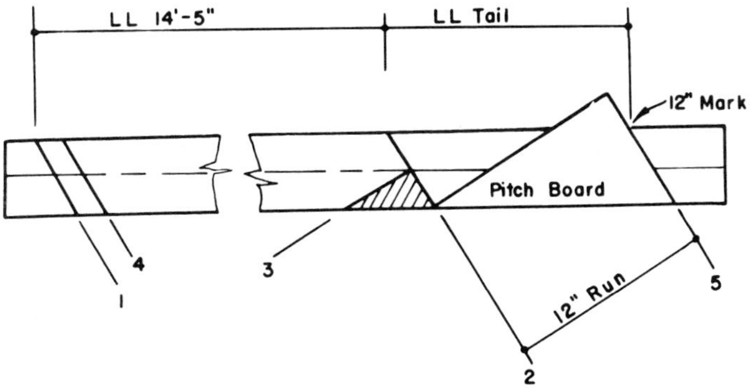

FIGURE 4.11.

17. Level cut of the tail

The plumb and level tail cut as in Figure 4.12 is drawn by placing the pitch board with the hypotenuse on the center line. The pitch board plumb cut is to the left. Slip the pitch board and draw in the level cut on the run of the pitch board. Draw in Line 6. See Figure 4.12.

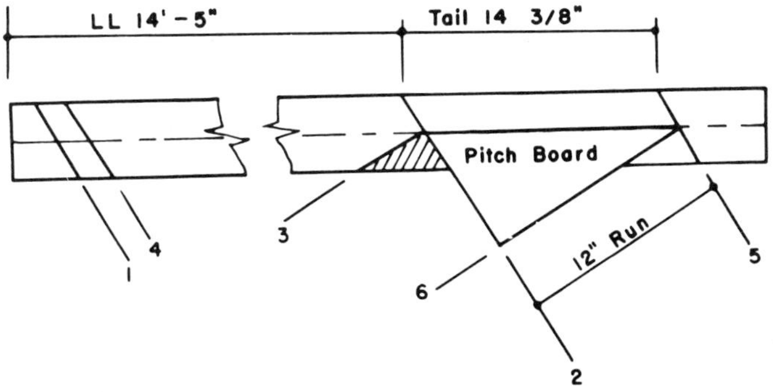

FIGURE 4.12.

18. Variations of tail end cuts

a. There are four variations of the tail end cuts of the rafters as shown in Figure 4.13.

b. In each case the tail LL must be determined before the shape of the tail may be drawn.

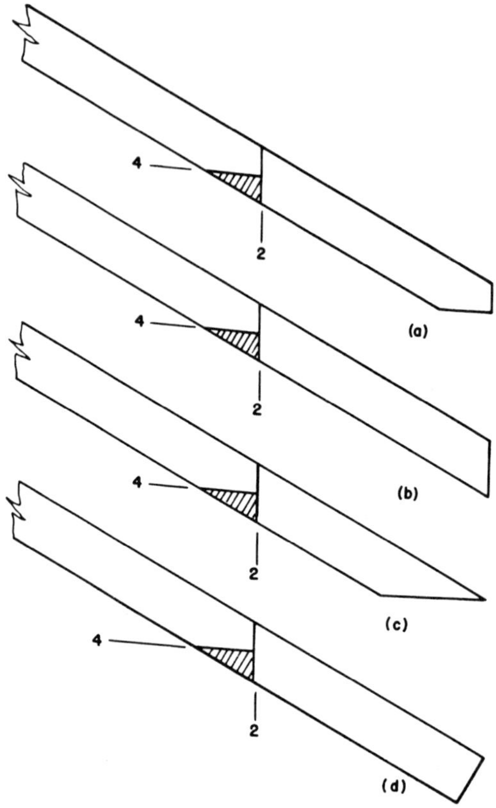

FIGURE 4.13. Variations of tail end cuts.

19. Complete layout of a common rafter

a. Chapter 3 covered the complete layout of Lines 1 to 6 inclusive and is detailed line by line.

b. Plate I, page 29, shows the complete layout. This should be very carefully studied.

20. The framing square

The graphic method involves the use of a pitch board for rafter layout. The same layout procedure may be used when laying out rafters with a framing square.

Review Questions

1. With respect to a piece of rafter stock, define the term *crown side.*

2. Describe a pitch board.

3. Supply the missing word:

   The base of a pitch board is ________.

4. If a pitch board is made for a $\frac{1}{3}$ pitch roof, the base is ________ and the rise is ________.

5. In building construction, the 30′, 40′, 50′ method of triangulation is used extensively in the layout of foundation building lines. Complete the following sequence of sets of numbers:

3	4	5
9	12	15
12	16	__
21	__	35
30	40	50

6. On your notepaper, draw two horizontal lines $3\frac{1}{2}''$ apart, representing rafter stock. Draw the position of a steel tape positioned to mark the center line of the stock. Now, show how you would position the tape to pencil spot for three equal width parts.

5

Common Rafter Layout with the Framing Square

There is much evidence that prehistoric races knew the use of the square. Numerous examples of stones that were dressed square by very early races may be seen on all continents.

It is recorded that a square was found in a kit of carpenter's tools discovered in an Egyptian tomb of some four or five thousand years ago.

The first steel square was probably forged by a blacksmith.

The precision-made framing square[1] with its rafter tables and fine graduations is of comparatively recent origin and is the outcome of the advancing technological age.

1. The framing square

a. The body of the framing square is 2 inches wide and 24 inches long.

b. The tongue is $1\frac{1}{2}$ inches wide and 16 inches long. In some countries the tongue is 18 inches long.

[1]The steel framing square with imperial system measurements is referred to in this book as "traditional."

c. The tool is two sides of a right-angle triangle. It usually has the rafter tables and the name of the manufacturer on the face side.

d. There are eight edges on the framing square. Each edge is divided into inches and fractions of an inch. Readings of $\frac{1}{16}''$, $\frac{1}{8}''$, $\frac{1}{4}''$, $\frac{1}{2}''$, $\frac{1}{12}''$, $\frac{1}{10}''$, $\frac{1}{5}''$, or $\frac{1}{3}''$ may be made on one or more edges. The use to which such graduations may be put in carpentry layout is indicated in the following example.

A $\frac{1}{12}$ scale-drawing is to be made of a rectangular building measuring 12′–3″ × 20′–6″. By use of the framing square edge having $\frac{1}{12}''$ graduations, these measurements can be scaled down immediately to read $12\frac{3}{12}''$ × $20\frac{6}{12}''$. The drawing may be done quickly and may just as easily be read.

2. Rafter tables

a. Plate II, page 42, shows in broken form a framing square for easy reading of the rafter tables during the study of the text.

First line: length of common rafter per foor run
Second line: length of hip or valley per foot run
Third line: difference in length of jacks 16″ centers
Fourth line: difference in length of jacks 24″ centers
Fifth line: side cut of jack rafter use
Sixth line: side cut of hip or valley use

b. In this chapter we will concern ourselves with the first line—the length of common rafter per foot run.

c. It must be carefully noted that the length of rafter per foot run refers to the number of feet of run in a common rafter (CR). The length of hip or valley per foot run also refers to the number of feet or run in the CR.

3. Relative pitches

a. Figure 5.1 shows graphically the relative pitches as used on the framing square. These are based on a unit of 12″ or 1′–0″ of run. The relative pitches are also shown in Table 5.1.

b. The framing square and the rafter tables have been designed and developed on the basis of 12″ run and 24″ span.

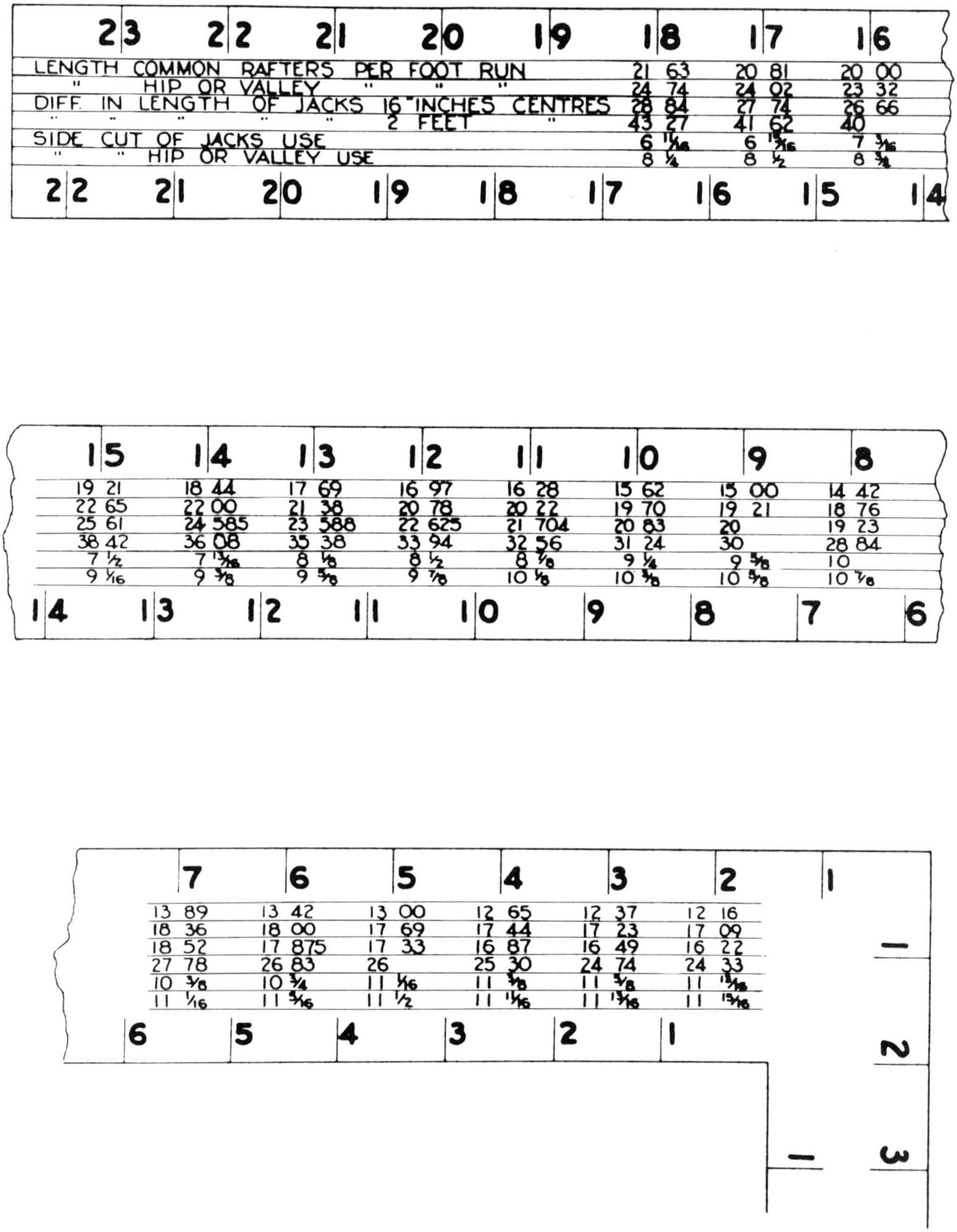

PLATE II. The Traditional Framing Square

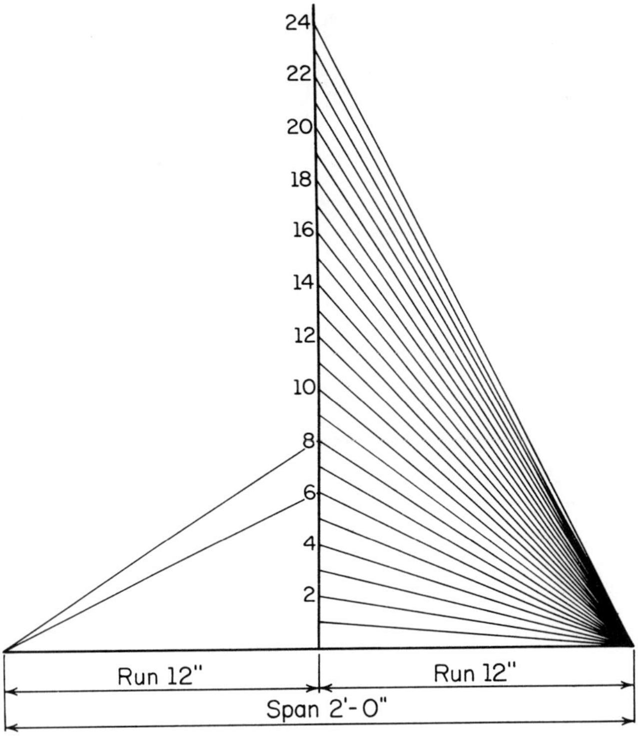

FIGURE 5.1. Relative pitches on the framing square.

c. Two common pitches are completely drawn in:

6″ rise per foot run—for a $\frac{1}{4}$ pitch

8″ rise per foot run—for a $\frac{1}{3}$ pitch

d. The slope of a roof is its pitch.

e. The pitch of a roof is rise divided by span, thus:

$$\frac{\text{Rise over}}{\text{Span}} \text{ (Remember if you can)}$$

Example 1: The pitch may be

$$\frac{\text{Rise (6)}}{\text{Span (24)}} \text{ or } \frac{1}{4}$$

Example 2: The pitch may be

$$\frac{\text{Rise (8)}}{\text{Span (24)}} \text{ or } \frac{1}{3}$$

TABLE 5.1

Relative Pitches

		Framing Square Read	
Rise over Span	*Pitch*	*Rise*	*Run*
$\frac{2}{24}$	$\frac{1}{12}$	2	12
$\frac{3}{24}$	$\frac{1}{8}$	3	12
$\frac{4}{24}$	$\frac{1}{6}$	4	12
$\frac{5}{24}$	$\frac{5}{24}$	5	12
$\frac{6}{24}$	$\frac{1}{4}$	6	12*
$\frac{8}{24}$	$\frac{1}{3}$	8	12*
$\frac{10}{24}$	$\frac{5}{12}$	10	12
$\frac{12}{24}$	$\frac{1}{2}$	12	12

*As drawn in at Figure 5.1.

4. Rise per foot run when the span and the rise are given

a. As has been stated, the framing square has been designed and developed as a unit of 12″ or 1′–0″ of run. The first consideration in understanding the rafter tables is the ability to find the number of inches of rise to every single foot of run in the CR.

b. Assume that a gable roof is to be erected on a building with a span of 36′–0″ and a rise of 9′–0″.

c. To find the number of inches of rise for every single foot of run of the CR, simply divide the total number of inches of rise in the roof by the actual number of feet of run of the CR. See Figure 5.2.

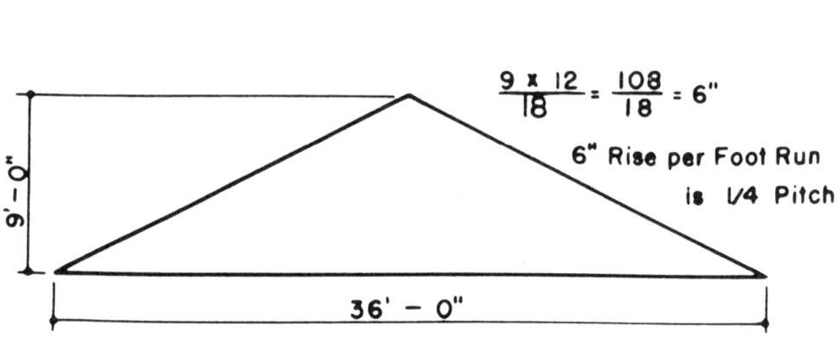

FIGURE 5.2.

5. Rise per foot run when the pitch is given

Example 1: Assume that a $\frac{1}{4}$ pitch roof is to be erected on a building, the span of which is 36′–0″.

a. To find the rise, multiply the pitch by the span thus: $\frac{1}{4} \times$ 36′–0″ equals 9′–0″. This is the rise of the roof over the span of 36′–0″.

b. The number of inches of rise is $9 \times 12 = 108''$.

c. Divide the number of inches of rise in the roof by the number of feet in the run of the CR.

d. Rise 9′–0″ equals 108″. Divide by the run of 18. $\frac{108}{18}$ equals 6″, the rise for every single foot of run of the CR. A $\frac{1}{4}$ pitch roof has a rise of 6″ for each foot of run of the CR. See the table of relative pitches on page 44.

Example 2: Assume that a $\frac{1}{3}$ pitch roof is to be erected on a building, the span of which is 36′–0″.

a. To find the rise, multiply the pitch by the span. $\frac{1}{3} \times$ 36′–0″ = 12′–0″. This is the rise of the roof over the span of 36′–0″.

b. Find the number of inches in the rise: $12 \times 12 = 144''$.

c. Divide the number of inches of rise of the roof by the number of feet in the run of the CR. Thus, $\frac{144}{18} = 8''$, the rise for every single foot of run of the CR. A $\frac{1}{3}$ pitch roof has a rise of 8″ for each foot of run of the CR. See the table of relative pitches, page 44.

d. Most architects and house designers show the rise per foot run on the drawings as in Figure 5.3.

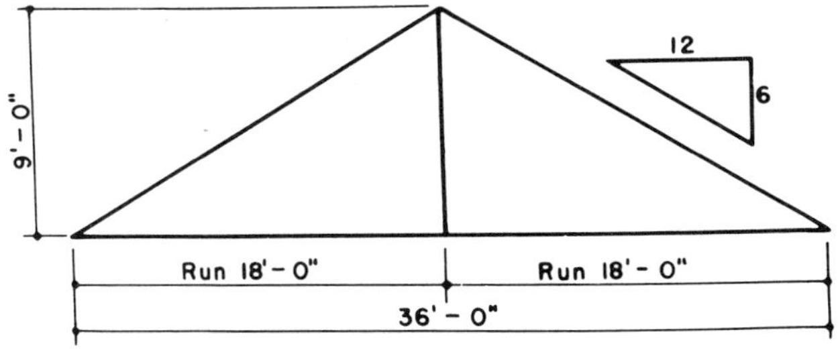

FIGURE 5.3.

e. The rise per foot run may be found when the span and the rise are given, when the pitch is stated, or when the pitch is given directly on the drawings as in Figure 5.3.

6. Line length per foot run by the graphic method

Example 1: Take a tape and measure very carefully across a framing square from the 12″ mark on the body to the 6″ mark on the tongue. It will be seen that the measurement is $13\frac{3}{8}''$ to the nearest one eighth of an inch. See Figure 5.4.

a. Turn the framing square over, and on the first line of the rafter tables (Plate II, page 42) read under the 6″ mark:

Length of Common Rafters per Foot Run . . . 13.42″

It will be seen that 13.42 is also $13\frac{3}{8}''$ to the nearest one eighth of an inch.

b. The line length per foot run for a $\frac{1}{4}$ pitch roof is 13.42″.

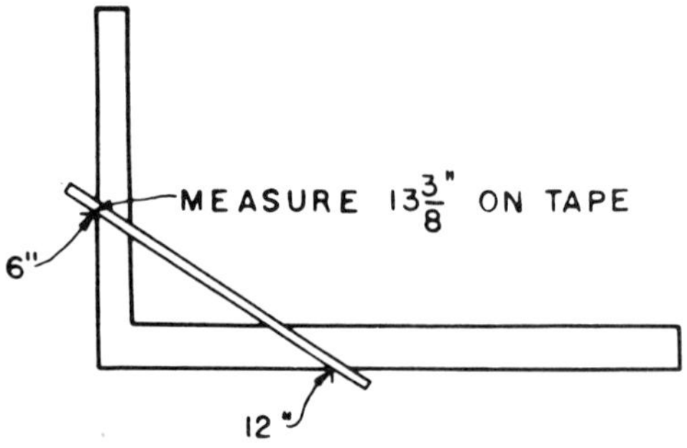

FIGURE 5.4.

Example 2: Measure across the framing square from the 12″ mark on the body to the 8″ mark on the tongue. Read $14\frac{3}{8}''$ to the nearest one eighth of an inch. Read on the rafter tables under the 8″ mark on the first line 14.42″. It will be seen that 14.42″ is $14\frac{3}{8}''$ to the nearest one eighth of an inch.

a. The LL per foot run for a $\frac{1}{3}$ pitch roof is 14.42″.

b. The rafter tables are ready reckoners for reading the line lengths per foot run of rafters for different pitches. The framing square also contains other information in ready reckoner form. In such ways man is continually making easily read tables of accepted "truths" and thus frees himself to accomplish more operations in less time.

7. Line length of common rafter by the rafter tables

Example 1: Assume that a $\frac{1}{4}$ pitch gable roof is to be erected on a building with a span of 36′–0″.

a. Find the rise per foot run.

b. Read on the first line of the rafter tables under the 6″ mark. (See rafter tables on the framing square—Plate II, page 42.)

Length of Common Rafters per Foot Run . . . 13.42

c. The figure 13.42″ is the ready-reckoned hypotenuse of a right-angle triangle with a base (run) of 12″ and a height (rise) of 6″.

d. The reading on the framing square for a $\frac{1}{4}$ pitch roof is 13.42. The length of the CR for every foot of run of the CR is 13.42″.

e. To find the LL of the CR for this roof, multiply the length of the CR per foot run (13.42″) by the actual number of feet of run in the CR (18′–0″).

f. 13.42″ × 18 = 241.46″. Reduce 241.46 inches to feet and inches, and express the result to the nearest one eighth inch. All rafters should be laid out to a tolerance of $\pm\frac{1}{8}''$.

g. Thus

$$\frac{12\overline{)241.56''}}{20' \text{ and } 1.56''} = 20'\text{–}0'' + 1'' + .56''$$

20′ 1″

To change .56″ to eighths of an inch:

$$.56'' = \frac{?}{8} \qquad .56'' \times 8 = 4.48''$$

$$.56'' = \frac{4.48''}{8}$$

$$.56'' = \tfrac{4}{8}'' \qquad \text{to the nearest } \tfrac{1}{8}''$$

Similarly, to change .5″ to eighths of an inch:

$$.5'' = \frac{?}{8} \qquad .5'' \times 8 = 4.0$$

$$.5'' = \tfrac{4}{8}''$$

To change .5″ to sixteenths of an inch:

$$.5'' = \frac{?}{16} \qquad .5'' \times 16 = 8.0''$$

$$.5'' = \tfrac{8}{16}''$$

Rule: *Multiply the decimal fraction of an inch by the denominator required.* This will give the numerator required.

(given decimal) times (given denominator) = required numerator

$$.5 \qquad \times \qquad 8 \qquad = \qquad 4$$

Example 2: Assume that a $\frac{1}{3}$ pitch gable roof is to be erected on a building with a 36′–0″ span.

a. Find the rise per foot of run.
b. Read under the 8″ mark on the rafter tables, page 42:

   Length of Common Rafters per Foot Run . . . 14.42

c. Multiply 14.42″ by the number of feet of run in the CR.
d. 14.42 × 18 expressed in feet and inches to the nearest one eighth of an inch is 21′–7$\frac{5}{8}$″.
e. The following set of decimal equivalents should be committed to memory:

Decimal Equivalents		
$\frac{1}{8}''$	equals	.125″
$\frac{1}{4}''$	equals	.25″
$\frac{3}{8}''$	equals	.375″
$\frac{1}{2}''$	equals	.50″
$\frac{5}{8}''$	equals	.625″
$\frac{3}{4}''$	equals	.75″
$\frac{7}{8}''$	equals	.85″

8. Layout of a common rafter with the framing square

Figure 5.5 shows a framing square with the stair gauges set for the layout of the CR for a $\frac{1}{3}$ pitch roof. The square is shown in position for the marking of the plumb cut. Stair gauges may be purchased in the tool section of most hardware stores, and they are almost indispensable in rafter framing with the square.

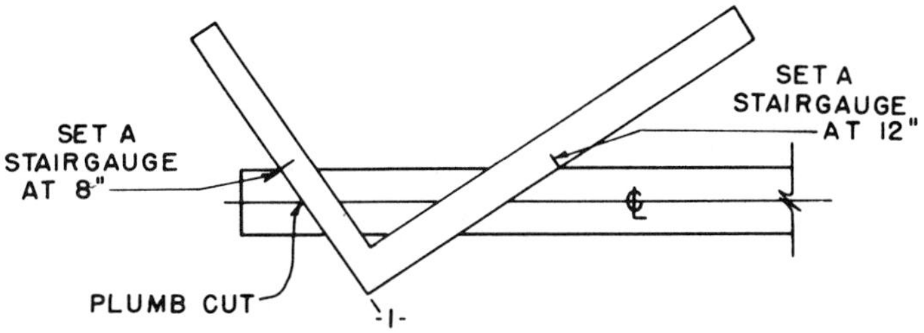

FIGURE 5.5. Stair gauges set for $\frac{1}{3}$ pitch.

It should be very carefully noted that all positions of the framing square for obtaining plumb and level cuts, as well as the line numbering, are the same as in Chapter 4 in the graphic method of layout.

Specifications

a. Assume that a $\frac{1}{3}$ pitch gable roof with a 1 × 6 ridge is to be erected on a building with a span of 36′–0″. The overhang is 16 inches.

b. Select a straight, clean piece of 2 × 4 of suitable length. Examine the stock for the crown side by holding it with the wide face flat in the palm of the hand and sighting along the narrow face. Place the stock on the sawhorses with the crown side away from you.

c. Draw a ℄ on the wide face of the 2 × 4.

d. Set the stair gauges for a $\frac{1}{3}$ pitch roof by attaching one gauge to the 8″ mark on the tongue of the square and the other on the 12″ mark on the body of the square. Be sure that both gauges are placed on the outside edges of the square.

e. The imaginary hypotenuse of the right-angle triangle is between the stair gauges.

Plumb cut

a. Place the square with the plumb cut to the left and the stair gauges snug to the back of the rafter stock as in Figure 5.5.

b. Draw in the plumb cut as in Figure 5.6. Study the complete CR layout in Plate I, page 29.

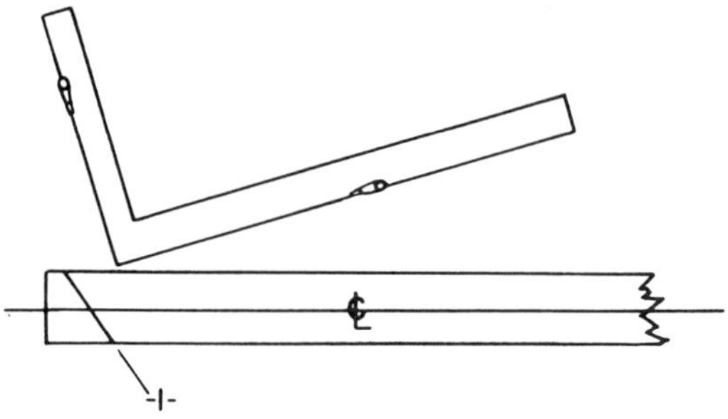

FIGURE 5.6.

Line length

a. A little to the right of the plumb cut and actually on the CR, work out mathematically the LL of the CR. Multiply 14.42 (as read on the rafter tables, page 42) by the actual number of feet of run in the CR, that is, $14.42'' \times 18 = 21'\text{-}7\frac{5}{8}''$.

b. With a steel tape, measure on the back of the rafter stock $21'\text{-}7\frac{5}{8}''$.

Checking the LL of the CR by the step-off method

a. An excellent additional checking method, and one that should always be used by carpenters is the step-off method. Place the stair gauges for the pitch of the roof. Set the framing square at Line 1 as in Figure 5.5.

Mark a short line on the run of the square to intersect with the top edge of the rafter stock.

Slip the square along the rafter stock until the plumb marking line intersects with the run line already drawn.

Repeat this operation until the number of steps taken is the number of feet of run in the CR.

The last application at the run end of the framing square should reach to within $\frac{3}{8}''$ of the LL as determined from the rafter tables.

When the step-off method checks to such a tolerance, accept the LL as reckoned from the rafter tables on the framing square and as measured by the tape.

The carpenter should use the step-off method for checking only—not for the original layout.

b. It must be carefully noted that, with the graphic method of layout, the pitch board run is $\frac{1}{12}$ the length of the run of the CR. The number of steps to be taken would therefore be twelve.

When using the framing square, the run is always 12″. The number of steps to be taken would therefore always be the number of times that 12″ (or one foot) is contained in the run of the CR—in this case 18 times for a run of 18 feet.

c. With the plumb cut of the framing square to the left, draw in another plumb cut at the 21′–7$\frac{5}{8}$″ mark as in Figure 5.7.

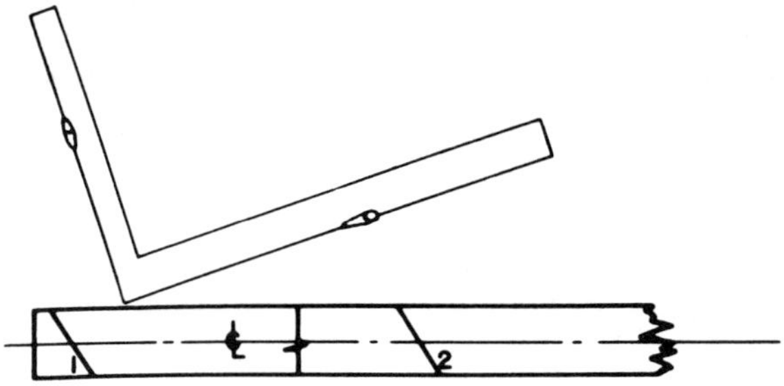

FIGURE 5.7.

The birdsmouth

a. Slip the framing square to the left until the level cut may be drawn to intersect with No. 2 plumb line at the ℄ as in Figure 5.8.

b. The meeting of Lines 2 and 3 at the ℄ lays out the birdsmouth. Lines 2 and 3 should be squared under the bottom edge of the rafter. After the birdsmouth has been removed, the level and plumb cuts fit to the wall plate.

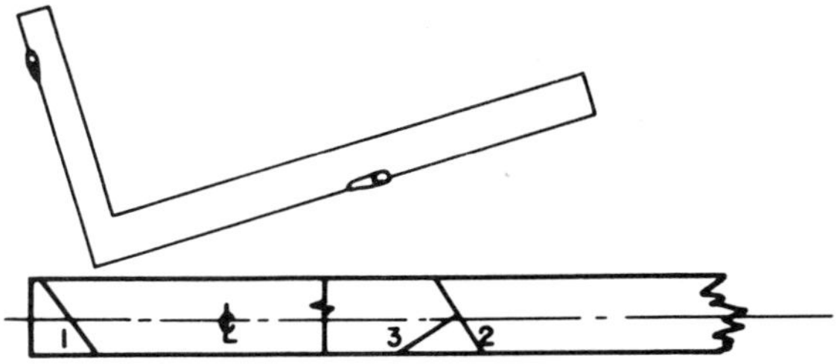

FIGURE 5.8.

Shortening of the CR

a. Referring to the specifications on page 49, you will see that the ridge board is 1 × 6. The actual size of this member would be $\frac{3}{4}'' \times 5\frac{1}{2}''$.

b. It is possible to erect the roof without using a ridge board, but a ridge board simplifies the actual erection of a roof. Apart from any other reason, an economy in labor time is effected.

c. It will be seen in Figure 5.9 that the introduction of a ridge board necessitates the shortening of the CRs. The amount of shortening necessary on each CR is half the thickness of the ridge board, measured at right angles to the plumb cut.

d. The shortening is measured at right angles (that is, on the run) to the plumb Line 1.

e. Place the framing square with the plumb marking line to the left on Line 1. Slip the square so that it is $\frac{3}{8}''$ back at right angles to Line 1 as in Figure 5.10. The tape may be seen placed in the position of checking the shortening. Draw in Line 4 as in Figure 5.10.

f. Square over Line 4 to the back of the rafter. This is the cutting line.

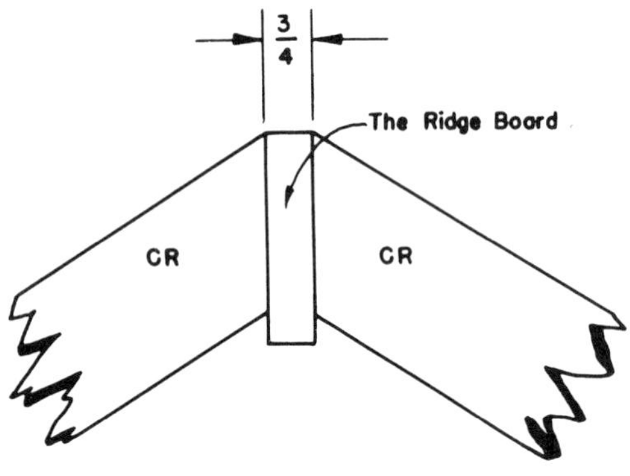

FIGURE 5.9. Shortening of CR at ridge board.

Tail LL of a CR

The tail length is reckoned in a manner similar to the LL of the CR.

a. Referring to the specifications, you will recall that the overhang is 16″ and that the pitch is $\frac{1}{3}$.

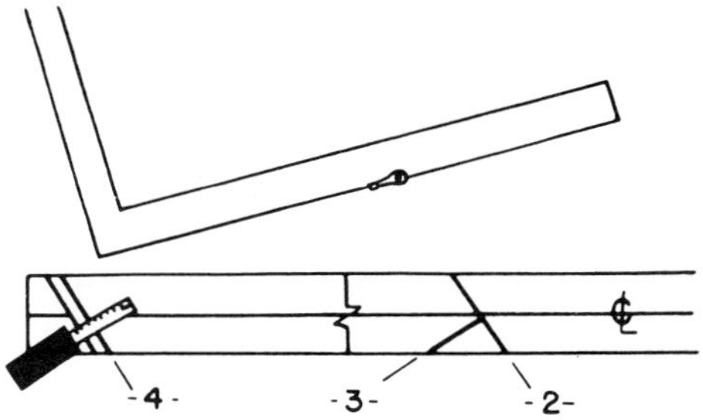

FIGURE 5.10.

b. On the CR, just to the right of Line 2, make the calculations for the LL of the tail for the overhang.

c. Reading on the framing square for a $\frac{1}{3}$ pitch roof, we see that the rise per foot of run is 8″. On the first line of the rafter tables, page 42, read 14.42 under 8″.

d. To find the LL of the tail of the CR, multiply the length of the rafter per foot run, 14.42, by the actual number of feet of run in the tail of the CR. This is 16″, or one and one third feet of run.

e. $14.42'' \times 1\frac{1}{3} = 19\frac{1}{4}''$, the LL of the tail. The tape may be seen in the position for checking the LL of the tail as in Figure 5.11 Draw in Line 5.

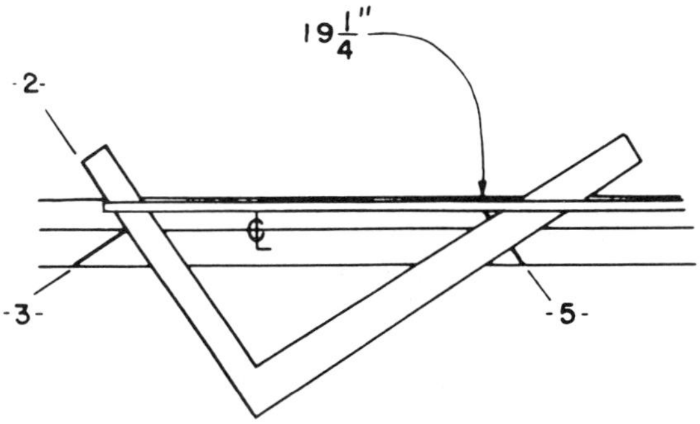

FIGURE 5.11.

Checking the LL of the tail by the step-off method

With the plumb marking side of the framing square to the left and placed on Line 2, measure 1′–0″ on the run. Slip the square to the right and measure a further 4″ as in Figure 5.12. The 12″ and the 4″ together equal 16″, the given overhang measured on the run.

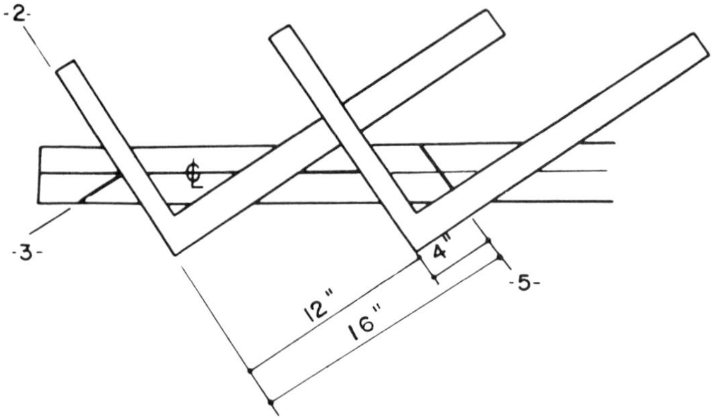

FIGURE 5.12.

It will be seen that the procedure for the layout of the LL of the tail of a CR from Line 2 is exactly the same as for the layout of the LL for the CR.

Plumb cut of the tail of the CR

a. When the rafter stock is of just sufficient length for the CR, it is necessary to turn the square over to draw Line 5. The square should be turned over so that the heel is away from the carpenter. The plumb side of the square should be to the right and the stair gauges should fit snug against the bottom of the rafter stock.
b. Figure 5.13 shows the measured LL as calculated from the rafter tables, and as checked by stepping off with the framing square from Line 2 to Line 5.

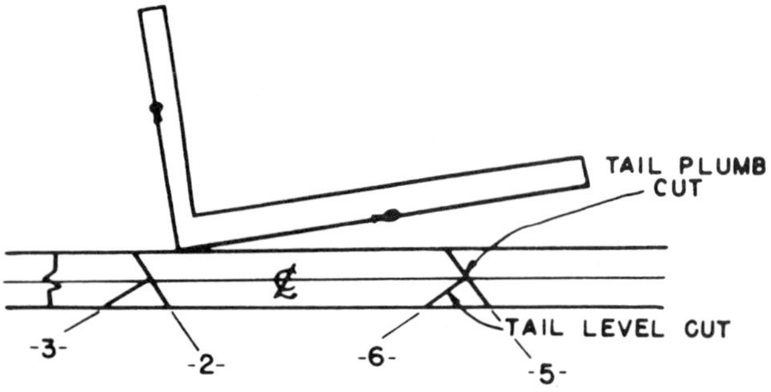

FIGURE 5.13.

Level cut of the tail of the CR

a. Where a plumb and level cut is to be made at the end of a CR, place the square with the plumb side to the left. Adjust the position of the square until it will intersect with Line 5 at the ℄.

b. The complete CR rafter layout with the line numbering as in this chapter is shown in Plate I, page 29.

Actual sawing lines of the CR

The actual cutting lines are on:

Line 4 for the shortened plumb cut

Line 3 and the lower part of Line 2 for the birdsmouth

Lines 5 and 6 for the plumb and the level tailcut.

Review Questions

1. Fill in the blanks of the questions:

   The body of the traditional steel framing square is ______ long.
   The tongue of the traditional steel framing square is ______ long.
   The body of the traditional steel framing square is ______ wide.
   The tongue of the traditional steel framing square is ______ wide.
   The edges of the steel traditional framing square are divided into ______ and ______.

2. On the first line of the rafter tables is shown ______ ______ ______.
   On the second line of the rafter tables is shown ______ ______ ______.
   On the third line of the rafter tables is shown ______ ______ ______ ______ ______.

3. The traditional steel framing square was designed to use ______ inches as the unit of the run of the roof.

4. If a ⅓ pitch roof has a span of 36′–0″, its rise is ______ feet, or ______ inches.
If a ¼ pitch roof has a span of 36′–0″, its rise is ______ feet, or ______ inches.
The number of inches of rise for a ½ pitch roof with a span of 36′–0″ is ______.

5. Complete the following table of relative pitches:

Rise over Span	*Pitch*	*Rise*	*Run*
$\frac{2}{24}$	$\frac{1}{12}$	2″	12″
$\frac{3}{24}$	$\frac{1}{8}$	3″	12″
$\frac{4}{24}$	—	—	—
$\frac{5}{24}$	$\frac{5}{24}$	5″	12″
$\frac{6}{24}$	—	—	—
$\frac{8}{24}$	—	—	—
$\frac{10}{24}$	$\frac{5}{12}$	—	—
$\frac{12}{24}$	—	—	—

6. Place a steel framing square before you with the rafter tables uppermost. Measure across the square from the 8″ mark on the tongue to the 12″ mark on the body and record it here. ______
Read on the framing-square rafter tables the measure given immediately under the 8″ mark; compare the tape measure recording with that given on the rafter tables and record it here. ______
What is the difference between the two readings?
______
______.

6

Common Rafter Layout with the Frederickson Metric Framing Square*

Where the measurement system in use is changing from imperial to metric, metric framing squares are available for rafter layouts. Two of the metric squares have tables based on different unit runs: one of 250 mm and one of 1000 mm (1m). The interested builder needs to decide which square to use.

Rafter tables of two metric squares are shown on Plates III and IV:

Plate III The Frederickson Metric Framing Square (CR unit run of 250 mm)—Chapter 6, page 59.

Plate IV The Stanley Metric Steel Square (CR unit run of 1000 mm)—Chapter 7, page 66.

1. The Frederickson Metric Framing Square, CR unit run of 250 mm

a. The body of the square is 50 mm wide and 600 mm long.

b. The tongue is 38 mm wide and 400 mm long.

---

*(Unit run of 250 mm)

c. The square, as a tool, is two sides of a right-angle triangle. It has rafter tables and the name of the manufacturer on the face side.

d. There are eight edges on the square, all graduated in millimeters (mm). Readings of 1 mm, 5 mm, 10 mm, 100 mm, and their multiples may be made on all edges.

The use to which such graduations may be put in carpentry layout is indicated in the following example:

Example: A one-tenth ($\frac{1}{10}$) scale drawing is to be made of a rectangular building measuring 3100 mm × 5800 mm. By use of the framing square, these measurements can be scaled down immediately to 310 mm × 580 mm. The drawing may be done quickly and be easily dimensioned. Division by 10 is readily accomplished.

2. Rafter tables

a. Plate III shows in broken form the Frederickson Metric Framing Square with rafter tables based on 250 mm as the unit run of the common rafter (CR).

First line:	length of CR per unit run of 250 mm
Second line:	length of HR per unit run of 353.5 mm
Third line:	difference in JRs at 400 mm OC
Fourth line:	side cut of hip or valley rafters
Fifth line:	side cut of JRs
Sixth line:	angle of CR at the plate

b. In this chapter discussion will be limited to the length of the CR per metric unit of a run of 250 mm.

3. Relative pitches and slopes for a unit run of 250 mm

a. Figure 6.1 shows graphically some relative pitches and one slope triangle with respect to a framing square with rafter tables based on a unit run of 250 mm and a span of 500 mm.

b. The rafter tables are considered to be developed on the basis of a 250 mm run, which may or may not imply a 500 mm span.

c. In Figure 6.1 two pitches are completely drawn:

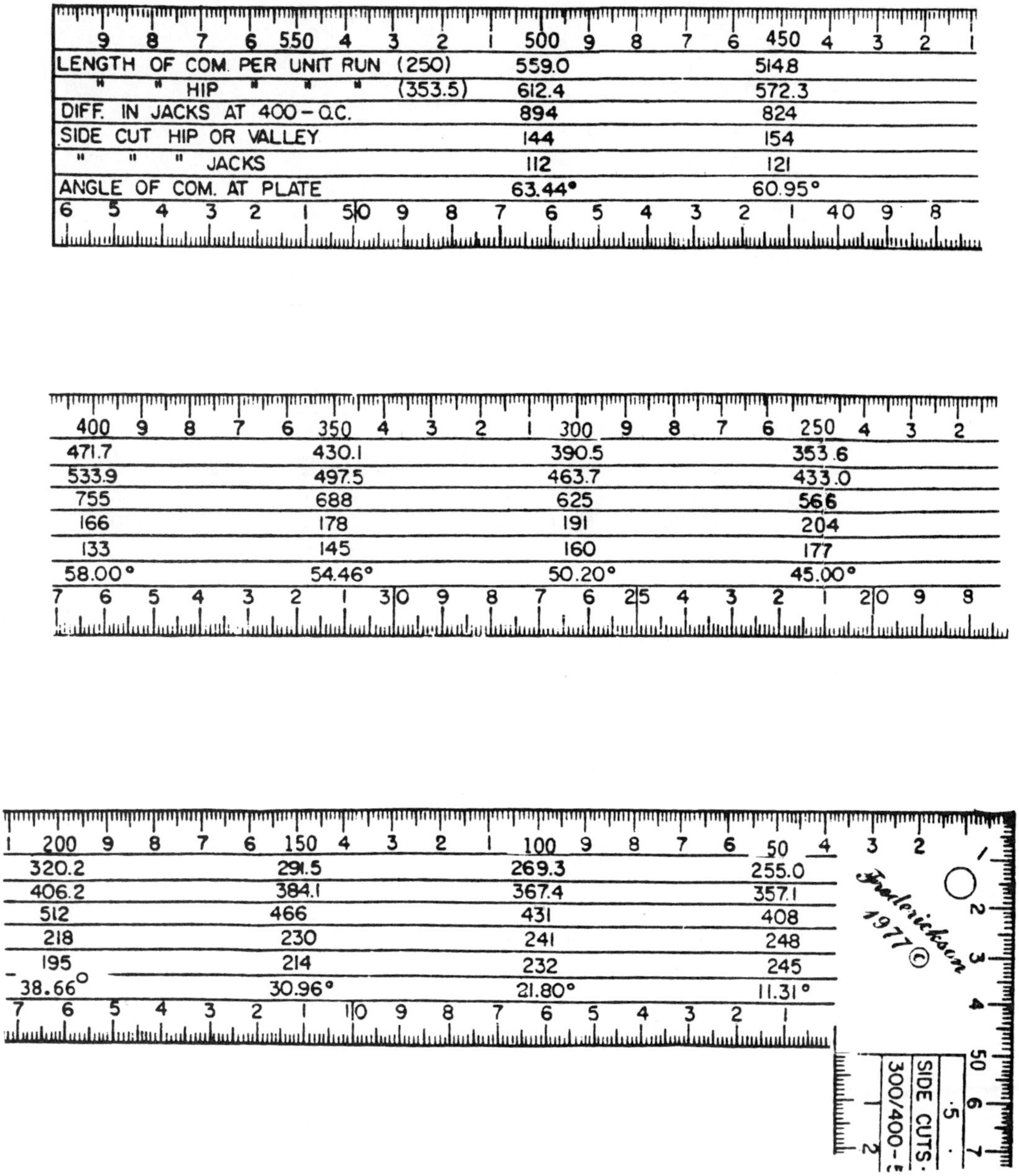

PLATE III. The Frederickson Metric Square[1], with a Unit Run of 250 mm.

---

[1]Through the kind permission of Paul H. Frederickson, 8878–213 Street, R.R. 10, Langley, B.C. Canada V3A 6X5.

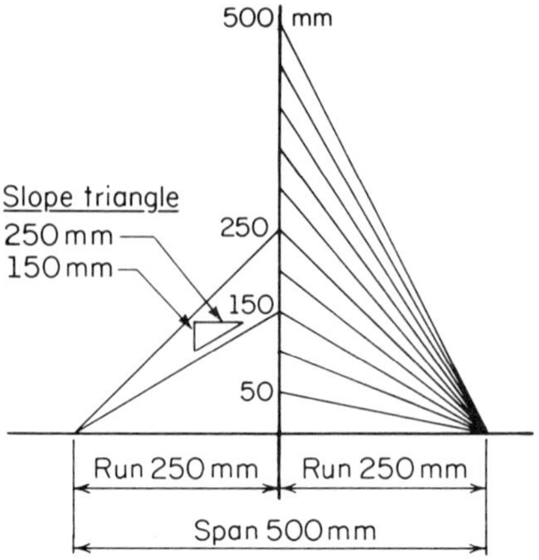

FIGURE 6.1. Pitches and one slope triangle for a 250 mm run.

250 mm rise for a run of 250 mm for $\frac{1}{2}$ (0.5) pitch
150 mm rise for a run of 250 mm for $\frac{3}{10}$ (0.3) pitch

For the latter, a slope triangle is also drawn.

d. The pitch of a roof may be indicated on a drawing as a slope triangle:  . The two sides of the triangle represent the unit rise and the unit run. The sides are dimensioned to help avoid errors in communication between the planner and the builder. For example, in Figure 6.1 and in more detail in Figure 6.2 which follows, the pitch $\frac{3}{10}$ (0.3) is represented by the slope triangle dimensioned 150 mm rise and 250 mm run, for an implied unit span measurement of 500 mm:

$$\frac{150}{500} = \frac{3}{10} \ (0.3)$$

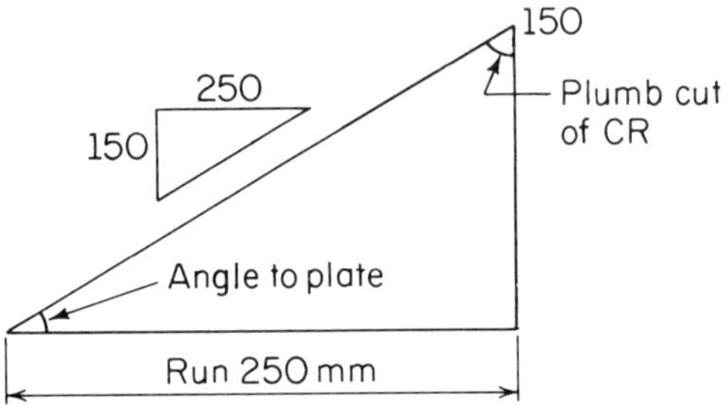

FIGURE 6.2. Dimensioning a slope triangle.

Observe that two similar triangles of different sizes (one upside down) are dimensioned the same. The slope triangle is not necessarily to scale and is merely a useful device for indicating the slope of a roof without reference to the span.

e. Where the span is an element in the description of the pitch, as in Figure 6.1, then the pitch is defined as the rise divided by the span (See first chapters).

Example 1:

$$\frac{\text{(Rise)}\ 250\text{ mm}}{\text{(Span)}\ 500\text{ mm}} \text{ or } \tfrac{1}{2} \text{ or } 0.5$$

Example 2:

$$\frac{\text{(Rise)}\ 150\text{ mm}}{\text{(Span)}\ 500\text{ mm}} \text{ or } \tfrac{3}{10} \text{ or } 0.3$$

Table 6.1 which follows relates some roof pitches to the Frederickson Metric Framing Square settings.

TABLE 6.1

Relative Pitches* for Unit Run of 250 mm (Frederickson Metric Framing Square)

		Framing Square Setting	
Rise over Span	*Pitch*	*Rise*	*Run*
500/500	1.0 (Full)	500	250
450/500	0.9	450	250
400/500	0.8	400	250
350/500	0.7	350	250
300/500	0.6	300	250
250/500	0.5	250	250
200/500	0.4	200	250
150/500	0.3	150	250
100/500	0.2	100	250
50/500	0.1	50	250

*Only ten pitches are shown in the table. There are many others. One, for example, is 125/500 or $\frac{1}{4}$ pitch.

4. Reading the line lengths of common rafters using the Frederickson Metric Framing Square, with tables based on a unit run of 250 mm:

a. Figure 6.3 indicates three readings of LL for a common unit run of 250 mm and three rise measurements: 200 mm, 250 mm, and 500 mm. The first line of the rafter table, the length of common rafter per a unit run of 250 mm, indicates the following LLs: (1) 320.2mm. (2) 353.6 mm. (3) 559.0 mm.

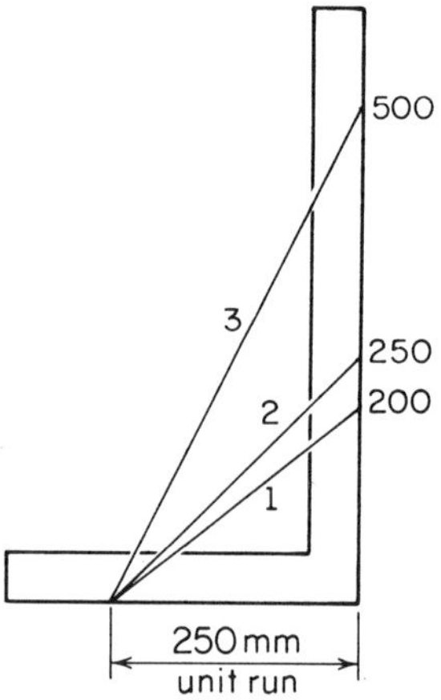

FIGURE 6.3. Line lengths for a unit run of 250 mm.

b. Figure 6.4 provides, in addition to readings of the LL of the CRs, the angle to the plate for a CR, the plumb cut angle of the same CR, and one slope triangle.

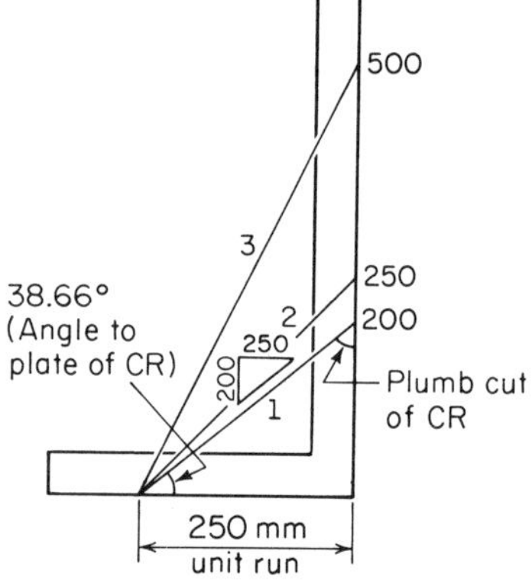

FIGURE 6.4. LL line and the slope triangle.

5. Pitch and slope triangle relationships in tabular form

Table 6.2 is a summary statement for the Frederickson Metric Framing Square, based on a CR unit run of 250 mm. Table 6.2 differs from Table 6.1 in that additional data is given in columns designated "angle to plate", and "slope triangle."

TABLE 6.2

Roof Framing Data with Respect to the Frederickson Metric Framing Square

Unit Run of 250 mm					
Angle to Plate (°)	*Slope Triangle Run / Rise* *Rise Run*	*Unit Run*	*Span*	*Rise*	*Pitch*
63.44	500 in 250	250	500	500	1.0
60.95	450 in 250	250	500	450	0.9
58.00	400 in 250	250	500	400	0.8
54.46	350 in 250	250	500	350	0.7
50.20	300 in 250	250	500	300	0.6
45.00	250 in 250	250	500	250	0.5
38.66	200 in 250	250	500	200	0.4
30.96	150 in 250	250	500	150	0.3
21.80	100 in 250	250	500	100	0.2
11.31	50 in 250	250	500	50	0.1

Data of Table 6.2 will be combined in Chapter 8 with similar data from Chapter 7 with respect to the Stanley Metric Steel Square.

7

Common Rafter Layout with the Stanley Metric Steel Square*

The tables of the Stanley Metric Steel Square are based on a unit run of 1000 mm (1m).

1. The Stanley Metric Steel Square, unit run of 1000 mm (1m)

a. The body of the square is 50 mm wide and 600 mm long.
b. The tongue is 40 mm wide and 400 mm long.
c. The rafter tables and the name of the manufacturer are on the face side.
d. There are eight edges on the square, all edges being graduated in 2 millimeter (2 mm) spaces. Readings of 2 mm, 10 mm, and 100 mm and their multiples may be made on all edges.
e. One-tenth ($\frac{1}{10}$) scale drawings may be readily accomplished.

*(Unit Run of 1000 mm)

2. Rafter tables

a. Plate IV shows in broken form the Stanley Metric Steel Square, with rafter tables based on 1000 mm (1m) as the unit of run of the CR.

First line:	mm rise per meter of run
Second line:	mm length of CR per meter of run
Third line:	mm length of hip or valley per meter of run
Fourth line:	mm difference in length of jacks, 400 mm OC
Fifth line:	mm difference in length of jacks, 600 mm OC
Sixth line:	side cut of jacks, use opposite 200 mm line
Seventh line:	side cut of hip or valley, use opposite 200 mm line.

b. In this chapter discussion will be limited to the length of the CR per metric unit of run of 1000 mm (1m).

3. Relative pitches for a unit run of 1000 mm (1m)

a. Figure 7.1 on page 67 shows graphically some relative pitches with respect to a framing square with rafter tables based on a unit run of 1000 mm (1m).

b. In Figure 7.1 three pitches are indicated with heavy lines:

1. $\frac{\text{(Rise)} \quad 600\text{ mm}}{\text{(Span)} \quad 2000\text{ mm}}$ or $\frac{3}{10}$ or 0.3

2. $\frac{\text{(Rise)} \quad 1000\text{ mm}}{\text{(Span)} \quad 2000\text{ mm}}$ or $\frac{1}{2}$ or 0.5

3. $\frac{\text{(Rise)} \quad 2000\text{ mm}}{\text{(Span)} \quad 2000\text{ mm}}$ or full or 1.0

4. Relative slopes for a unit run of 1000 mm (1m)

a. Figure 7.2 shows graphically the set of measurements using slope triangles which would produce the same results as the pitch measurements of Figure 7.1.

b. Span is not shown in Figure 7.2 as the term is not always applicable, such as with a lean-to roof or a roof of two different slopes.

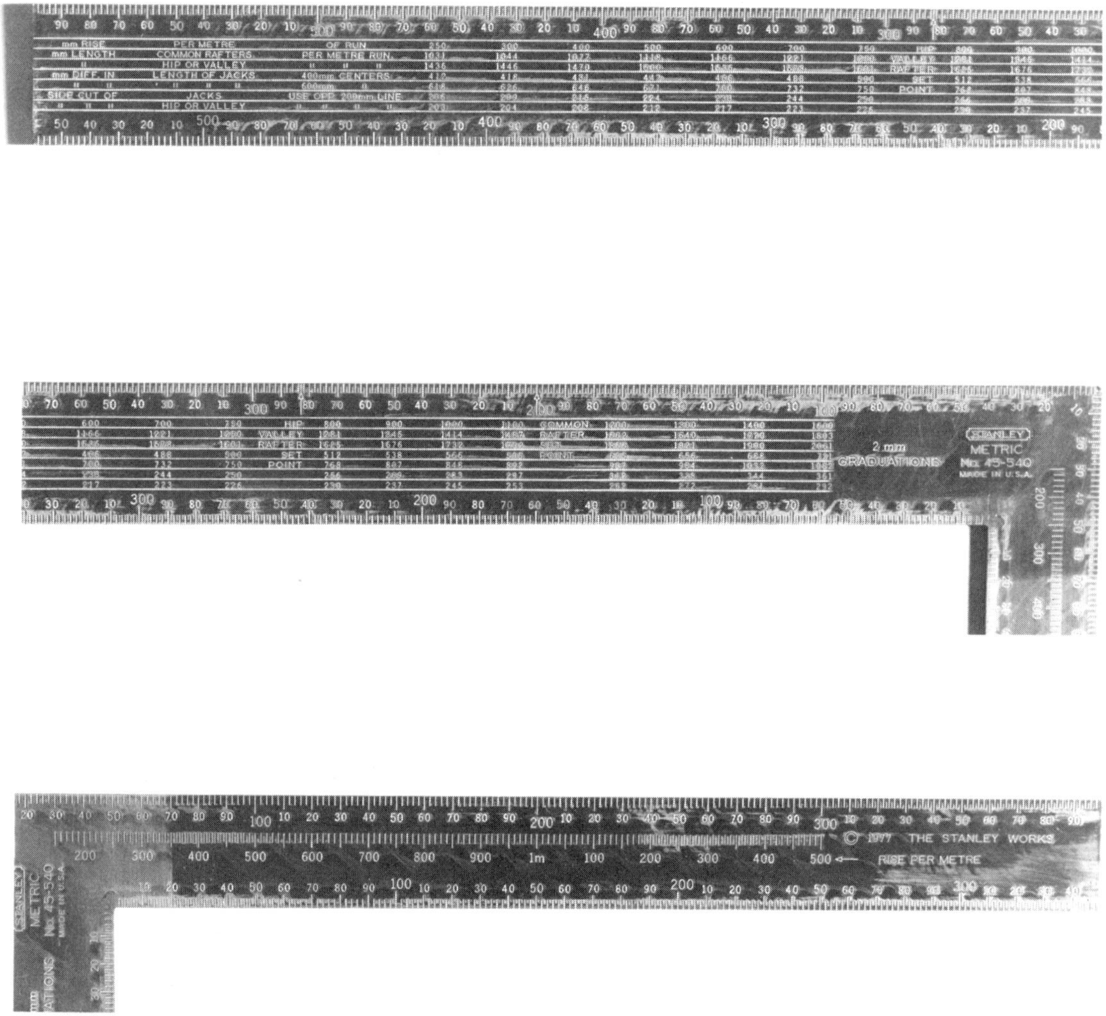

PLATE IV The Stanley Metric Steel Square No. 45–540 with a Unit Run of 1000 mm (1m).*

---

*Through the kind permission of the Stanley Tool Company, Inc., 600 Myrtle Street, New Britain, Connecticut, USA 06050

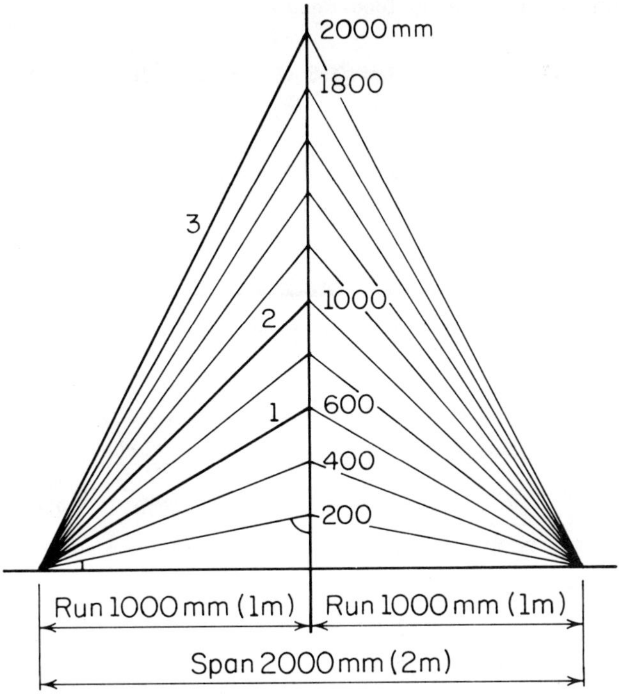

FIGURE 7.1. Some relative pitches for a CR unit run of 1000 mm (1m) and a span of 2000 mm (2m).

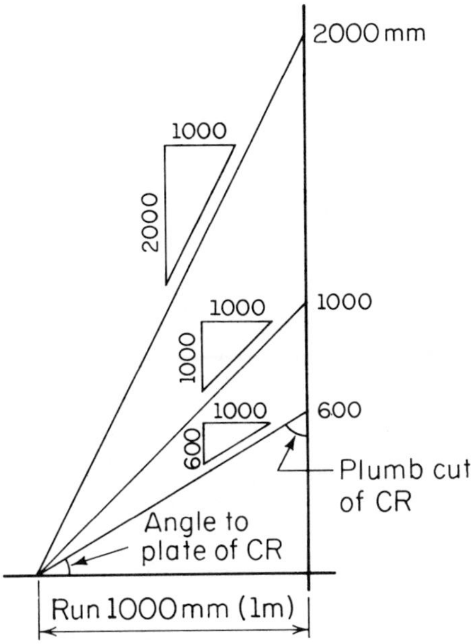

FIGURE 7.2. Some slope triangles for a CR unit run of 1000 mm (1m).

5. Pitch and slope triangle data relationships in tabular form

Table 7.1 is a summary statement for the Stanley Metric Steel Square based on a CR unit run of 1000 mm (1m).

TABLE 7.1

Roof Framing Data with Respect to the Stanley Metric Steel Square

	Unit Run of 1000 mm (1m)				
Pitch[1]	*Rise*	*Span*	*Unit Run*	*Slope Triangle Run / Rise — Rise Run*	*Angle to Plate (°)*[2]
1.0 (Full)	2000	2000	1000	2000 in 1000	
0.9	1800	2000	1000	1800 in 1000	
0.8	1600	2000	1000	1600 in 1000	
0.7	1400	2000	1000	1400 in 1000	54.5
0.6	1200	2000	1000	1200 in 1000	50.2
0.5 (Half)	1000	2000	1000	1000 in 1000	45.0
0.4	800	2000	1000	800 in 1000	38.7
0.3	600	2000	1000	600 in 1000	31.0
0.2	400	2000	1000	400 in 1000	21.8
0.1	200	2000	1000	200 in 1000	

[1]Only ten pitches are shown in the table. There are many others. One, for example, is 500/2000 or $\frac{1}{4}$ pitch.

[2]Only selected angular measurements are shown on the square.

6. The set point

a. As the body of the Stanley Metric Steel Square is 600 mm in length, the actual unit run of 1000 mm is not measurable on the square. Actual measurements are scaled down to one-fifth full size by the use of a set point at the 200 mm mark on the edge of the square. Thus, 200 mm is $\frac{1}{5}$ of 1000 mm.

b. The slope triangle helps one understand the use of the set point and the corresponding markings on the square which relate to the rafter tables of the square. See Figure 7.3.

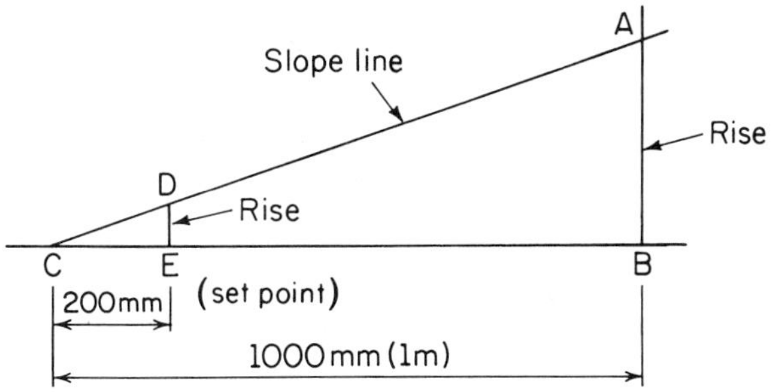

FIGURE 7.3. The set point and the slope triangle.

On the basis that triangles DEC and ABC are similar, the rise measurements DE and AB are in the same proportion as the corresponding run measurements EC and BC. Slopes DC and AC are of the same degree. Corresponding measurements of rise and slope line lengths are in the ratio of 200 mm to 1000 mm or of 1 to 5.

c. On the Stanley Metric Steel Square, the reference point for the unit run of the CR is 200 mm rather than 1000 mm. The scale is 1 to 5.

d. The set point is fixed. Rise measurements vary according to the wishes of the designer/carpenter.

e. Triangle DEC as applied to the square represents the slope triangle ABC.

Example: To draw the slope line for a rise of 800 mm and a run of 1000 mm, use the set point 200 and the 800 mark on the rise-per-meter markings on the tongue of the square, as in Figure 7.4. Extend the slope line CD to A and the unit run line CE to B.

f. The actual length of the slope line CA, for a unit run of 1000 mm, may be found in the rafter tables. Locate 800 on line 1, as the mm rise per meter of run. Under the 800 mark, find 1281 mm on line 2, the mm length of CRs per meter of run.

g. The slope line may be drawn by using the set point and the appropriate mark on the "rise-per-meter" scale.

h. The corresponding measurement of the LL of the CR may be found in the rafter tables.

i. The 800 mark of the "rise per meter" markings is at the 160 mm mark on the edge of the square. Use this 160 mm mark when drawing the slope line.

j. The mathematical relationships of the preceding example are described in the following equation:

$$\frac{160 \times 5}{200 \times 5} = \frac{800}{1000}$$

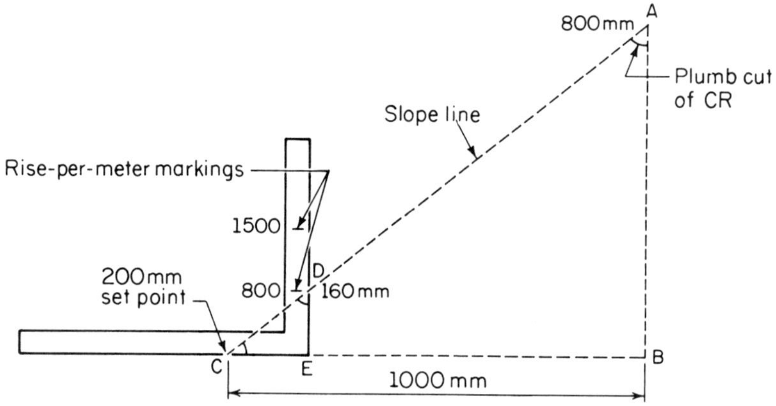

FIGURE 7.4. Relationship of the slope triangle to the square.

The data of Chapters 6 and 7 are related in Chapter 8. Some suggestions for further study are also provided.

8

Design Relationships of Two Metric Framing Squares: Frederickson and Stanley

As described in the two preceding chapters, the Frederickson Metric Framing Square is designed on the basis of a 250 mm unit run of the CR, and the Stanley Metric Steel Square is designed on the basis of a 1000 mm (1m) unit run of the CR, with a set point of 200 mm.

1. Comparison of the two unit-run systems of measurement

Figure 8.1 helps describe the relationship for a slope of 0.8.

$$\text{Slope of line 1: } \frac{200}{250} \text{ or } \frac{800}{1000}$$

$$\text{Slope of line 2: } \frac{160}{200} \text{ or } \frac{800}{1000}$$

2. Assumptions

a. If slope lines 1 and 2 of Figure 8.1 are parallel, they are of the same slope.

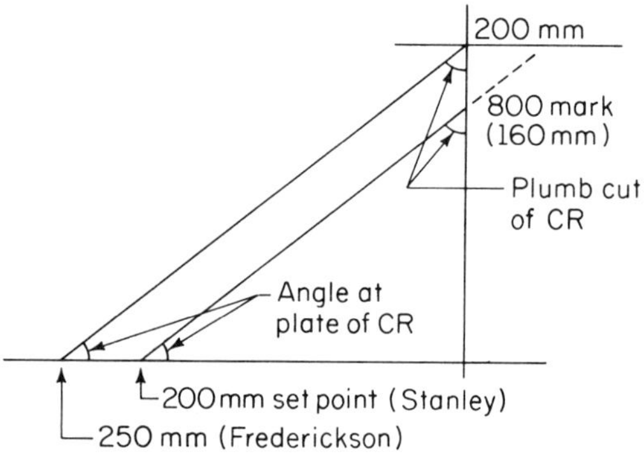

FIGURE 8.1. Unit run/set point relationship of two metric squares: Frederickson and Stanley, for a slope of 0.8.

b. If the right triangles formed by lines 1 and 2 are similar, then slope lines 1 and 2 are of the same slope.

3. Conclusions

a. Conditions of a) and b) above are satisfied with the mathematical statement describing the two right triangles:

$$\frac{\text{(Rise)}}{\text{(Run)}}\ \frac{200}{250} = \frac{160}{200} \text{ that is } \frac{200 \times 4}{250 \times 4} = \frac{160 \times 5}{200 \times 5} = \frac{800}{1000} \text{ (mm unit run)}$$

b. Slope lines 1 and 2 are LL lines for specific unit runs: 250 mm and 1000 mm (using the set point).

c. Ratios of unit runs and of LLs are the same. The readings of LLs in the tables on the squares are respectively:

Line 1. 320.2, for unit run 250 mm
Line 2. 1281 mm, for unit run 1000 mm (1m)

As the unit runs are in the ratio 1:4 (250 to 1000), then the LL measurements should be in the same ratio, 320.2 to 1281.

$$4 \times 320.2 \text{ mm} = 1280.8 \text{ mm, which is approximately } 1281 \text{ mm.}$$

$$\frac{250 \text{ mm}}{1000 \text{ mm}} = \frac{320.2 \text{ mm}}{1281 \text{ mm}}$$

4. Pitch and slope data relationships in tabular form

Table 8.1 is a summary statement for two squares: one based on a unit run of 250 mm (Frederickson); and one based on a unit run of 1000 mm

TABLE 8.1

Roof framing data[1] related to two metric framing squares: one based on a unit run of 250 mm; and one based on a unit run of 1000 mm (1m)

Unit Run of 250 mm						*Unit Run of 1000 mm (1m)*				
Angle to Plate (°) of CR	*Slope Triangle* (Run, Rise; Rise/Run)	*Unit Run*	*Span*	*Rise*	*Pitch*	*Rise*	*Span*	*Unit Run*	*Slope Triangle* (Run, Rise; Rise/Run)	*Angle to Plate (°) of CR[2]*
63.44	500 in 250	250	500	500	1.0	2000	2000	1000	2000 in 1000	[3]
60.95	450 in 250	250	500	450	0.9	1800	2000	1000	1800 in 1000	
58.00	400 in 250	250	500	400	0.8	1600	2000	1000	1600 in 1000	
54.46	350 in 250	250	500	350	0.7	1400	2000	1000	1400 in 1000	54.5
50.20	300 in 250	250	500	300	0.6	1200	2000	1000	1200 in 1000	50.2
45.00	250 in 250	250	500	250	0.5	1000	2000	1000	1000 in 1000	45.0
38.66	200 in 250	250	500	200	0.4	800	2000	1000	800 in 1000	38.7
30.96	150 in 250	250	500	150	0.3	600	2000	1000	600 in 1000	31.0
21.80	100 in 250	250	500	100	0.2	400	2000	1000	400 in 1000	21.8
11.31	50 in 250	250	500	50	0.1	200	2000	1000	200 in 1000	

[1]Most of the data is based on the definition that pitch is rise divided by span.
[2]Angle to plate may not be the same as angle of rafter cut shown on a framing square.
[3]Some angular measurements may not be shown on the square.

(1m) (Stanley). The pitch column is common to each set of roof-framing data. Additional data is given in columns designated "slope triangle" and "angle to plate."

The metric system of measurement seems to simplify roof framing calculations. However, the builder needs to decide which subsystem, which unit of run, he or she prefers when choosing a metric framing square.

5. For further study

a. Relate your experiences with imperial framing squares to data given in the three chapters about metric framing squares (Chapter 6, 7, and 8). If you have worked with residential roof pitches such as $\frac{1}{4}$, $\frac{1}{3}$, or 5 in 12, expand the data for each, and insert the sets in Table 8.1 at the appropriate places.

b. Prepare for your use a set of comparable pitches, and answer questions such as:

   What is the metric pitch nearest to $\frac{1}{4}$? To $\frac{1}{3}$? To 5 in 12?
   Which metric pitches do I intend to use?

c. Which term should I use: pitch? or slope? or "other"?
   What advantages are there in using the term *slope*?
   Which term, *pitch* or *slope*, is used by draftspersons and architects to describe a roof?

d. When describing a roof, how is measurement expressed? As a common fraction? A decimal fraction? As a dimensioned slope triangle? Or other?

e. In terms of roof framing, what is the meaning of each of the following expressions?

   $\frac{1}{2}$ pitch

   0.5 pitch

   45 degree slope

   1:1 slope

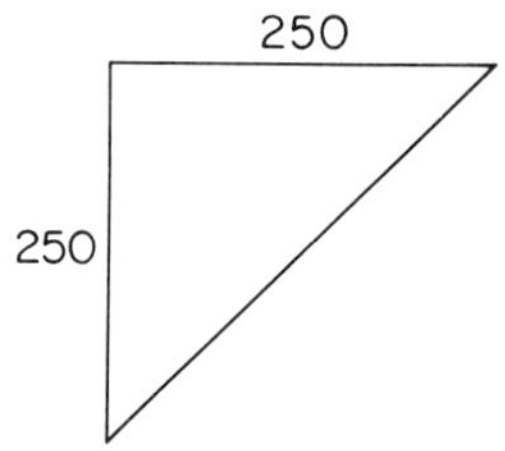

9

Gable Studs*

1. The line length of the gable studs

a. The LL of the longest gable stud when centrally placed is equal to the rise of the roof. The other gable studs are in pairs on either side of the longest. They are spaced at regular intervals along the plate. See Table 9.1.

b. The LL is also equal to the pitch multiplied by the span. Assume that a $\frac{1}{4}$ pitch roof has a span of 36′–0″. The LL of the longest gable stud is $\frac{1}{4}$ of 36′–0″ which is 9′–0″.

2. Rise per foot run

a. Since the framing square tables are based on a unit of one foot of run, they are also based on two feet of span.

b. To find the rise per foot of run, multiply the pitch of the roof by 24 inches. This is the unit of span and also the length of the blade of the framing square. For a $\frac{1}{4}$ pitch roof, the rise per foot of run is $\frac{1}{4}$ of 24″ which is 6.

*Note: The traditional framing square is used throughout this chapter.

3. Differences in line lengths of gable studs

Example 1: A $\frac{1}{4}$ pitch roof has gable studs placed at 12″ OC. The difference in the length of adjacent studs is $\frac{1}{4}$ of 2 × 12″, that is, one quarter of the unit of span. One stud will be 6″ longer (or shorter) than its neighbor.

Example 2: A $\frac{1}{4}$ pitch roof has gable studs placed at 16″ OC. The difference in the length of adjacent studs is $\frac{1}{4}$ of 2 × 16″, that is, one quarter of the unit span. One stud will be 8″ longer (or shorter) than its neighbor.

Example 3: A $\frac{1}{3}$ pitch roof has gable studs placed at 16″ OC. The difference in the length of adjacent studs is $\frac{1}{3}$ of 2 × 16″, that is, one third of the unit span. One stud will be $10\frac{2}{3}$″ longer (or shorter) than its neighbor. It will be recalled that thirds of inches may be read on the twelfth-scaled edge of the traditional framing square.

4. Differences in lengths of gable studs

Rule: *To find the difference in lengths of the gable studs for any regular gable roof, multiply the pitch by twice the spacing of the gable studs.*

TABLE 9.1

Differences in Lengths of Gable Studs

Roof Pitches	*Gable Studs Spacings*	*Twice Gable Stud Spacings*	*Differences in Lengths of Successive Gable Studs*
$\frac{1}{4}$	12″	24″ $\frac{1}{4}$ of 24″	6″
$\frac{1}{4}$	16″	32″ $\frac{1}{4}$ of 32″	8″
$\frac{1}{3}$	16″	32″	$10\frac{2}{3}$″
$\frac{3}{8}$	16″	32″	12″

5. Mathematical differences in the lengths of the gable studs for a ⅓ pitch roof

a. The difference in the lengths of the gable studs for a $\frac{1}{3}$ pitch roof where the studs are placed 12″ OC is 8″.

b. The difference in the length from the wall plate to the ℄ of the gable studs, for every inch of run on the wall plate for a $\frac{1}{3}$ pitch roof, is $\frac{2}{3}$″, that is, 8 ÷ 12 inches.

c. It follows that the difference in the lengths of the gable studs for a $\frac{1}{3}$ pitch roof where the studs are placed 16″ OC is 16 times $\frac{2}{3}$″ or $10\frac{2}{3}$″.

6. Differences may be shown graphically

It should be carefully noted that the pitch does not change when the OC measurement changes. See Figure 9.1.

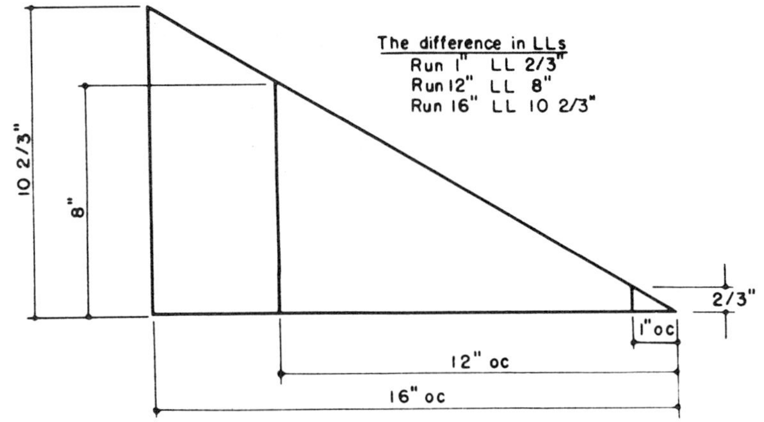

FIGURE 9.1. The pitch does not change when the OC measurement changes.

7. Layout of the gable studs at 16″ OC for a ⅓ pitch roof

a. The layout of the gable studs is usually made on the wall plate from the center gable stud.

b. The position of the longest gable stud should be located first.

c. The remaining gable studs should be laid out each way from the center gable stud.

d. Figure 9.2 shows a center line drawing of the CRs and the gable studs.

e. The distance *a* to *b* is the ℄-to-℄ distance between a pair of studs. This is 32″, which is twice the spacing or run distance.

f. At *C*–*C*2 is shown the LL of the longest gable stud.

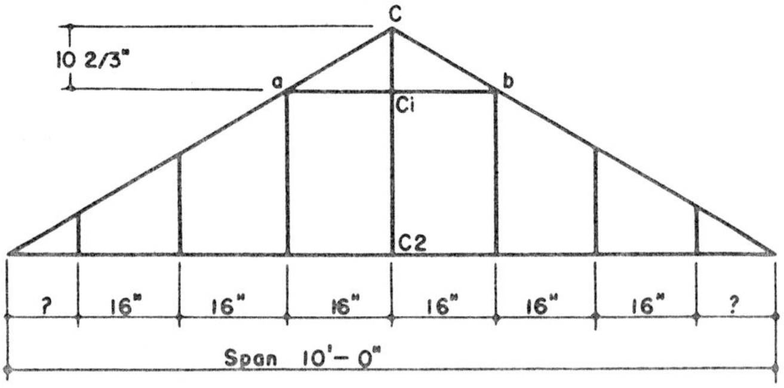

FIGURE 9.2. Center line drawing of the CRs and the gable studs.

g. At C–$C1$ is shown the rise of the center gable stud over line a–b, $10\frac{2}{3}''$.

h. The pitch of the roof is $\frac{1}{3}$. Pitch times span is the rise. One third of 32″ (a to b) is $10\frac{2}{3}''$. The latter measurement is the common difference in the length of the gable studs for a $\frac{1}{3}$ pitch roof with studs placed at 16″ OC.

8. Line length of the longest gable stud

a. The LL of the longest gable stud is equal to the rise of the roof from the wall plate to the center line of the CRs.

b. The LL is shown in Figure 9.3. It is from Line 1 to Line 2.

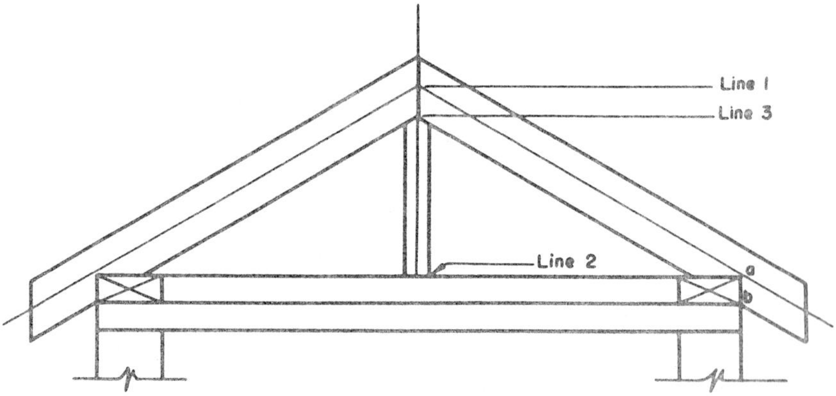

FIGURE 9.3. Shortening of the longest gable stud.

9. Shortening of the longest gable stud

a. The LL of the longest gable stud should be laid out on the edge of the gable stud stock. Secondly, the shortening should be laid out.

b. The shortening is equal to the vertical measurement of the birdsmouth from the bottom edge of the CR to the level cut. The vertical measurement from Line 1 to Line 3 is equal to measurement *a–b* at the birdsmouth.

10. Shop drawing

A full-scale shop drawing should be made. This should show the upper portions of the longest gable stud and the upper portions of each stud on either side of the longest stud.

11. Layout of a pattern gable stud for an equal-pitch roof

a. Select a straight, sound piece of 2 × 4 of convenient length.

b. Place the stock on the sawhorses with the narrow face uppermost.

c. Draw a ℄ about eight or ten inches long from the left hand end as in Figure 9.4. See Line *x–x*.

d. Lay off the LL of the longest gable stud (Lines 1–2).

e. Lay off from Line 1 a measurement equal to the vertical distance between Lines 1 and 3. See Figure 9.5.

f. Place the pitch board with its hypotenuse on the edge of the gable stud stock with plumb cut end towards the top of the stud.

g. Place the plumb cut end of the pitch board at the intersection of Lines *x–x* and 3.

h. Draw in the rake angle of the cuts. See Lines 4–4 and 5–5 in Figure 9.6.

i. When using the framing square, first place the stair gauges for the required pitch.

j. Set the framing square with the plumb cut towards the top of the gable stud and the stair gauges fitting snug along the bottom edge of the stud stock. Draw in the angles for the rake cuts.

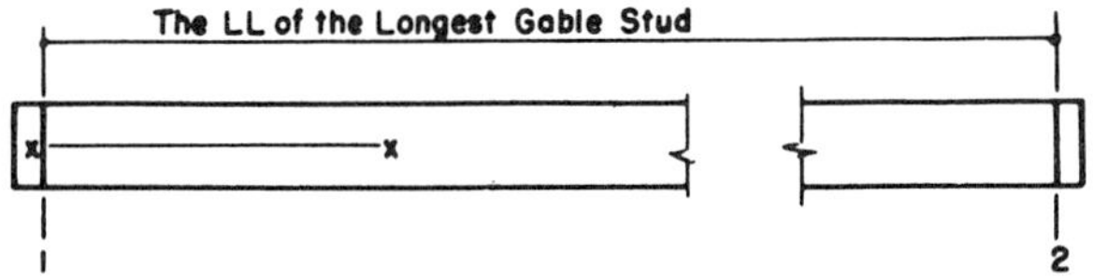

FIGURE 9.4.

FIGURE 9.5

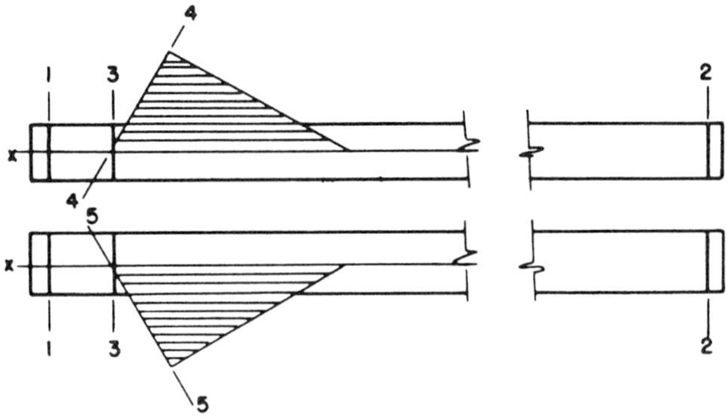

FIGURE 9.6

12. Position of the framing square on the gable stud for a ⅓ pitch roof

a. Figures 9.7 and 9.8 show the framing square in the same positions as the pitch board for laying out the rake cut on a gable stud.

b. The stair gauges should be set at the 8″ mark on the tongue of the square and at the 12″ mark on the body.

c. Figure 9.11 shows the pitch board and the framing square placed in the positions in which the rake cuts were drawn.

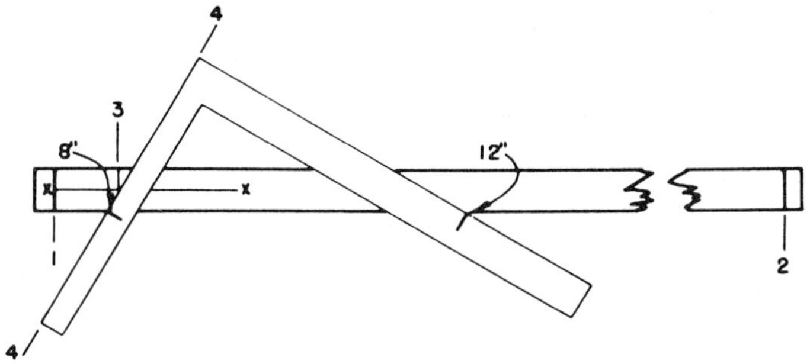

FIGURE 9.7

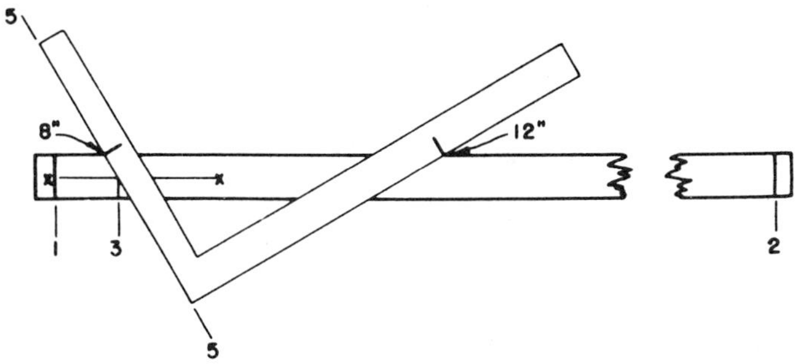

FIGURE 9.8

13. Layout of the gable studs from the pattern stud ⅓ pitch

a. From the outside edge of the gable stock at Line 4, complete the layout of all the remaining gable studs.

b. The common difference in the length of each succeeding gable stud is determined from the pitch of the roof and the spacing of the studs as at Lines 5–6 and 6–7 in Figure 9.9. Measurement 2–6 is the LL of the second longest pair of studs. Measurement 2–7 is the LL of the third longest pair of gable studs.

c. Pieces of stock of convenient length may now be squared off in readiness for the layout from the pattern stud. See Figure 9.10.

d. Set the stair gauges for the pitch of the roof. Lay off the cutting lines for each pair of studs.

e. The cutting lines should be drawn in from the outside edge of the stud, and the rake cut should be towards the peak of the roof.

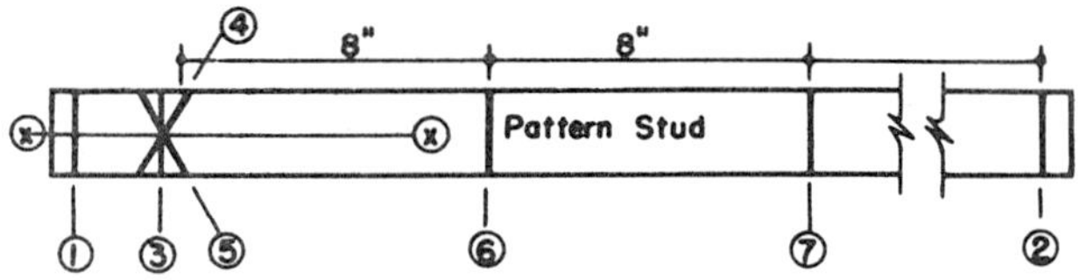

FIGURE 9.9

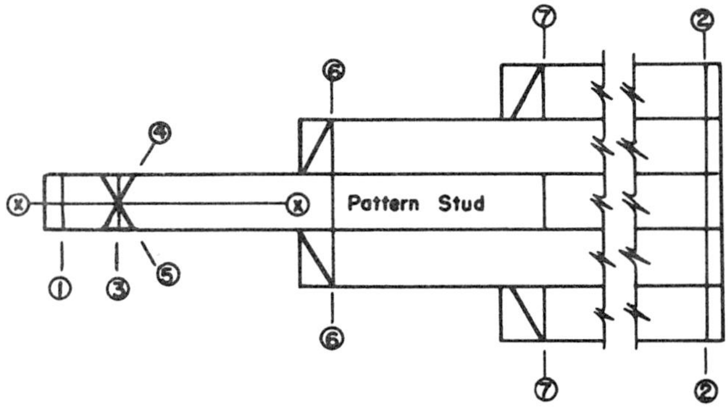

FIGURE 9.10

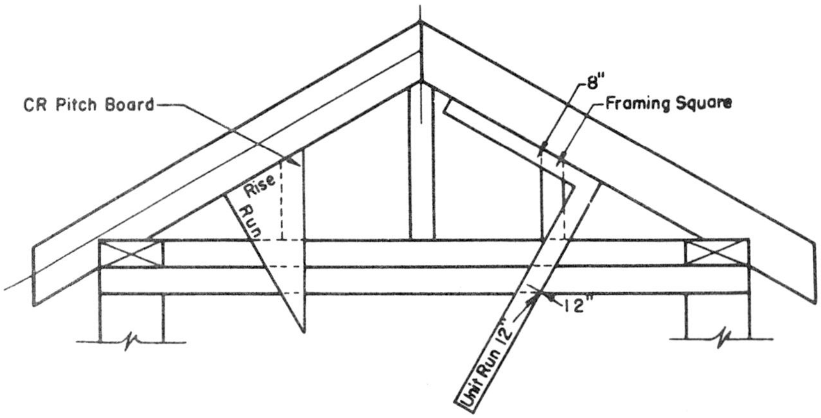

FIGURE 9.11

Review Questions

1. Fill in the blanks:

   a. The line length of the longest gable stud when centrally placed on a regular roof is equal to the ______ of the roof.

   b. The line length of the longest gable stud when centrally placed on a regular roof is equal to the ______ multiplied by the ______.

   c. Assume that a $\frac{1}{4}$ pitch roof has a span of 36′–0″, the line length of the longest stud ______ 36′–0″, which is ______ inches.

2. Make a scale drawing of a regular gable roof showing the center lines of the common rafters and the gable studs, with the gable studs placed at 16″ on centers.

   Make a scale drawing of a regular gable roof with a 32′–0″ span showing the center lines of the common rafters.

   Make a scale drawing of a section of a regular gable roof with a 32′–0″ span showing the center lines of the members.

   Make a scale drawing of a section of a gable roof with a 32′–0″ span, showing the center lines of the gable studs and the common rafters.

   Measure the difference in lengths of the gable studs which is ______ inches.

   Make a scale drawing of the section of a regular gable roof with a 32′–0″ span. Show the center lines of the gable studs at 16″ OC and of the two common rafters. Measure the difference in the lengths of the successive gable studs which is ______ inches.

3. The shortening for successive gable studs is equal to the ______ distance from the bottom of the common rafter to its center line. The shortening for successive gable studs is equal to the ______ distance from the bottom of the common rafter to its center line.

4. Using the traditional steel square to mark the rake cut for the gable studs for a $\frac{1}{3}$ pitch roof, set the stair gauges at ______ inches and ______ inches.

10

Ceiling Joists and Collar Ties

Most authorities state, "Where possible, ceiling joists shall be secured to opposite rafter feet." This forms a truss effect. See Figure 10.1.

1. Layout of ceiling joists

a. Select a sound piece of stock upon which to lay out the pattern ceiling joist.

b. Set the stair gauges on the framing square to the pitch of the roof (CR setting).

2. Rake cut of the ceiling joists

a. The rake cut of the ceiling joists is the angle cut at the end of each joist. This is made to match the angle of the pitch of the roof.

b. Place the framing square with the stair gauges set snug along the bottom edge of the ceiling joist. The 12″ setting should be at the left hand end. Draw the rake cut, Line 1.

c. The framing square and the pitch board are both shown in position at Figures 10.2 and 10.2(a), respectively.

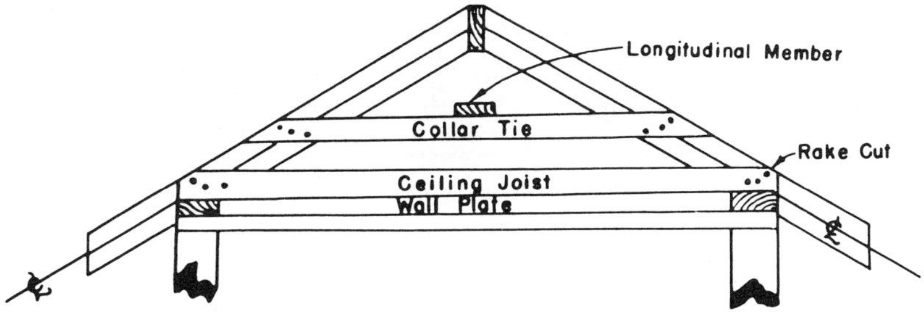

FIGURE 10.1. Ceiling joist and collar tie in gable construction.

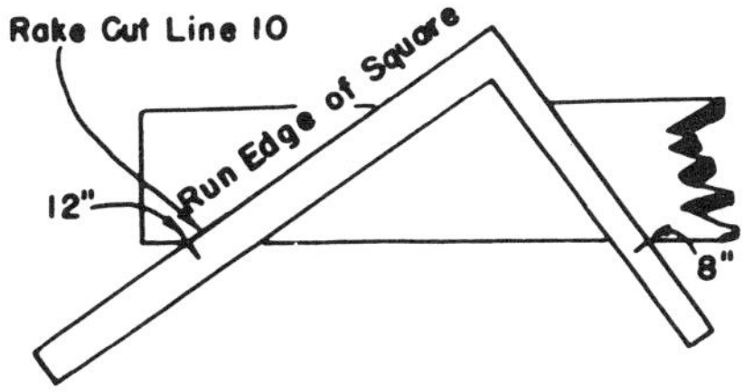

FIGURE 10.2

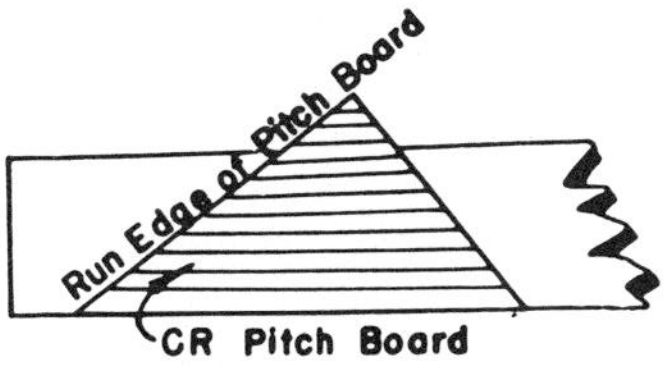

FIGURE 10.2(a)

3. Plumb cut of the ceiling joists

a. The ceiling joists should be cut plumb at the wall plate end. Set a carpenter's combination square equal to the vertical measurement *a–b* on Line 2 above the birdsmouth of the CR. See Figures 10.3 and 10.3(a).

b. With the square setting *a–b* of the CR, draw a plumb line on the ceiling joist to intersect with rake cut Line 1 as in Figure 10.3(a).

c. When the ceiling joist is in position on the building, Line 2 in Figure 10.3(a) should be plumb over the outside edge of the wall plate. The rake cut should be flush with the back of the CR as in Figure 10.4.

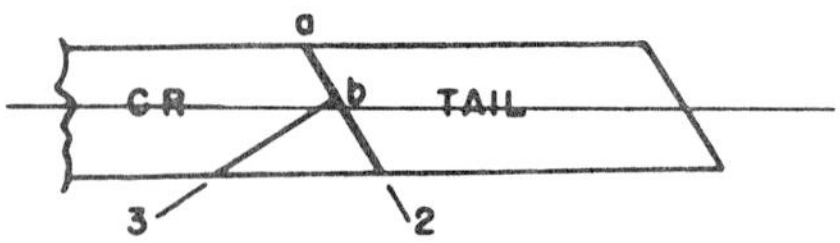

FIGURE 10.3

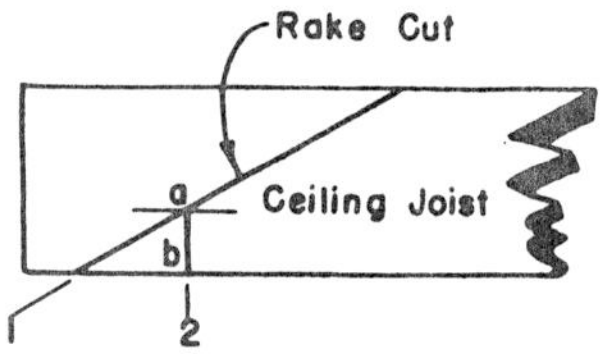

FIGURE 10.3(a)

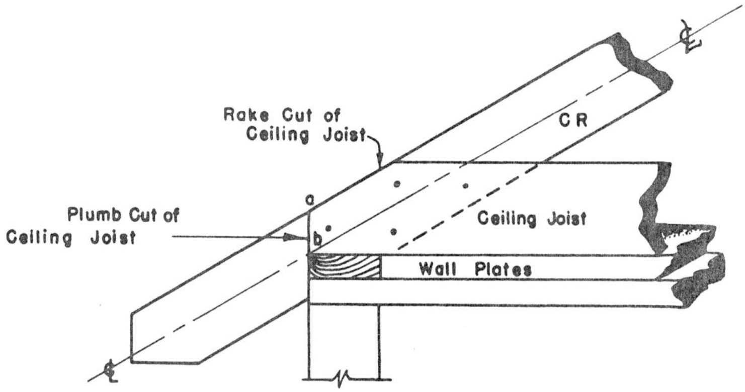

FIGURE 10.4. Ceiling joist in position at CR.

4. Ceiling joist template

a. A template should be made of the wall plate end of the ceiling joist. This template may be made out of a piece of material about 18″ in length with exactly the same width as the ceiling joist.

b. Where portable hand power saws are used, the template may be made into a jig. If a stationary power saw is used, the angle of the saw may be set by using the jig, the framing square, or the CR pitch board.

5. Layout of the collar ties

a. Read the plans and the specifications for the size and the positions of the collar ties. See Figure 10.1.

b. Many authorities state that, for residential construction, the collar ties (where there is no attic) shall be placed in the upper third of the LL of the CR.

c. The length of the collar ties may be measured on the drawings. They are laid out in a manner similar to that for the ceiling joists. The layout may be made by using the template, the framing square, or the pitch board.

d. The actual nailing position of the collar tie should be marked on the pattern common rafter. The nailing position should be transferred to all the CRs during the layout of these members.

e. It cannot be too strongly emphasized that for exactness of layout and economy of time, all lines and cuts on roof framing members should be made on the ground before the members are passed up for nailing on the structure.

f. A longitudinal member is usually secured at right angles to the center of the collar ties. See Figure 10.1. Read the drawings and specifications for the dimensions of this member. Lay off the rafter spacings on this member before it is passed up for nailing.

Review Questions

1. Fill in the blanks:

   Most authorities state, "Where possible, ceiling joists shall be secured to ______ feet." This forms a ______ effect.

2. The rake cut of the ceiling joists is at the ______ of the common rafter which matches the angle of the ______ of the roof.

3. Using a ruler, make a neat (no scale) sketch of a section of a regular gable roof showing the features listed on the following page.

a. the top of the wall framing studs
b. the wall plate
c. the cap plate
d. the common rafter
e. the ridge board
f. the collar tie
g. the longitudinal member over the collar tie.

4. The rake cut at the end of the collar tie may be drawn by setting the stair gauges at ______ inches, and ______ inches, on the traditional framing square.

5. To draw the end cut of ceiling joists and collar ties, it will save time to make a ______ which may be used over and over again.

6. Many authorities state that for residential construction the collar ties (where there is no attic) shall be placed in the ______ of the length of the common rafter.

11

Regular Hip Roof

1. Specifications

Plan 28′–0″ × 40′–0″. Pitch $\frac{1}{3}$. Ridge 1 × 6. Overhang 16″. Rafters 16″ OC. CRs 2 × 4. Hips 2 × 6.

2. Carpenter's working drawing

a. The plan of a carpenter's working drawing for a roof is usually drawn to the scale of 1″ to 1′–0″. For simple roofs a working drawing is not necessary, but for roofs with many features such as hips and valleys a working drawing is indispensable.

b. A regular hip roof has an equal pitch from each wall plate towards the ridge.

3. Hip rafter

a. The rise of a hip rafter (HR) is equal to the rise of the CR against which it fits.

b. The run of an HR is equal to the diagonal of a square whose sides are equal to the run of the CRs. See Figure 11.1.

c. The run of the HR tail is equal to the diagonal of a square whose sides are equal to the run of the CR tail.

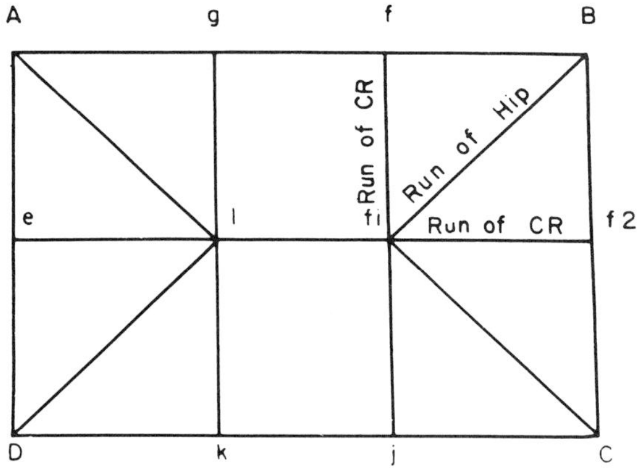

FIGURE 11.1. Skeleton outline of a hip roof.

4. Plan

a. Figure 11.1 shows the skeleton outline of a regular hip roof.

b. The extremities of the wall plates are shown at *A, B, C,* and *D.*

c. The runs of the CRs (six) are shown at *g–l, f–fi, fi–f*2, *fi–j, l–k,* and *l–e.*

d. The ridge is shown at *l–fi.*

e. The runs of the hips are shown at *A–l, D–l, B–fi,* and *C–fi.*

f. Altogether Figure 11.1 shows the outline of the wall plates, six CRs (all the same length), the ridge, and four hips.

g. The most important feature is that the HR run is the diagonal of a square whose sides are equal to the run of the CR.

5. Development of the line length of the common rafter

a. To find the LL of the CR, project the rise of the roof from *l* to *n* on the line of the ridge. See Figure 11.2.

b. Joint *g–n*, which is the LL of the CR. To find the LL of the CR, measure it.

c. The run of the CR is *g–l*. The run of the CR also forms one side of a square *A–g–l–e*. The diagonal of this square is the run of the hip.

d. The rise of the roof is projected at *l–n*. **The rise of the CR is the same as the rise of the hip.**

e. If we assume that the drawing is made to the scale of 1″ to 1′–0″, the right-angle triangle *g–l–n* would be the outline of the No. 1 pitch board.

f. If a scaled drawing is made and the lines *g–n* and *n–l* are cut through with a sharp razor, the triangle *g–l–n* may then be folded up on its run *g–l*.

g. It may then be clearly seen that the angle at *n* is the plumb cut angle and that the angle at *g* is the seat-cut angle.

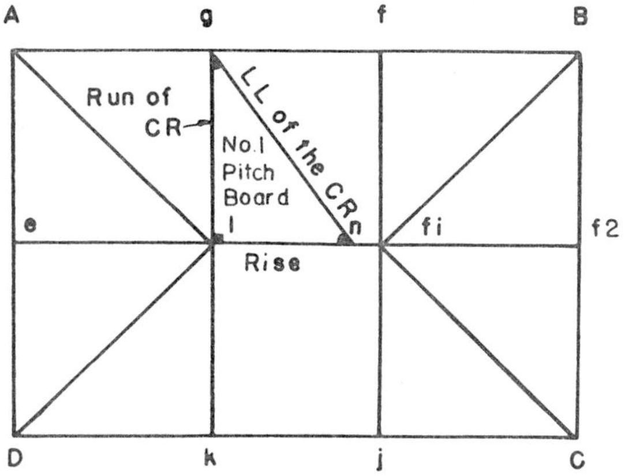

FIGURE 11.2. The LL of the CR.

6. Development of the line length of the hip rafter

a. **The run of an HR is equal to the diagonal of a square whose sides are equal to the run of the CR.** The run of the HR in Figure 11.3 is equal to the diagonal of the square *A–g–l–e*.

b. The angle formed at *l* by the intersection of *A–l* and *l–d* is an angle of 90°.

c. It will be recalled that on the rafter tables of the framing square the "length of rafter per foot run" refers to the number of feet and portions of feet of run in the common rafter.

d. The "length of hip or valley per foot run" read on the second line of the rafter tables refers to the number of feet and portions of feet of run also in the common rafter.

e. For every 12″ of run in the CR, the length of run of the hip would be the diagonal of a square whose sides are 12″. The diagonal of a square with 12″ sides is 17″. Measure across the framing square from 12″ to 12″ on the outside edges and read 17″ as a check.

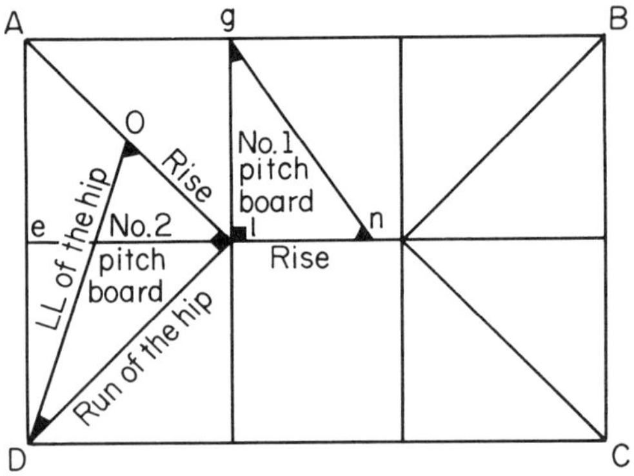

FIGURE 11.3. The LL of the hip rafter.

f. *D–l–o*, Figure 11.3, shows:

the run of the hip rafter at *D–l*

the rise of the hip rafter at *l–o*

the LL of the hip rafter at *D–o*

the plumb cut angle is at *o*

the seat cut is at *D*.

g. If No. 2 pitch board is made from a working drawing (scale 1″ equals 1′–0″), the LL of the hip rafter may be stepped off twelve times with its hypotenuse. The graphic method of layout may be used. This would be done in a manner similar to that used for the CR (graphic method) in Chapter 4.

h. If the hip development lines *D–o* and *l–o* are cut through with a sharp razor, the triangle *D–l–o* may be turned up on its run *D–l*. If both the CR triangle and the HR triangle are turned up on their runs, then *n* and *o* will be vertical over *l*.

i. The length of any roof member may be determined by making a scale drawing and measuring.

7. Development of the line length and the side cuts of the jack rafters

a. The runs of the jack rafters (also referred to in the text as JRs) are shown at *g–gi* and *gi–g*2 and at *h–hi* and *hi–h*2. See Figure 11.4.

b. The JR spacings should always be laid out from the CRs (on the wall plates) towards the corners of the building as at *f* through *g* and *h* to B; also as at *f*2 through *g*2 and *h*2 to *B*.

c. It should be very carefully noted that *g–gi–g*2*–B* and *h–hi–h*2*–B* both form squares on the plan.

d. **Rule:** *The run of any jack rafter on a regular hip roof is equal to its distance from the outside corner of the wall plates.* (See Figure 11.4.)

e. Project *f*2–*f*3 equal to the common rafter LL.
Project *B–f*3 which is the LL of the hip rafter.
Project *g*2–*g*3 which is the LL of the longest JR shown.
Project *h*2–*h*3 which is the LL of the shortest JR shown.

f. The triangle *B–f*2–*f*3 is revolved around *B–f*2 from a position where *f*3 was in a position directly vertical over *fi.*

g. The projected LLs of the JRs are shown at *g*2–*g*3 and at *h*2–*h*3. They are assumed to have revolved with the triangle *B–f*2–*f*3 from vertical points over *gi* and *hi*, respectively.

h. The angles *f*3, *g*3, and *h*3 are all equal. They are the angles for the side cuts of the JRs and the face cuts for the sheathing;

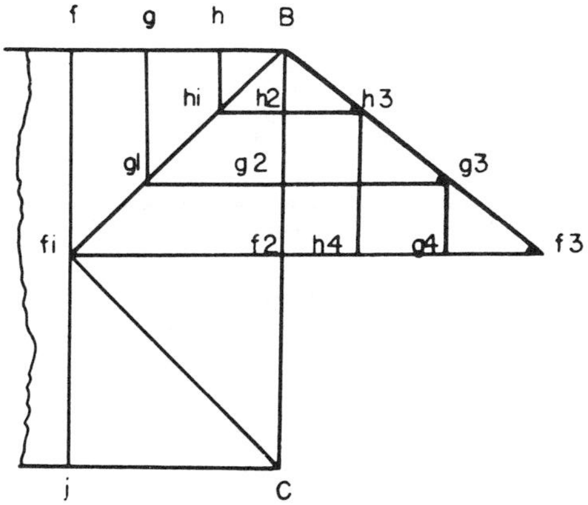

FIGURE 11.4. The run of a JR = its distance from the outside corner of the wall plates.

and they are also the purlin miter cuts. Purlins are cross timbers resting on trusses or posts and are used to support the rafters when these have excessive spans.

i. The projected lines *g*3–*g*4 and *h*3–*h*4 to the CR LL at *f*2–*f*3 show the common difference in lengths of the JRs. The common difference is measured from *f*3 to *g*4 and from *g*4 to *h*4. It will be seen that these two measurements are equal.

8. Layout of the hip rafter

Make a pitch board for the HR out of similar material and in a similar manner to the pitch board for the CR. It will be recalled that the development drawing should be made to the scale of 1″ to 1′–0″.

9. Hip rafter stock

a. The HR stock is usually of wider material than the CR stock. In house construction 2 × 4s are often used for CRs and jack rafters; 2 × 6s are usually used for the hip rafters.

b. Select a straight clean piece of 2 × 6 of a suitable length. Place the stock on the sawhorses with the crown side away from you.

10. Plumb line for a $\frac{1}{3}$ pitch roof

a. For a $\frac{1}{3}$ pitch roof, the stair gauges are set on the framing square at 8″ on the rise and at 12″ on the run (for the common rafter).

b. The rise of an HR is equal to the rise of the common rafter against which it fits. For a $\frac{1}{3}$ pitch roof, one stair gauge is set at 8″ on the framing square for the rise of the hip rafter.

c. The run of an HR is equal to the diagonal of a square, the sides of which are equal to the run of the CR. For every single foot of run of the CR, the run of the HR is the diagonal of one foot square. Check this by measuring across from 12″ to 12″ on the outside edges of the framing square. See Figure 11.5.

d. For a $\frac{1}{3}$ pitch roof the stair gauges are set on the framing square at 8″ for the rise and at 17″ for the run.

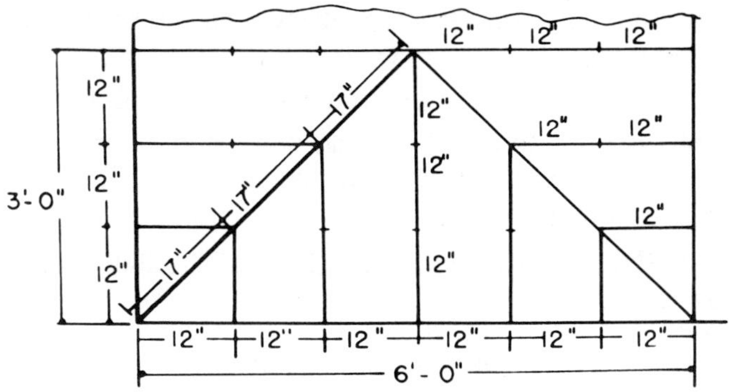

FIGURE 11.5. Run of a hip rafter = a diagonal whose side = run of the CR.

11. Plumb line

Lay the hypotenuse of the pitch board (with the plumb marking edge to the left) on the face of the 2 × 6 as in Figure 11.6, or use the framing square with the stair gauges set at 8″ and 17″ for this pitch as in Figure 11.6(a). Draw in Line 1.

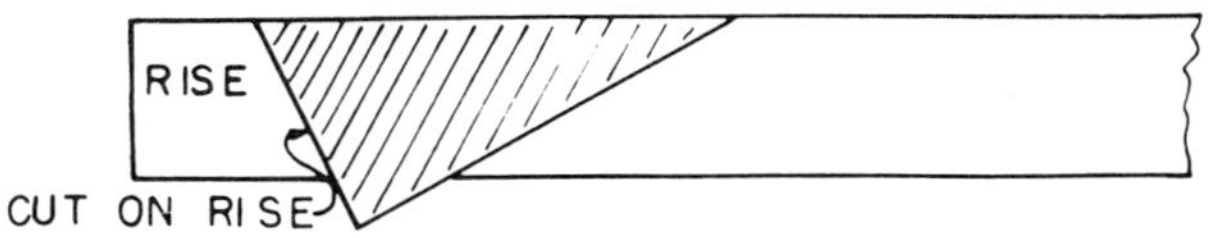

FIGURE 11.6

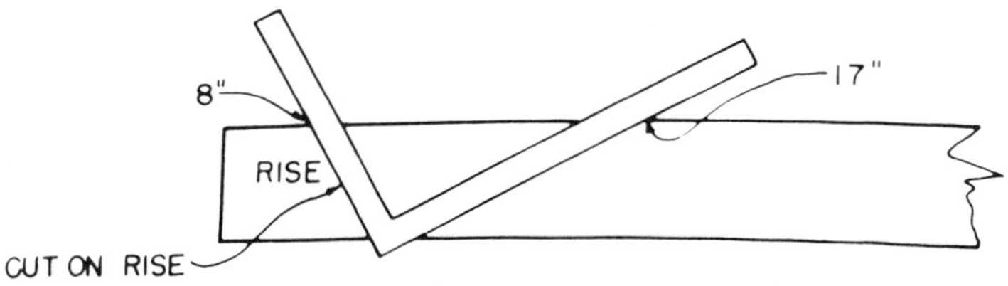

FIGURE 11.6(a)

12. Line length of the hip rafter

Accept the LL measurements in the following order:

a. The mathematical length as calculated by using the framing square rafter tables. This method is used by carpenters.

b. The mathematical length as calculated by the extraction of the square root.

c. The scaled measurement from the development drawing.

d. The step-off length by using the pitch board. Step off the hypotenuse of the pitch board twelve times. This should check to a tolerance of $\frac{3}{8}''$.

e. The step-off method by using the framing square with the stair gauges set at 8″ and 17″ for a $\frac{1}{3}$ pitch. Step off as many times with the framing square as there are numbers of feet of run to the CR. This should check to a tolerance of $\frac{3}{8}''$.

13. Method of layout of the line length of the hip rafter

a. It should be carefully noted that throughout this section all positions of the framing square for obtaining plumb and level lines are in exactly the same relative positions as those for the pitch board. Study Plate V along with the text.

b. Make all LL calculations actually on the face of the rafter stock. These calculations should be made a little to the right of the plumb Line 1 for the LL of the hip and a little to the right of the birdsmouth for the LL of the HR overhang.

c. Read on the rafter tables (page 42) under 8″ and on the second line:

Length of Hip or Valley per Foot Run . . . 18.76

(The specifications for this roof, at the beginning of this chapter, are span 28′–0″, pitch $\frac{1}{3}$).
Multiply 18.76 by the number of feet of run in the CR, which is fourteen. Then, 18.76″ × 14 equals 21′–10$\frac{5}{8}''$. This is the LL of the hip rafter.

d. Measure with the tape from Line 1 on the back of the hip rafter 21′–10$\frac{5}{8}''$, and draw Line 2 as in Figure 11.7.

e. Step off the hypotenuse fourteen times (with the framing square), and check to a tolerance of $\frac{3}{8}''$.

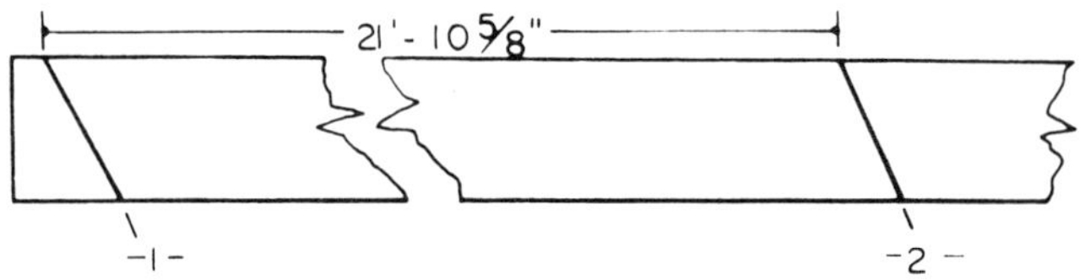

FIGURE 11.7

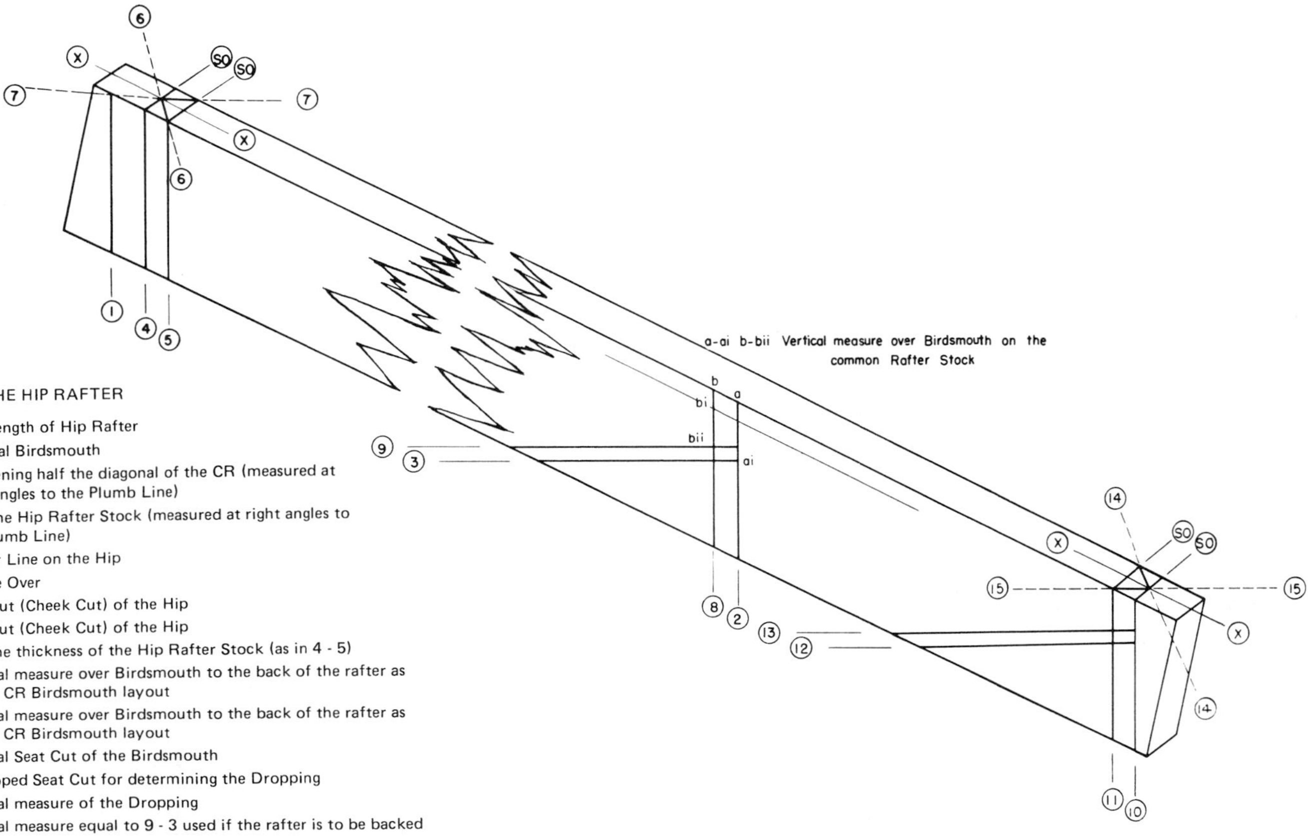

LAYOUT OF THE HIP RAFTER

Lines		
	1 - 2	Line length of Hip Rafter
	2 - 3	Original Birdsmouth
	1 - 4	Shortening half the diagonal of the CR (measured at right angles to the Plumb Line)
	4 - 5	Half the Hip Rafter Stock (measured at right angles to the Plumb Line)
	X - X	Center Line on the Hip
	SO	Square Over
	6 - 6	Side Cut (Cheek Cut) of the Hip
	7 - 7	Side Cut (Cheek Cut) of the Hip
	8 - 2	Half the thickness of the Hip Rafter Stock (as in 4 - 5)
	a - ai	Vertical measure over Birdsmouth to the back of the rafter as on the CR Birdsmouth layout
	b - bii	Vertical measure over Birdsmouth to the back of the rafter as on the CR Birdsmouth layout
	3 - ai	Original Seat Cut of the Birdsmouth
	9 - bii	Developed Seat Cut for determining the Dropping
	9 - 3	Vertical measure of the Dropping
	b - bi	Vertical measure equal to 9 - 3 used if the rafter is to be backed
	2 - 10	LL of the Hip Rafter Tail
	2 - 10	Run of the overhang when measured at right angles to the Plumb Line
	10 - 11	Half the Hip Rafter Stock (measured at right angles to the Plumb Line)
	10 - 12	The Layout of the Original Birdsmouth for the Level Cut, used when the Hip Rafter is to be Backed
	10 - 13	The Layout of the Level Cut used when the Hip Rafter is to be Dropped
	X - X	Center line on the back of the Hip
	14 - 14	Side Cut (Cheek Cut) for the Tail of the Hip Rafter
	15 - 15	Side Cut (Cheek Cut) for the Tail of the Hip Rafter

PLATE V. Layout of the Hip Rafter

14. Hip rafter original level line at the birdsmouth

a. Figure 11.8 shows the pattern CR. On Line 2 at *a–ai* is shown the perpendicular height of the CR from the seat cut, Line 3, to the back of the rafter.

b. Figure 11.9 shows the pattern HR. On Line 2 at *a–ai* is shown the amount of rafter above the plate as measured from the CR and reproduced on the HR. It will be seen that, since a 2 × 6 is used for the HR, the latter will require a larger birdsmouth to be removed.

c. Figure 11.10 shows the framing square in position for drawing the level line. Draw the level Line 3.

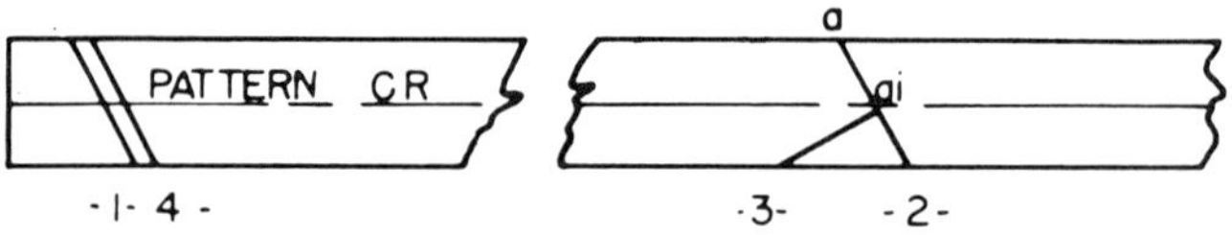

FIGURE 11.8

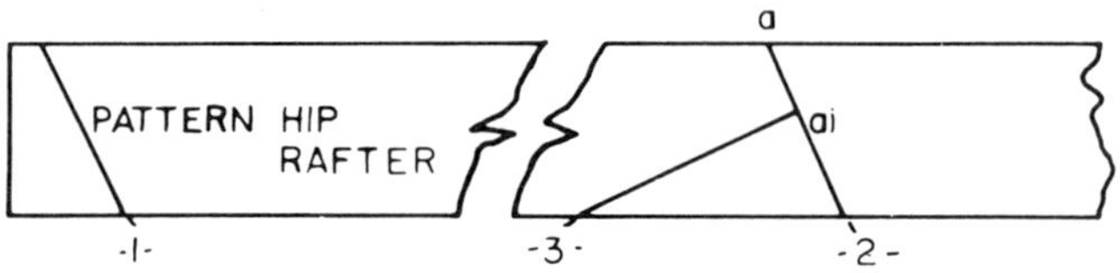

FIGURE 11.9

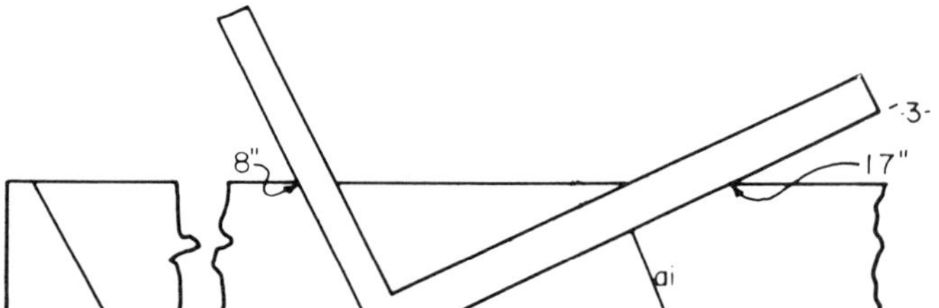

FIGURE 11.10

15. Shortening of the hip rafter

a. From an inspection of Figure 11.11, it will be seen that the run of the shortening of the HR is equal to half the diagonal thickness of the CR as at Lines 1 and 4.

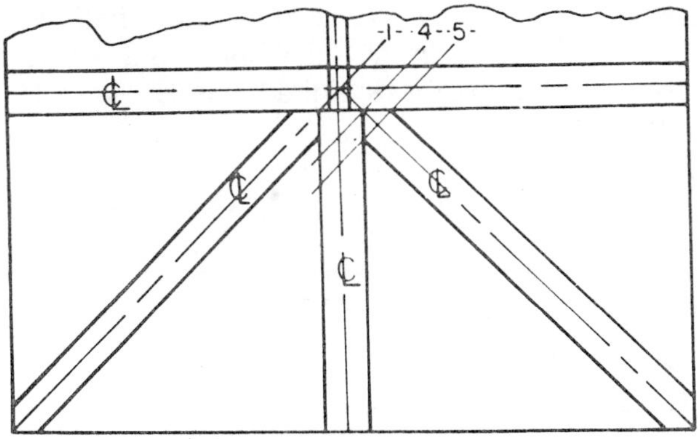

FIGURE 11.11. Shortening of the hip rafter.

b. It will be recalled that it is possible to erect a gable roof without a ridge board. It is equally possible to erect a hip roof without a ridge board. A ridge board is placed between the CRs to facilitate the erection of the roof.

c. It should be constantly kept in mind that the HR is shortened against the roof member at which it fits. It fits against the CR.

d. Measure back, at right angles to Line 1, half the diagonal thickness of the CR. See Figure 11.12. In this case as a 2 × 4 is specified for the CR, half the diagonal of this member is $1\frac{1}{8}''$. Draw in Line 4 measured at right angles to Line 1.

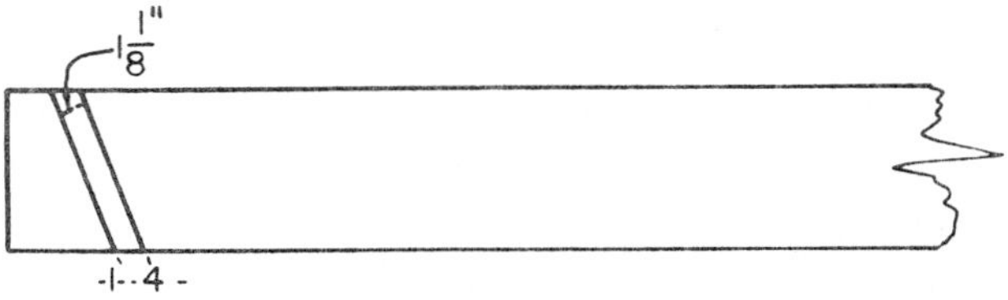

FIGURE 11.12

16. Side cuts (cheek cuts) of the hip rafter

a. Upon close inspection of related plates and figures it will be seen that the run of the side cuts of the hip rafter is equal to half the thickness of the hip rafter stock. The run of the side cut is measured on the plan between Lines 4 and 5 as in Figure 11.11.

b. Measure back at right angles from Line 4 half the thickness of the hip rafter stock. In this case, since a 2 × 6 is specified for the hip, half the thickness of the hip is $\frac{13}{16}''$. Draw in Line 5 as in Figure 11.13.

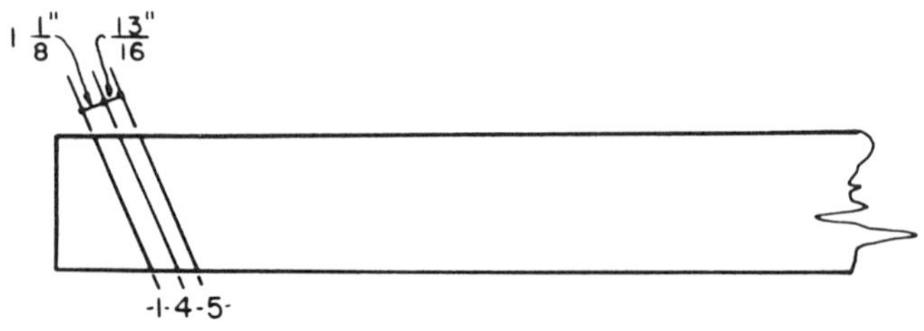

FIGURE 11.13

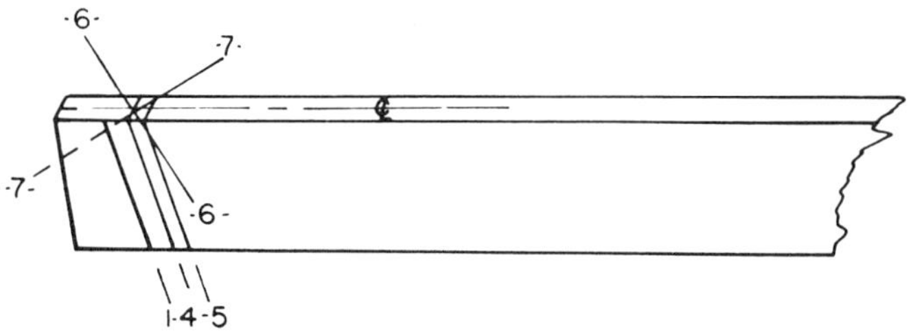

FIGURE 11.13(a)

17. Backing (beveling) of the hip rafter

a. The sheathing of a roof is the material which is secured to the backs of the rafters. Usually roofs are sheathed with common boards, shiplap, or plywood.

b. The sheathing of adjacent sides of the roof meets at a point above the center of the back of the HR. The steeper the pitch of the roof, the more acute will be the angle at the meeting place of the adjacent pieces of sheathing.

c. Theoretically, and sometimes in actual practice for special projects, the back of the HR must have the edges bevelled to receive the sheathing. Unless some provision is made, the roof sheathing will not lie flat where it comes in contact with the edges of the HR.

d. Closely examine Figure 11.14 at Line 2. A perpendicular line through the center of the back of the hip on Line 2 at ***a*** would intersect with the inside intersection of the wall plates.

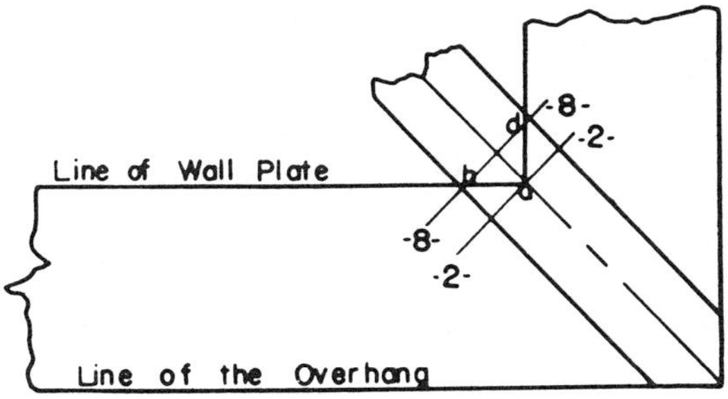

FIGURE 11.14

e. This perpendicular height corresponds to the perpendicular height of the CR from the seat cut Line 2 of the CR.

f. The length of the run between Lines 2 and 8 on the plan is equal to half the thickness of the HR stock. The length of the run is also equal to the run between Lines 4 and 5. The length of the run is shown between Lines 2 and 8, Figure 11.15.

g. On the pattern hip rafter, draw Line 8 forward from Line 2, a distance of $\frac{13}{16}''$ as at Line 8, Figure 11.16.

h. The edges of the HR are vertical over the wall plates as on the plan of the hip at *b* and *d* on Line 8, Figure 11.14. Points *b* and *d* are the end views of plumb lines from the wall plate to the top edges of the HR.

i. Upon very close examination of the plan in Figure 11.14, it will be seen that *b* and *d* on Line 8 and *a* on Line 2 are on the same level plane (when the rafter is dressed) as corresponding points on the JRs and the CRs.

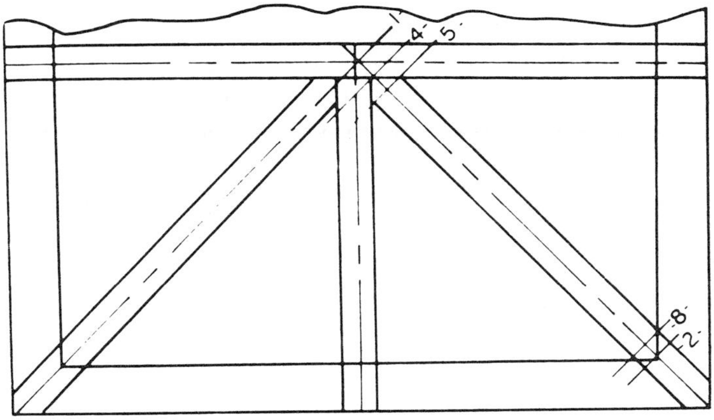

FIGURE 11.15. Backing of the hip rafter.

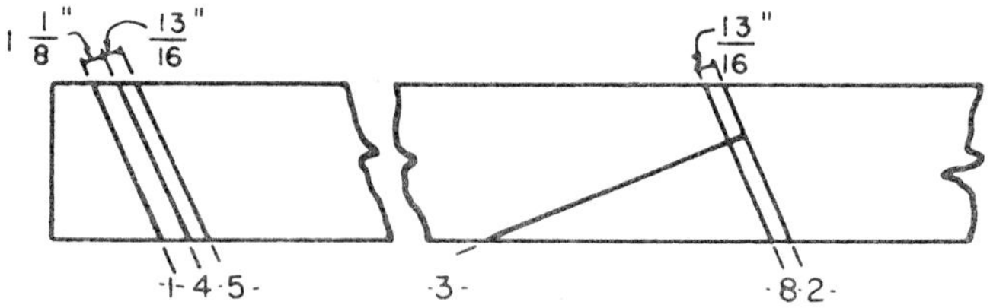

FIGURE 11.16

j. The height of the center of the hip rafter above the wall plate was located as at *a–ai* on Line 2 of the pattern hip rafter layout. This is the same height as is measured from the back of the CR on Line 2 of the common rafter layout.

k. The measurement from the wall plates to the outside edges of the HR, where the HR passes directly over the wall plate, is *a–ai* and must be measured from the back of the rafter on Line 8.

l. Measure down from the back of the hip rafter on Line 8 the same distance as on Line 2 at *a–ai*. Spot *bi* on Line 8 as in Figure 11.17.

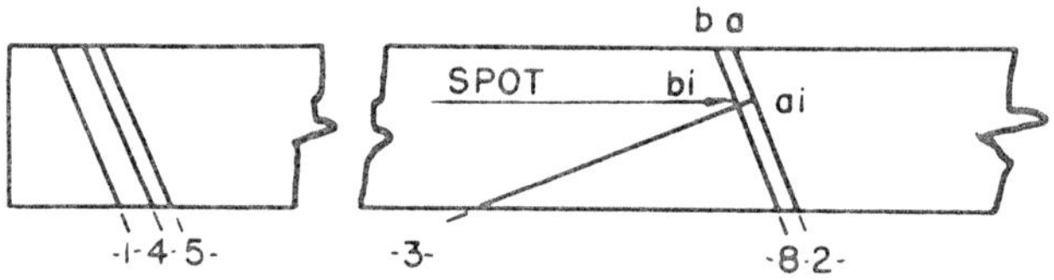

FIGURE 11.17

m. Place the framing square with the run to intersect Line 8 at *bi*. Draw a level line as at Line 9, Figure 11.18.

n. The measurement between *bi* and *b*2 (measured vertically) on Line 8 is the amount of bevel required on the HR. From the back of the rafter on Line 8, gauge the amount of backing (bevelling) required. See Figure 11.19.

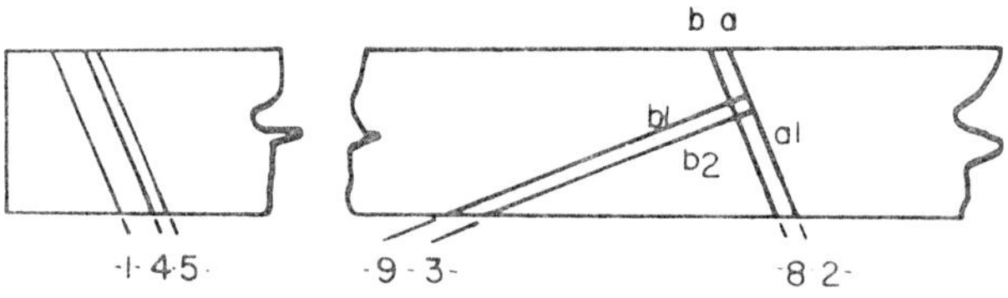

FIGURE 11.18

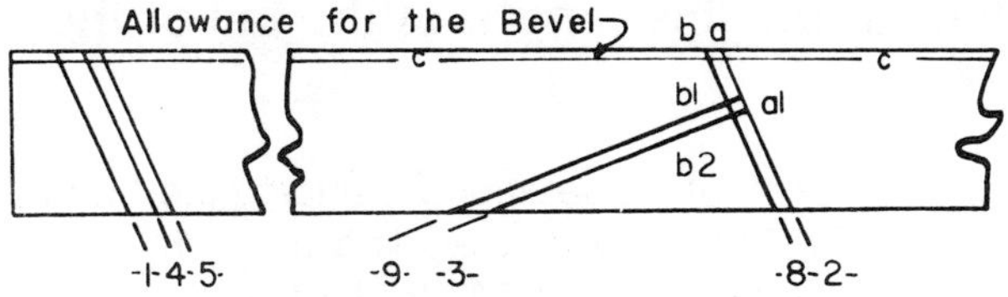

FIGURE 11.19

o. With the carpenter's combination square, gauge from the back of the rafter stock an amount equal to *bi–b*2 measured as at *c–c* on Line 8. Draw a ℄ down the back of the rafter. The amount to be backed is shown in Figure 11.20. Note *bi–b*2 equals *b–c*.

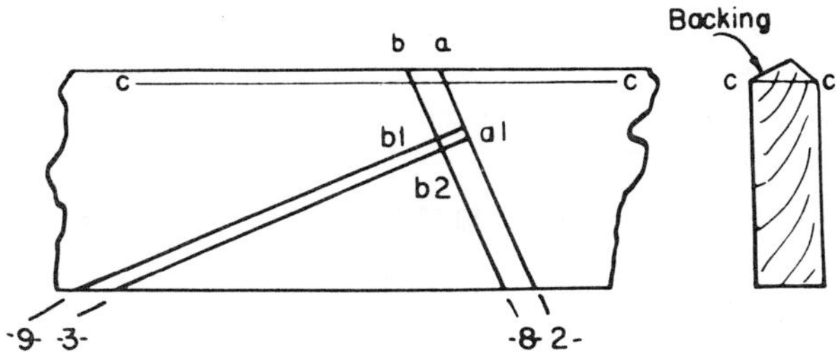

FIGURE 11.20

18. Dropping the hip rafter

a. In general building practice, unless the HR is of very large dimensions, it is the usual practice to drop the rafter, instead of backing it.

b. **Rule:** *A hip rafter must be either backed or dropped.* The amount that an HR is dropped is the amount that it would otherwise be backed.

c. Since the amount to be dropped is equal to the amount that it would otherwise be backed, it follows that if the rafter were to be dropped, the seat cut would be on Line 9. See Figure 11.18.

d. If the rafter is to be backed, the seat cut should be on Line 3.

e. An HR is either backed or dropped so that its edges will line up with the main plane of the roof.

19. Review of the line length of the tail of the common rafter

a. Upon reviewing the operation for finding the LL of the tail of the CR, it will be recalled that reading under 8″ on the first line of the rafter tables we find 14.42. To find the LL of the CR tail, multiply 14.42 by the number of feet and portions of feet of run in the CR overhang.

b. The number of feet and portions of feet of run in the CR overhang is 1′–4″, or $1\frac{1}{3}$ feet, or 1.33 feet. 14.42″ × $1\frac{1}{3}$ is $19\frac{1}{4}$″, which is the LL of the tail of the common rafter to the nearest $\frac{1}{8}$″.

20. The line length of the tail of the hip rafter

a. The LL of the tail of the HR is laid off from the plumb Line 2 of the original birdsmouth.

b. Make the calculations on the face of the rafter a little to the right of Line 2.

c. Read on the rafter tables of the framing square, page 42, under 8″ on the second line:

Length of Hip or Valley per Foot Run . . . 18.76

(It will be recalled that this roof is specified to have a span of 28′–0″, the pitch $\frac{1}{3}$, and the overhang 16″.)

d. Multiply 18.76″ by the number of feet of run in the *common rafter tail.* The LL of the tail is therefore 18.76″ × $1\frac{1}{3}$ which equals 25″ to the nearest $\frac{1}{8}$″.

e. Measure 25″ along the back of the HR from Line 2, and draw the plumb Line 10. This is the LL of the tail from Line 2. See Figure 11.21.

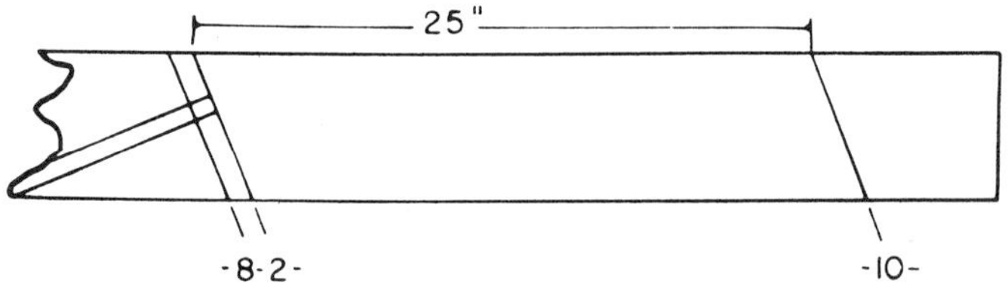

FIGURE 11.21

21. Check the line length of the hip rafter by the step-off method

a. The run of the overhang of the tail of the HR is the diagonal of the run of the overhang of the common rafter.

b. For every 1′–0″ of run of the CR tail, the run of the HR tail will be 17″. (17″ is the diagonal of 12″ and 12″.) Set the stair gauges to 8″ and 17″ for a $\frac{1}{3}$ pitch HR layout.

c. For every portion of 1′–0″ of run of the CR tail, the HR tail will require the diagonal of that portion of the CR tail. See Figure 11.22.

d. Apply the framing square to the rafter stock with the plumb marking side of the square to the left, and placed on Line 2. Measure 17″ on the run. This is the first step. Slip the square to the right so that the 8″ setting of the tongue is at the 17″ mark on the rafter. Draw a fine plumb line. See Figures 11.23 and 11.24.

e. Slip the framing square further to the right from the marked plumb line, and measure $5\frac{5}{8}$″ on the run as in Figure 11.24. The measuring edge of the framing square should now be within $\frac{1}{8}$″ of Line 10. See Section 20 and Figure 11.21.

f. The LL of the tail of the HR (25″) is measured on the back of the hip rafter stock from Line 2 to Line 10. The run of the overhang is measured on the blade of the square and is $22\frac{5}{8}$″.

22. The tail end side cuts of the hip rafter

a. Where the eaves are to be boxed in, the tail end side cuts of the HR will be the reverse of the side cuts at the ridge end of the HR.

b. From Line 10 on the plan, lay off half the thickness of the rafter stock as at Line 11, Figure 11.25.

c. On the HR, measure back, at right angles from Line 10 towards the birdsmouth, half the thickness of the rafter stock. See Figure 11.26. In this case, as a 2 × 6 is specified for the HR, half the thickness of this member is $\frac{13}{16}$″. Draw Line 11.

d. Draw a ℄ on the back of the rafter as at *x–x*. From Line 11 draw Lines 14 and 15 to intersect at the ℄ on Line 10.

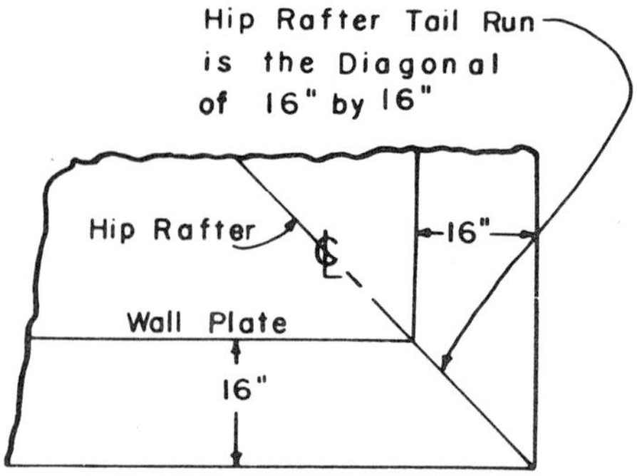

FIGURE 11.22

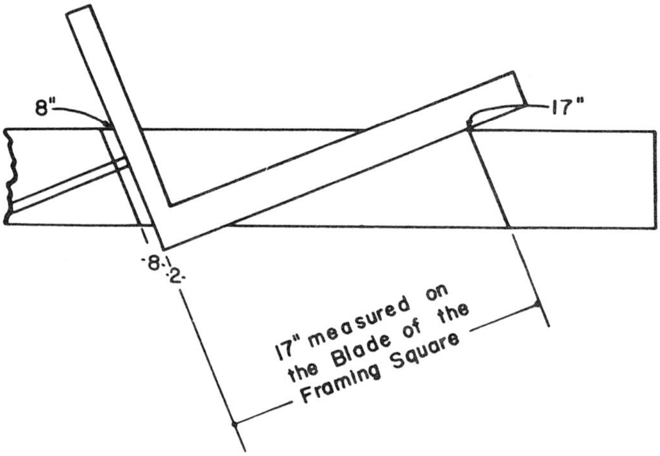

FIGURE 11.23

e. Draw from Line 10 two level lines in the same relative positions as are Lines 3 and 9 from the back of the rafter at the birdsmouth. The final cutting level line at the end of the rafter will correspond to the cutting line at the birdsmouth. If the rafter is to be dropped, cut on Line 9 at the birdsmouth and on Line 13 at the tail end. If the rafter is to be backed, cut on Line 3 and on Line 12.

f. Study Plate V.

g. It is common practice in building operations to allow the HR tail end to run out past the tail ends of the CRs and then to mark the actual length with a straight edge lined up with the tail ends of the CRs, from both sides of the roof.

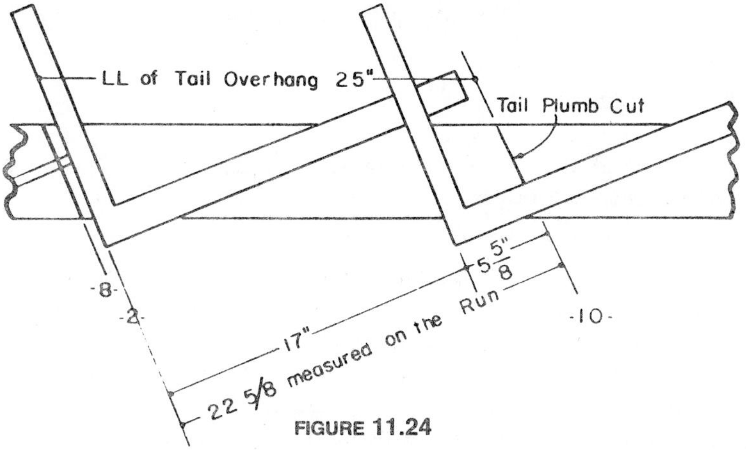

FIGURE 11.24

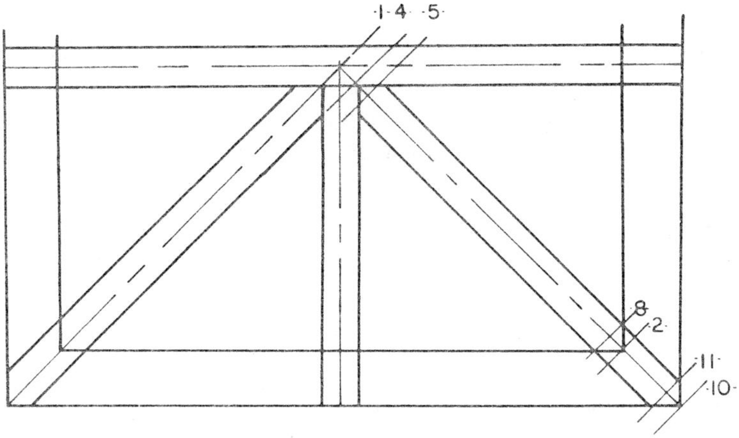

FIGURE 11.25. Tail end side cuts of hip rafter.

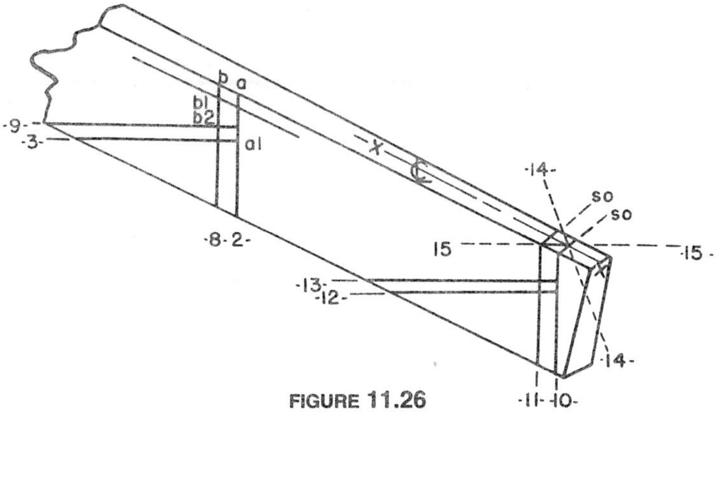

FIGURE 11.26

Review Questions

1. Fill in the blanks:

   The slope of a regular hip roof is ______ ______ from each wall plate towards the ridge.

2. Place a piece of tracing paper on Figure 11.3, and reproduce the line drawing. Using a sharp instrument, cut on the line length and the rise of the hip. Now cut on the line length and the rise of the common rafter. Turn up both triangles on their runs and examine the result. Both triangles have the same ______. The run of the hip is the diagonal of a square whose sides are equal to the ______ of the common rafter.

3. When using the traditional framing square, to draw the plumb cut of a common rafter for a $\frac{1}{3}$ pitch regular roof, set the stair gauges at ______ inches and ______ inches. Draw the plumb cut on the ______ side of the framing square.

4. The base of a CR pitch board for a $\frac{1}{3}$ regular pitch roof is ______ inches long, and the rise is ______ in length. To draw the plumb cut, place the ______ of the pitch board parallel with the ______ of the rafter, and draw the plumb cut on the ______ of the pitch board.

5. The shortening of the hip rafter is equal to ______ the thickness of the common rafter.

6. The seat cut of a common rafter is drawn on the ______ of the traditional framing square.

   The seat cut of the common rafter is drawn on the ______ of the pitch board.

7. The perpendicular height of the hip rafter at its birdsmouth is equal to ______ ______ ______ of the common rafter above its wall plate.

12

Layout of Jack Rafters for a Regular Hip Roof

1. Graphic method

a. Figure 11.4, page 93, shows the graphic method of establishing the LL and the side cut of the JRs.

b. With the exception of the side cuts of the JRs, all other cuts are identical with those of the CR for the same regular hip roof.

c. It should be carefully noted that the runs of pairs of the JRs and the wall plate lines form perfect squares on plan.

d. **Rule:** *The run of any jack rafter on a regular hip roof is equal to its distance from the outside corner of the wall plates.* See Figure 12.1.

2. Common differences in lengths of jack rafters

a. Following is shown the rafter table for a $\frac{1}{3}$ pitch roof as it appears on the framing square.

	8″
Length of common rafter per foot run	14.42
Length of hip or valley per foot run	18.76
Difference in lengths of jacks 16″ centers	19.23
Difference in lengths of jacks 24″ centers	28.84
Side cuts of jacks: Use	10″
Side cuts of hip or valley: Use	$10\frac{7}{8}$″

b. Assume that a regular hip roof is to be erected with a $\frac{1}{3}$ pitch and the rafters placed 16″ OC. Read under 8″ on the rafter tables, page 42. On the third line find:

Difference in Lengths of Jacks, 16″ Centers . . . 19.23

The common differences in LLs between succeeding JRs is 19.23″, or $19\frac{1}{4}$″ to the nearest $\frac{1}{8}$″. Note that the difference in the length of the pattern common rafter from Line 1 to the longest JR is also $19\frac{1}{4}$″ to the nearest $\frac{1}{8}$″. See Figure 12.2, and Plate VI, page 111.

c. Assume that a regular hip roof is to be erected with a $\frac{1}{3}$ pitch and the rafters placed 24″ OC. Read under 8″ on the rafter tables. On the third line find:

Difference in Lengths of Jacks, 24″ Centers . . . 28.84

The common differences in LLs between succeeding JRs is 28.84″ or $28\frac{7}{8}$″ to the nearest $\frac{1}{8}$″.

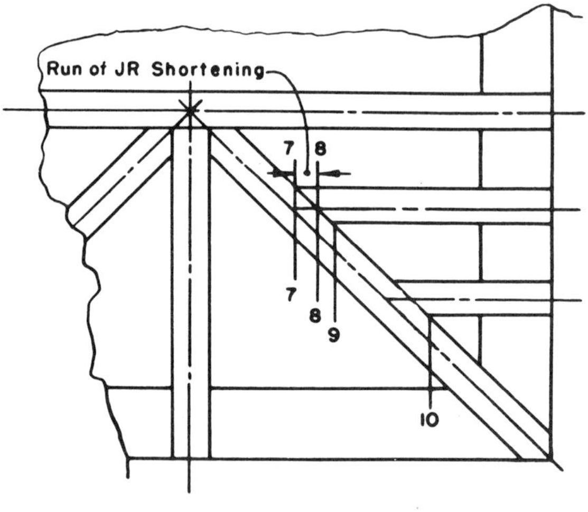

FIGURE 12.1. Run of any JR = its distance from the outside corner of the wall plates.

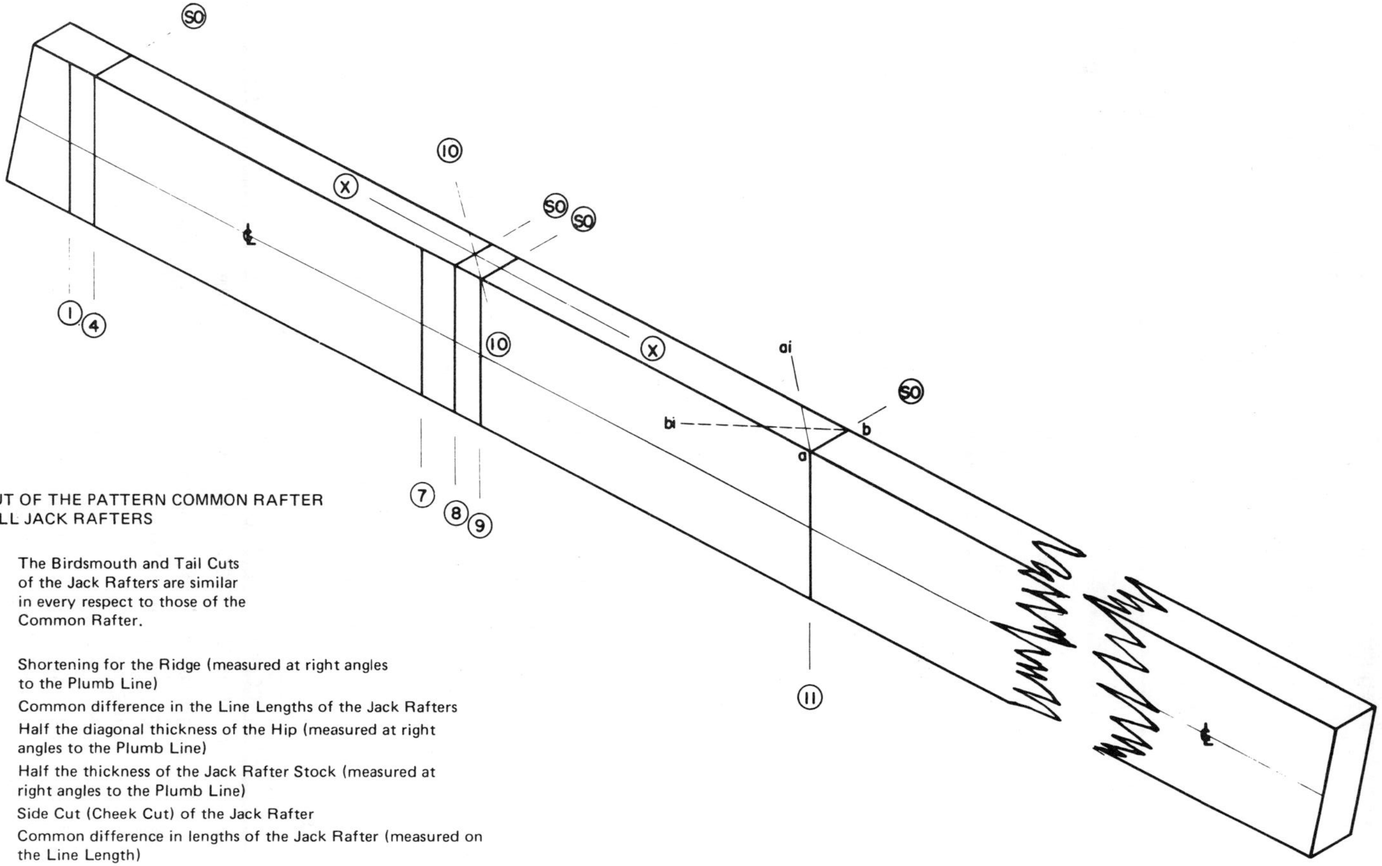

LAYOUT OF THE PATTERN COMMON RAFTER
AND ALL JACK RAFTERS

Note: The Birdsmouth and Tail Cuts of the Jack Rafters are similar in every respect to those of the Common Rafter.

Lines	1 - 4	Shortening for the Ridge (measured at right angles to the Plumb Line)
	1 - 7	Common difference in the Line Lengths of the Jack Rafters
	7 - 8	Half the diagonal thickness of the Hip (measured at right angles to the Plumb Line)
	8 - 9	Half the thickness of the Jack Rafter Stock (measured at right angles to the Plumb Line)
	10 - 10	Side Cut (Cheek Cut) of the Jack Rafter
	9 - 11	Common difference in lengths of the Jack Rafter (measured on the Line Length)
		Note: Line II has already allowed for the shortening. The Side Cut of the Jack Rafter would be drawn in from a or b according to which hand of Jack Rafter is required
	SO	Square Over
	X - X	Center Line on the back of the Rafter

PLATE VI. Layout of the Jack Rafter
NOTE: Jack Rafters are laid out on the Pattern Common Rafter

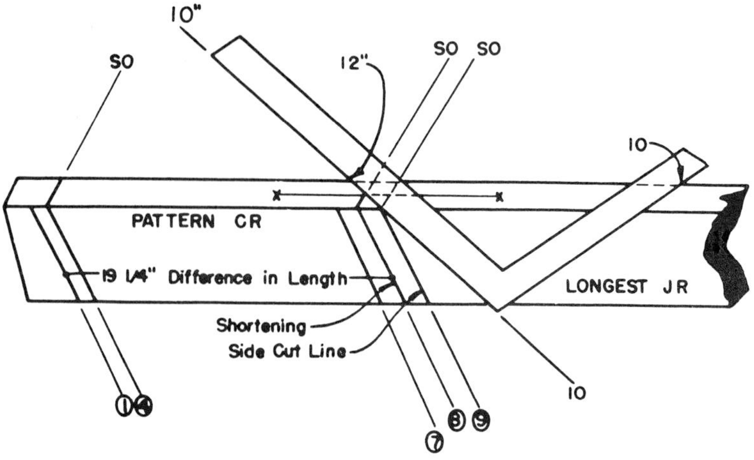

FIGURE 12.2

d. **Rule:** *The difference in the lengths of succeeding jack rafters on a regular hip roof is equal to the number of feet and portions of feet in the center-to-center spacings of the jack rafters multiplied by the LL of the common rafter per foot of run.*

e. *Example 1:* For a $\frac{1}{3}$ pitch roof with JRs placed at 16″ OC.
The common difference in the runs of the JRs placed at 16″ OC is $1\frac{1}{3}$ feet.
The LL per foot run of the CR is 14.42.
The common difference in the length of the JRs is $14.42 \times 1\frac{1}{3}$ which is 19.23 as read on the rafter tables.

f. *Example 2:* For a $\frac{1}{3}$ pitch roof with JRs placed at 24″ OC.
The common difference in the runs of JRs placed at 24″ OC is 2 feet.
The LL per foot run of the CR is 14.42
The common difference in the lengths of the JRs is 14.42×2, which is 28.84 as read on the rafter tables.

g. *Example 3:* For a $\frac{1}{3}$ pitch roof with JRs placed at 14″ OC.
The common difference in the runs of JRs placed at 14″ OC is $1\frac{1}{6}$ feet.
The LL per foot run of the CR is 14.42.
The common difference in the lengths of the JRs is $14.42 \times 1\frac{1}{6}$ which is 16.83 or $16\frac{7}{8}$ to the nearest $\frac{1}{8}$″.
This is an odd OC placing of JRs and is not shown on the rafter tables, but any problem may be solved in a similar manner.

3. Layout of the jack rafters on the pattern common rafter

a. The JRs are laid out on the pattern CR. This is then used for the layout of the JRs for the entire roof.

b. The example given is for a regular hip roof of $\frac{1}{3}$ pitch and rafters placed at 16″ OC. Figure 12.1 shows the plan of the JRs (not to scale).

4. The line length of the longest jack rafter for a roof of $\frac{1}{3}$ pitch and rafters 16″ OC

a. The pattern JRs are all laid off on the pattern common rafter.

b. The LL of the longest JR is $19\frac{1}{4}$″ shorter than the LL of the pattern common rafter.

c. The LL of the longest JR is measured from its ℄ intersection with the ℄ of the hip rafter through to the birdsmouth. See Figure 12.1.

d. From Line 1 on the face of the CR lay off $19\frac{1}{4}$″. This is the difference in the LLs of the CR and the JR. Draw the plumb Line 7. Do not square this line over. See Plate VI, page 111.

5. Shortening of the longest jack rafter

a. All rafters are shortened for the roof members against which they fit. The JRs fit against the HRs. Since the JRs meet the hip at 45° on plan, they must be shortened half the diagonal thickness on the run of the HR stock.

b. From Line 7 measure at right angles half the diagonal thickness of the HR stock and draw the plumb Line 8. Square this line over. See Figure 12.2.

6. Side cut of jack rafters: graphic method

a. Measure at right angles from Line 8 half the thickness of the JR stock. Draw the plumb Line 9 and square over. Draw *x–x* on the back of the rafter. Use pitch board No. 1.

b. Draw the side cut Line 10. This is drawn as shown on Plate IV from the top edge of the stock at Line 9, to intersect with the ℄ *x–x* on the back of the rafter at the center of the squared-over Line 8.

c. The actual length of the longest JR as prepared for nailing in position on the roof is from the cutting Lines 9–10 to the tail end of the rafter at the eaves.

d. On every regular hip roof without intersections there are four pairs of JRs. The four pairs of JRs are all laid off at one time from the pattern jack rafter length.

e. It will be recalled that the method of the layout of the gable studs from the pattern gable stud is somewhat similar to the layout of the JRs from the pattern JR.

7. Side cut of jack rafters: framing square method

a. Read on the fifth line of the framing square rafter tables, page 42, under the 8″ mark for a $\frac{1}{3}$ pitch roof:

Side Cut of Jacks Use . . . 10

b. The angle of the side cut is determined in the following manner:
Set one stair gauge on the outside edge of the blade on the framing square at the 12″ mark.
Set the other stair gauge on the outside edge of the tongue of the framing square at the 10″ mark.
Place the framing square on the back of the rafter with the 10″ setting towards the ridge end as at Figure 12.2.
Check the graphically drawn side cut line, Line 10.
Note that after this checking of the side cut, all other side cut lines are drawn with the use of the framing square.

c. After the longest JR has been laid off on the pattern common rafter, the second and succeeding JRs are all laid off (also on the pattern CR).
The common difference in the LLs of all the JRs for this roof is $19\frac{1}{4}$″ as read on the framing square.

d. From the squared-over Line 9 on the back of the rafter, lay off all the remaining differences in lengths of the JRs. Square over the spaces of $19\frac{1}{4}$″ on the back of the rafter as at Figure 12.3.

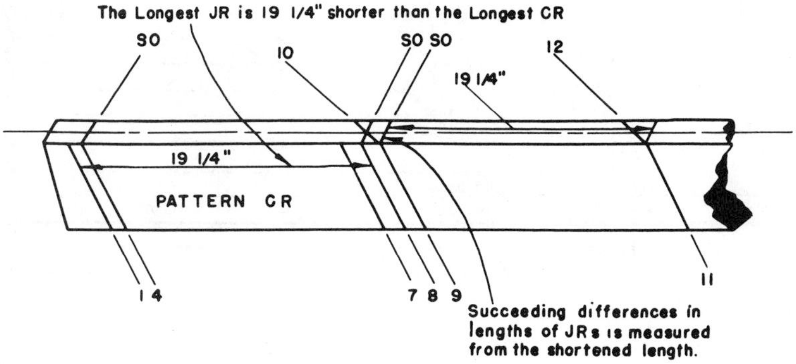

FIGURE 12.3

8. Layout of pairs of jack rafters from the pattern common rafter

a. From the squared-over Line 9 on the back of the pattern common rafter, lay off all the remaining common differences of $19\frac{1}{4}''$ the line lengths of the JRs. See Figure 12.4.

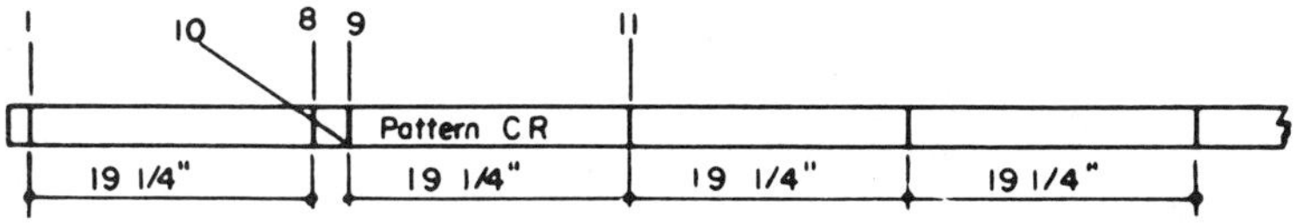

FIGURE 12.4

b. Pieces of stock of convenient length may now be placed in pairs on either side of the pattern common rafter. See Figure 12.5.

c. Square over four pairs of JRs from the pattern CR. Each set of four pairs will be $19\frac{1}{4}''$ different in length from tne next set.

d. With the stair gauges set on the framing square at 12″ and 10″, draw in the bevel side cut lines on the backs of the JRs as at Figure 12.5.

e. Reset the stair gauges to 8″ and 12″ for the plumb line of a $\frac{1}{3}$ pitch roof. Draw in the plumb cutting lines on all the pairs of JRs.

f. The completed pair of JRs ready for nailing into position on the roof would appear as at Figures 12.6(a) and 12.6(b).

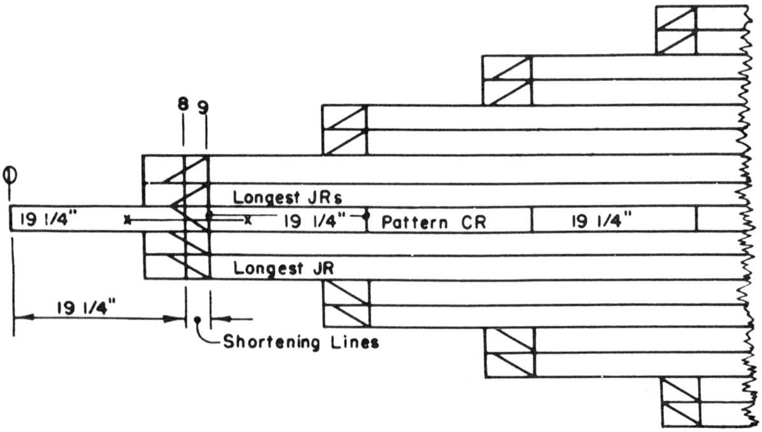

FIGURE 12.5

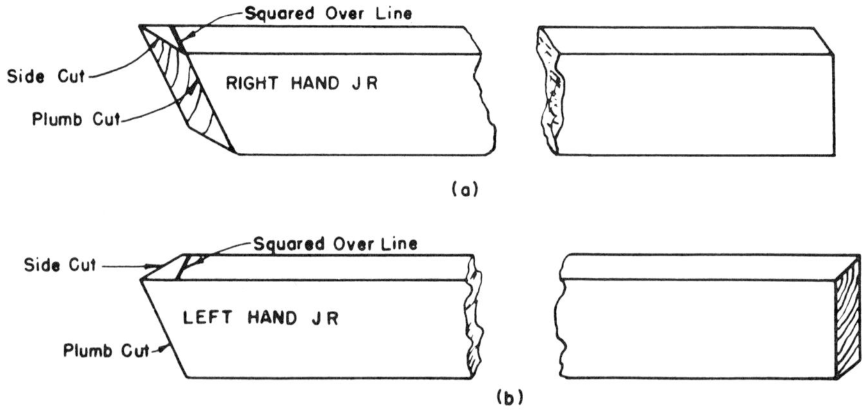

FIGURE 12.6

Review Questions

1. Fill in the blanks:

   The run of the first common rafter on a regular hip roof is equal to its ______ from the outside intersection of the wall plate.

2. The run of each jack rafter is equal to its ______ from the outside intersection of the wall plates. The birdsmouth of the jack rafter is ____.

3. Draw to the scale of $\frac{1}{4}''$ = 1′–0″ a pitch board for a regular $\frac{1}{3}$ pitch roof; then draw a sketch showing how you would apply it to a common rafter to draw the plumb cut.

4. Under which figure on the traditional framing square would you read the differences in lengths of jack rafter placed at 16″ on centers? ______.

5. Jack rafters placed at 24″ on centers for a $\frac{1}{3}$ pitch regular hip roof are shortened by ______ inches.

6. The plumb height of the hip rafter above the wall plate is ______ ____________________ of that of the common rafter.

13

Sheathing, Purlin, and Windblock Cuts

Roof sheathing may be shiplap, tongue and groove, common boards, plywood, or composition boards. It gives rigidity and provides insulation to the structure and also forms the base for the roof covering.

1. Specifications

The only parts of the specifications required by a carpenter to lay out a regular hip roof sheathing cut are the span and the pitch. In this chapter it will be assumed that the span is 28′–0″ and the pitch $\frac{1}{3}$.

2. Angle of the sheathing face cut

a. Figure 13.1 shows a single line drawing of a regular hip roof with one portion projected to a horizontal plane.

b. The right-angle triangle *a–b–c* has a base equal to the run, the height equal to the LL of the CR, and the hypotenuse equal to the LL of the HR.

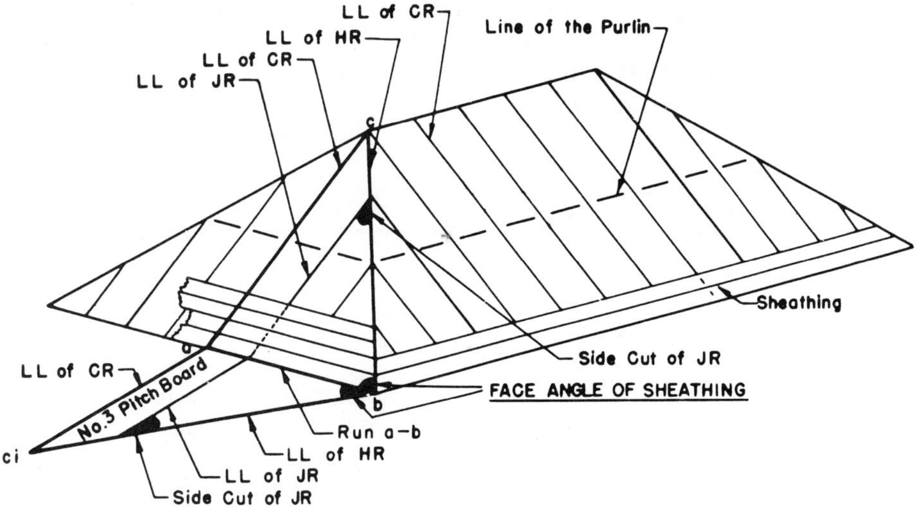

FIGURE 13.1. Regular hip roof.

c. This triangle is shown in horizontal plane at *a–b–ci*. The face angle of the sheathing is shown on both triangles at *b*. The sheathing face cut is the same angle as the side cut of the purlin and the windblock. See Figure 13.2.

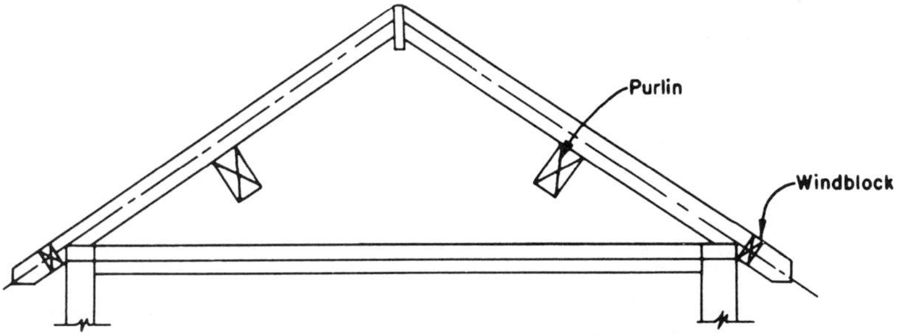

FIGURE 13.2. Section showing purlin and windblock.

3. Sheathing face cut (graphic method)

a. Make a pitch board to the scale of 1″ to 1′–0″. In this case the run is 14″, the height is the LL of the CR which is $16\frac{5}{8}$″±, and the hypotenuse is the LL of the HR which is $21\frac{11}{12}$″±.

b. Figure 13.3 shows two different positions of the pitch board as applied to a piece of common board to obtain the face angle to the sheathing. It must be carefully noted that in each case the angle of the sheathing cut is formed by the angle *b* which is at the intersection of the base and the hypotenuse of the pitch board.

c. Number 3 pitch board may be used to draw the cut of the sheathing face cut, the purlin side cut, the windblock side cut, and the JR side cut.

d. In each case apply the hypotenuse of the pitch board to the stock and lay off the cut as follows:

> for the sheathing face cut, mark and cut on the run;
> for the purlin side cut, mark and cut on the run;
> for the windblock side cut, mark and cut on the run;
> for the JR side out, mark and cut on LL of the CR.

This should be memorized.

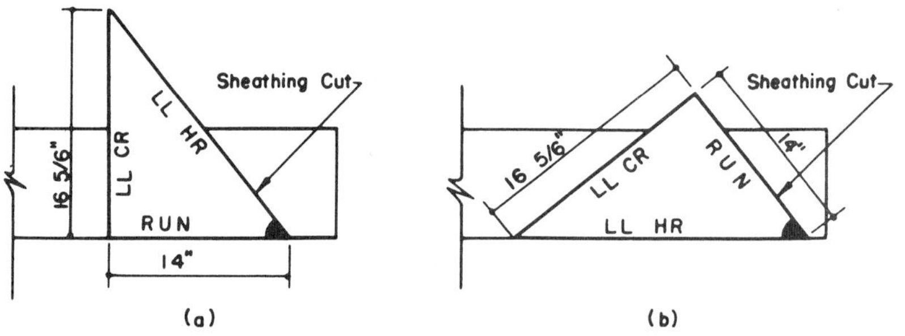

FIGURE 13.3

4. Sheathing face cut (framing square method)

a. With an understanding of the graphic method, it would follow that if the stair gauges are set at 14″ for the run and at 16$\frac{5}{6}$″ for the LL of the CR, the square may be used. Lay the square on a piece of sheathing stock as in Figure 13.4(a), and draw a line along the blade of the square.

b. Slip the framing square along the drawn line until the 12″ mark may be read at the bottom edge of the stock. See Figure 13.4(a). Notice very carefully that, with the framing square in this position on the stock, although 12″ is read on the blade, 10″

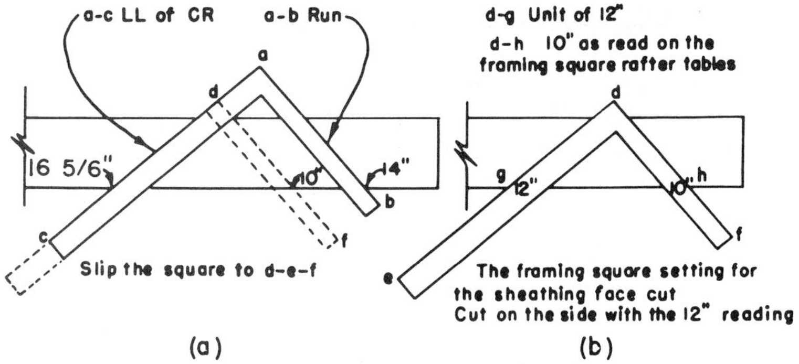

FIGURE 13.4

is read on the tongue. Lay off the sheathing cut on the 10″ side. Stair gauges are always set (for a regular hip roof) with one on 12″ for the layout of the sheathing face cut, the purlin side cut, and the windblock side cut.

c. Study the rafter tables (page 42) for a $\frac{1}{3}$ pitch roof. On the fifth line under the 8″ mark, read:

Side Cut of Jacks Use . . . 10

This refers to the 10″ in the previous paragraph. Set one stair gauge to 12″ and the other to 10″, and cut on the 10″ side. Use the same figures on the framing square for the face cut of the sheathing as are used for the side cut of the JRs. See Figure 13.4(b).

d. The framing square is based on a unit of 12″. In this case the unit is related to the LL of the CR which gives a relative figure of 10″ on the run side.

e. When the framing square is used to find the face cut of the sheathing for any regular hip roof, the only information required is the pitch.

Example: The pitch of a roof is $\frac{1}{4}$. What figures should be used on the framing square for the sheathing cut face angle? The rise per foot run of the roof is 6″. Read on the fifth line of the rafter tables (page 42) under 6″:

Side Cut of Jacks Use . . . $10\frac{3}{4}$

Set the stair gauges to 12″ and $10\frac{3}{4}$″. Mark and cut on $10\frac{3}{4}$″.

5. Sheathing edge cut (with the hand electric saw)

On many roofs the sheathing is nailed in place with the ends left running untrimmed past the hips. A chalk line is then struck from the ridge to the tail, over the ℄ of the hip. A hand power saw is used to cut through the sheathing from the ridge to the tail. This method is not practical for the valley sheathing.

6. Purlin and windblock cuts for a regular hip roof

a. The miter cuts of both the purlin and the windblock are the same.

b. The composite drawing, Plate VII, page 123, shows a section and partial plans of a hip roof. Both the purlin and the windblock are shown in section in Figure 13.2. The purlin is very important, giving support to the rafters, and, providing that it is mitered and fitted snug at the hips, giving strength to the roof as a whole. In many areas, slate, concrete, or clay tiles weighing many tons are used as roof covering. The roof must be engineered to meet the load. The windblock, as its name implies, is secured in place to close the open top of the wall frame to the underneath side of the roof sheathing.

c. The surfaces of the purlin that may be seen on the plan view are at *a–b* and *a–c* in section (Plate VII).

d. The center portion of the drawing shows a single line plan view of the miters of the purlin at *ai–bi–ci.* On the plan they appear as at 45°. The side cut angles of the purlin and the windblock are the same angle as the sheathing cut.

e. In the bottom portion of the drawing is shown a single line plan view of a portion of the hip roof with a developed true miter of the purlin face cut shown at *aii–aiii–biii.*

f. To establish the true miter cut of the purlin, the face and the edge must be projected to a horizontal plane. The face *a–b* of the purlin in section is assumed to have revolved about the point *a* to a horizontal plane *x–x* at *a–bii.* The true face width is now projected to the bottom portion of the drawing at *bii–biii* (Plate VII).

g. The shaded portion of the purlin on the plan is assumed to have revolved about *aii–aiii.* Study the right-angle triangle *aii–aiii–biii.*

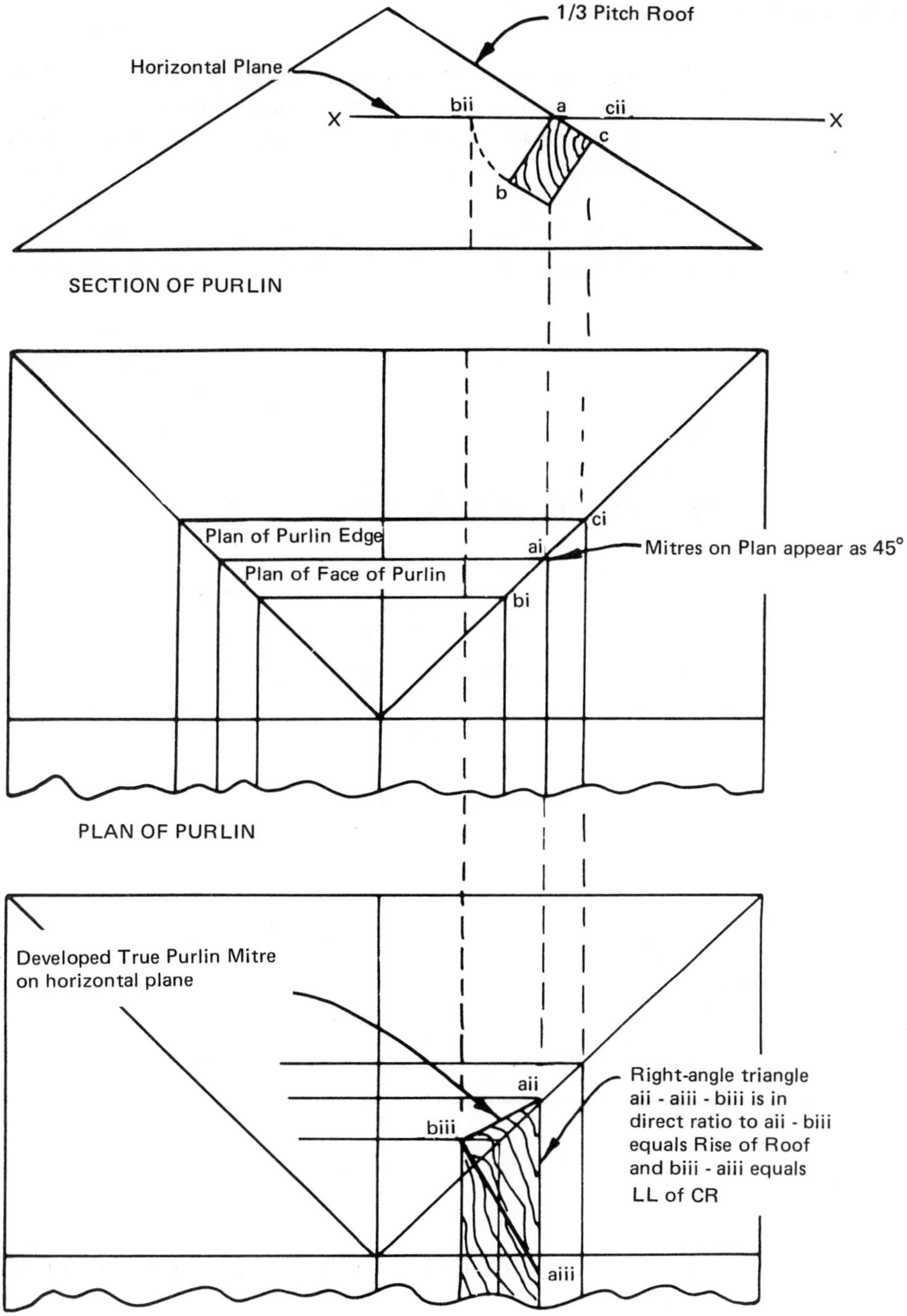

PLATE VII. Miter Cut of the Purlin

h. The developed right-angle triangle is in direct ratio to a right-angle triangle with a base equal to the rise of the roof and a height equal to the LL of the CR. Try it! On a piece of building paper or composition board, make a full-size layout of a section of a purlin for a $\frac{1}{3}$ pitch roof. Develop the face angle of the purlin. Check by placing a framing square with the 8″ mark of the tongue on *aii* and the 14.42″ ($14\frac{3}{8}''$) mark of the blade at *aiii*. Draw and cut the miter on the 8″ side.

i. For all (regular hip roof) purlin face cuts, set one stair gauge to the rise per foot run and the other to the LL of CR per foot run, in this case 8″ and $14\frac{3}{8}''$. Cut on the rise.

7. Purlin and windblock face cut (graphic method)

a. Assume a $\frac{1}{3}$ pitch roof with a span of 28′–0″. Make a No. 4 pitch board with a base of $9\frac{1}{3}''$ and a height of $16\frac{5}{6}''\pm$. These figures represent the run and the LL of the CR respectively.

b. Apply the hypotenuse of the pitch board to the top edge of the purlin stock and draw the face angle for the miter along the rise ($9\frac{1}{3}''$) side of the pitch board. See Figure 13.5(a).

c. It will be recalled that the side cut of the purlin is the same angle as the face cut of the sheathing.

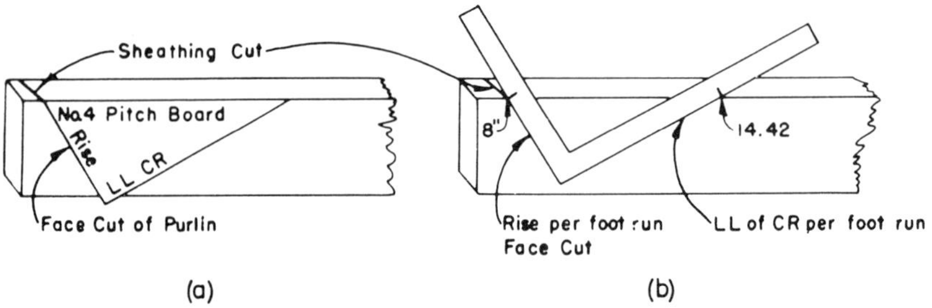

FIGURE 13.5

8. Purlin and windblock face cut (framing square method)

a. Consider a $\frac{1}{3}$ pitch roof. Set the stair gauges on the square, one to the rise per foot run of the roof (8″) and the other to the LL of the CR per foot run (14.42″) or $14\frac{3}{8}''$.

b. Apply the framing square to the top edge of the purlin and draw the face angle on the rise side (8″) as in Figure 13.5(b).

c. Figure 13.6 shows a complete single line drawing of a regular hip roof with the four pitch boards.

When sheathed, a regular hip roof is like a closed inverted hopper, as in Figure 13.7. Every cut on a regular hip roof may be developed by a close study of this section. Study Chapters 22 and 23.

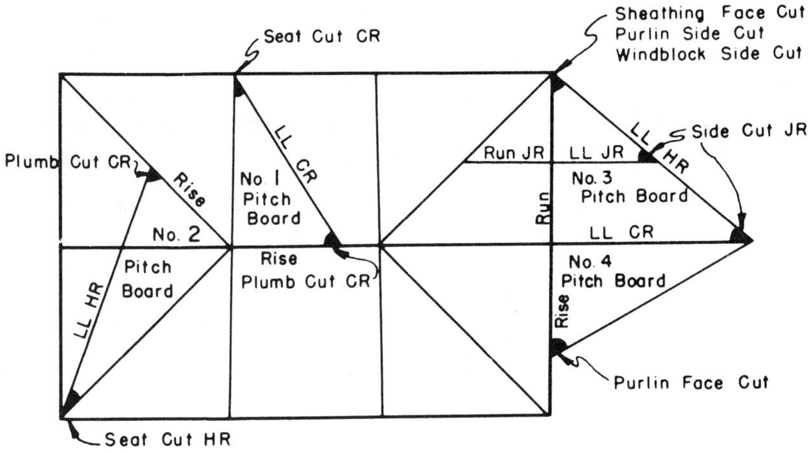

FIGURE 13.6

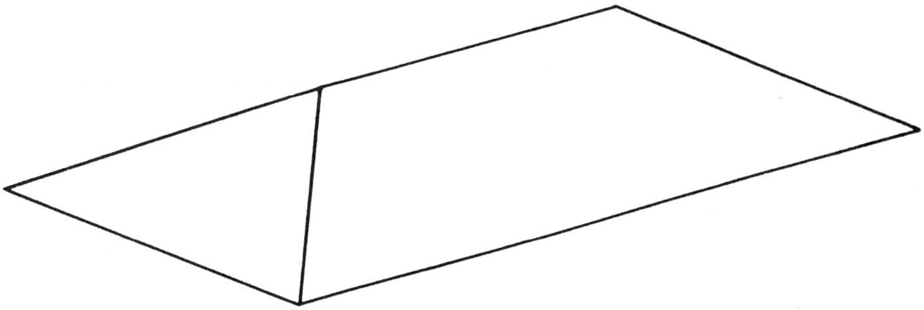

FIGURE 13.7. Closed inverted hopper.

Review Questions

Before answering the following questions, review the chapter.

1. The base of the No. 3 pitch board is equal (in scale) to the _______ of the roof.

2. The hypotenuse of the No. 3 pitch board is equal (in scale) to the _______________ of the _______________.

3. To draw the roof sheathing face cut, place the _______ of the No. 3 pitch board on the bottom edge of the sheathing board, and draw and cut on the _______ of the pitch board.

   Make a neat sketch showing the position of the No. 3 pitch board as placed on a roof sheathing board to mark the cut at the hip line.

4. When the traditional framing square is used to find the face cut of the sheathing for a regular hip roof, the only information needed is the pitch of the roof and a reading of the rafter tables.

   The pitch of a roof is $\frac{1}{4}$, read on the fifth line of the rafter tables under _______.

   For the side cut of jacks use _______.

   Set the stair gauges to _______ inches and _______ inches.

   Mark and cut on _______.

5. Make a line tracing of Figure 13.7, and name all the edges of the four different pitch boards.

14

Intersecting Roof of Equal Pitch

To obtain a pleasing effect and functional advantage, architects and house designers often draw roofs with an intersecting portion of equal pitch. The part of the roof covering the greatest area is called the major roof. The remaining part is called the minor roof. A Valley Rafter (VR), unless made into a half truss, is the weakest part of any roof. This chapter deals with the intersecting roof of equal pitch. See Plate VIII.

1. Specifications

MAJOR ROOF. Outside plate measurements 28′–0″ × 46′–0″. Pitch $\frac{1}{4}$. Ridge 1 × 6 (nominal). Cornice run 32″. CRs 16″ OC. Hip rafters 2 × 8. Valley rafters 2 × 8. Rafter tails to have plumb and level cuts to receive a fascia board and soffit.

MINOR ROOF. Span 28′–0″. Projection 10′–0″. Pitch $\frac{1}{4}$. Ridge 1 × 6 (nominal). Rafter tails to have plumb and level cuts to receive fascia board and soffit.

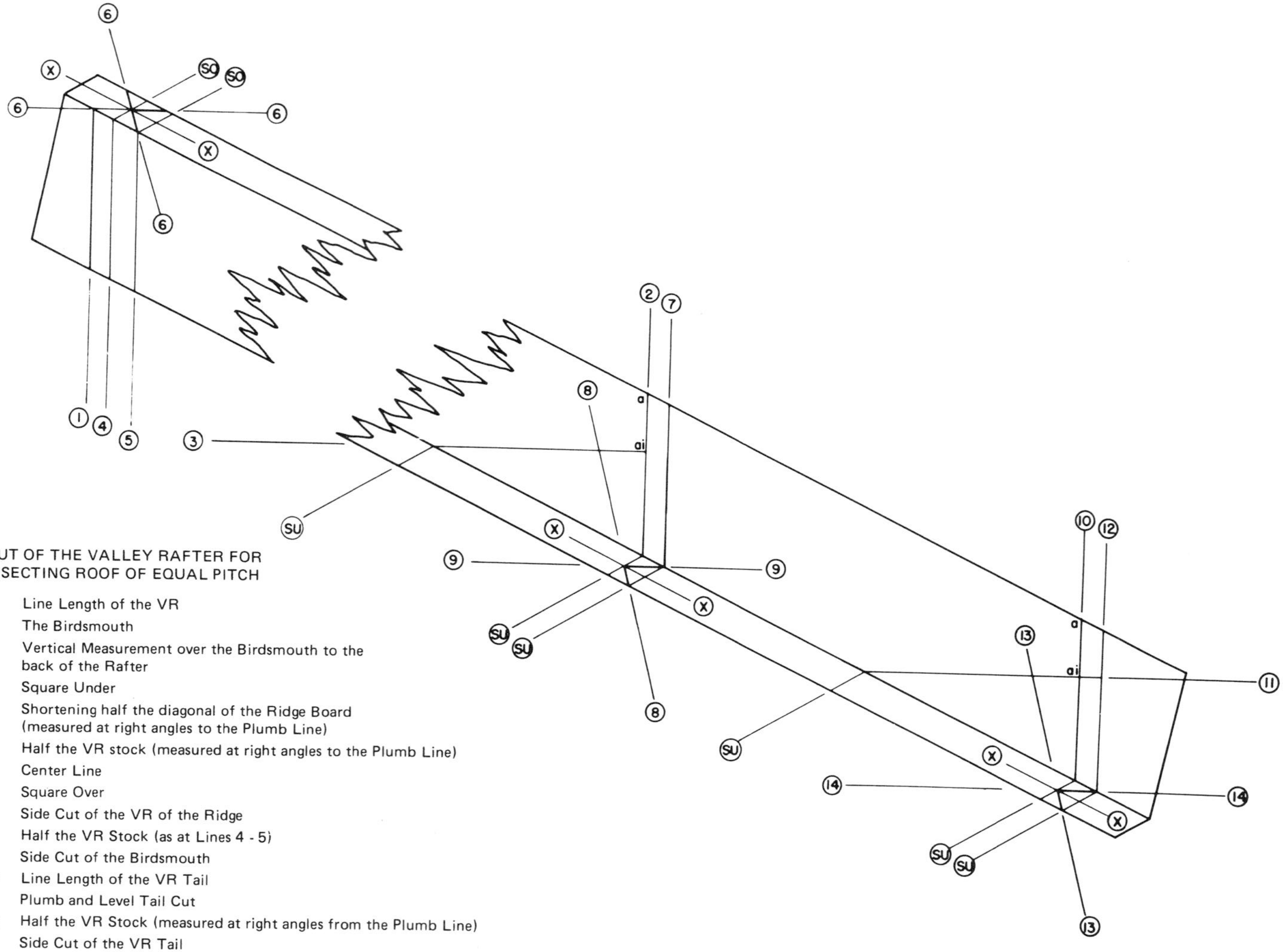

LAYOUT OF THE VALLEY RAFTER FOR
INTERSECTING ROOF OF EQUAL PITCH

Lines		
	1 - 2	Line Length of the VR
	2 - 3	The Birdsmouth
	a - ai	Vertical Measurement over the Birdsmouth to the back of the Rafter
	SU	Square Under
	1 - 4	Shortening half the diagonal of the Ridge Board (measured at right angles to the Plumb Line)
	4 - 5	Half the VR stock (measured at right angles to the Plumb Line)
	X - X	Center Line
	SO	Square Over
	6 - 6	Side Cut of the VR of the Ridge
	2 - 7	Half the VR Stock (as at Lines 4 - 5)
	8 - 8	Side Cut of the Birdsmouth
	2 - 10	Line Length of the VR Tail
	10 - 11	Plumb and Level Tail Cut
	10 - 12	Half the VR Stock (measured at right angles from the Plumb Line)
	13 - 13	Side Cut of the VR Tail
	14 - 14	Side Cut of the VR Tail

PLATE VIII. Layout of the Valley Rafter for Intersecting Roof of Equal Pitch

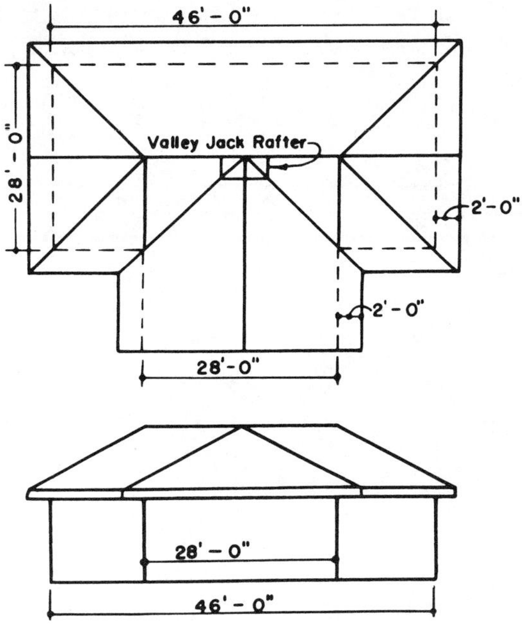

FIGURE 14.1. Intersecting roof of equal pitch.

2. Carpenter's working drawing

a. In building practice a partial plan of the roof is drawn to the scale of 1″ to 1′–0″ as shown in Figure 14.2.

b. Each roof member is later drawn to full-scale thickness on the single line $\frac{1}{12}$ scale drawing. This is very important and must be remembered.

c. The scaled drawings, layout, and erection of the CRs, JRs, and HRs for the main roof are the same as for those of an ordinary hip roof. In addition the CRs for the minor roof are the same as those of the major roof. The run and the rise for measurements of the VR are identical to those of the HR.

3. Run of the valley rafter

a. The run of the VR is the diagonal of a square, one side of which is equal to the run of the major roof and the other side equal to the run of the minor roof.

b. The VR of an equal pitch roof intersects both roofs at the center lines of their ridges and also at their wall plates.

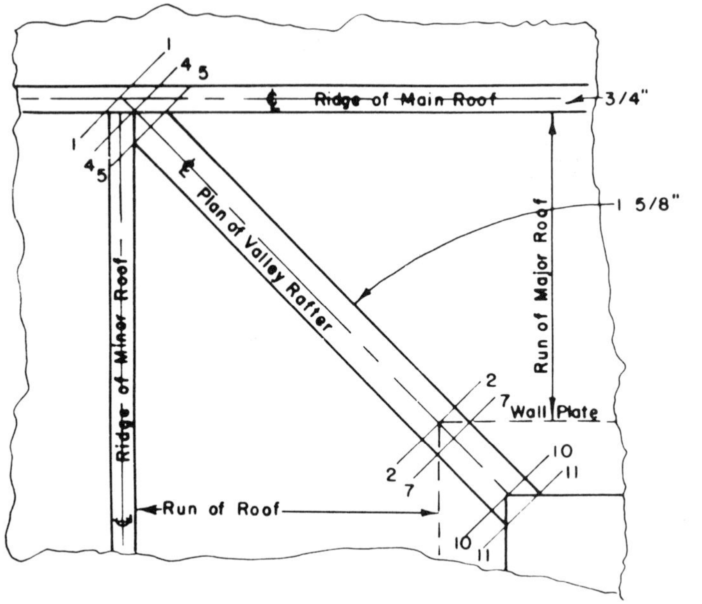

FIGURE 14.2. Partial plan.

4. Layout of the valley rafter

a. If the VR is to be laid out by the graphic method, use the same pitch board that was made for the HR.

b. If the roof is being laid out by the framing square method, set the stair gauges to 6″ and 17″ for a $\frac{1}{4}$ pitch roof. Near the left-hand end of the rafter stock, draw Line 1. See Figure 14.3.

c. The calculations for the LL of the VR should be made on the face of the VR a little to the right of plumb Line 1. The LL is the same as the LL of the HR.

d. Read on the rafter tables (page 42) under 6″ and on the second line:

Length of Hip or Valley per Foot Run . . . 18.00

This is the LL in inches of the VR per foot run of the CR.

e. Multiply 18″ by the number of feet of run in the CR which is fourteen. Then 18″ × 14 = 21′–0″ which is the LL of the VR.

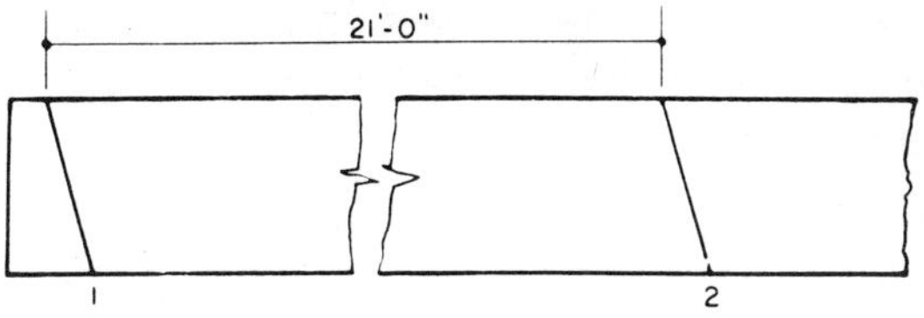

FIGURE 14.3

f. On the back of the rafter measure the LL 21′–0″ from Line 1 and draw Line 2. Square this line under. See Figure 14.3.

5. Valley rafter seat cut

a. Unlike the seat cut of the HR, there is only one level line at the birdsmouth.

b. Figure 14.4 shows the pattern CR. On Line 2 at *a–ai* is shown the perpendicular height of the CR from the seat cut to the back of the rafter.

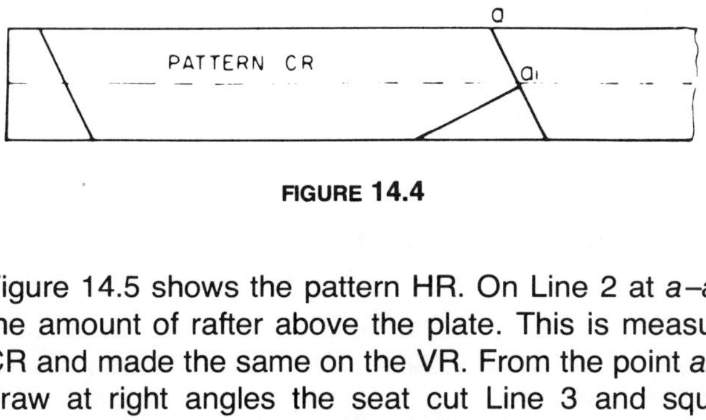

FIGURE 14.4

c. Figure 14.5 shows the pattern HR. On Line 2 at *a–ai* is shown the amount of rafter above the plate. This is measured on the CR and made the same on the VR. From the point *ai* on Line 2, draw at right angles the seat cut Line 3 and square under. Since a 2 × 8 rafter is to be used, a very large birdsmouth will be removed.

d. When the VR is in place on the roof, the ℄ on the back of the rafter is in the same plane as the backs of the CRs of both the major and the minor roofs.

e. The VR is neither dropped, dubbed, nor backed. The center of the back of the rafter is in the same plane as the backs of the CRs of both the major and the minor roofs.

f. On VRs with very large dimensions (especially for steep roofs), the edges of the rafter may have to be built up with angled strips as shown in Figure 14.6. The amount required to be built up would equal the amount that the edges of the HR are dropped.

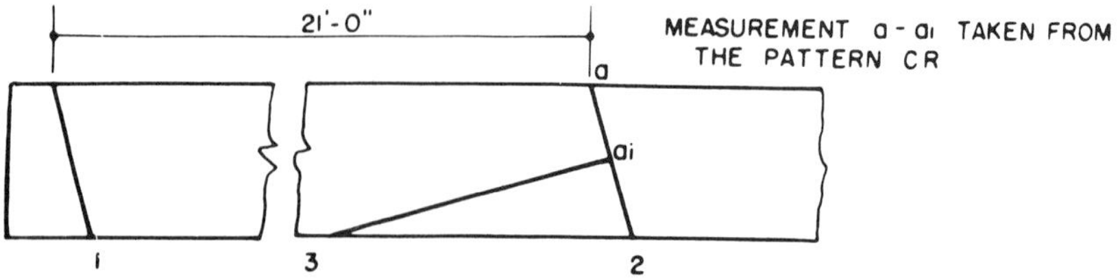

FIGURE 14.5

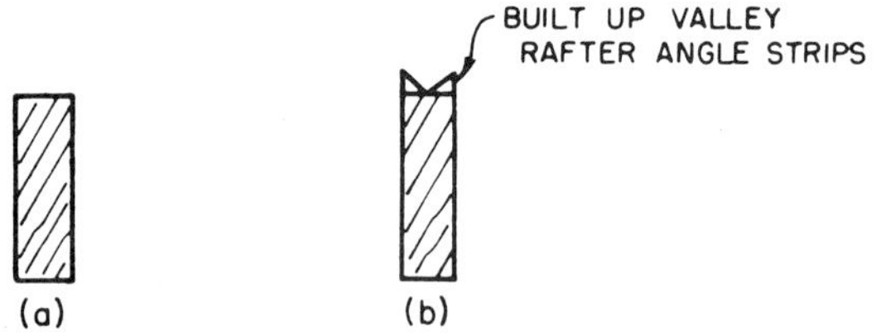

FIGURE 14.6

6. Shortening of the valley rafter

a. Lay out a full-scale drawing of the upper part of the VR and both ridges. See Figure 14.7. Measure the run of the shortening between Lines 1 and 4. The run is $\frac{1}{2}''$ for all practical purposes.

b. Measure the run of the shortening ($\frac{1}{2}''$) at right angles from Line 1 on the VR and square this line over. Lay off Line 4. See Figure 14.8.

7. Side cuts of the valley rafter at the ridge

a. The run of the side cuts of the VR is equal to half the thickness of the VR stock, in this case $\frac{13}{16}''$. Check this measurement on the drawing, Figure 14.7, between Lines 4 and 5.

b. Measure the run of the side cut ($\frac{13}{16}''$) at right angles from Line 4 on the VR, and draw Line 5. Square-over Line 5 to the opposite face of the VR. See Figure 14.8.

c. Draw a ℄ on the back of the stock as at *x–x*. Join the outside edges of the squared-over Line 5 to the ℄ on Line 4. Cut on Lines 5 and 6, from both sides.

d. The run of the shortening and the run of the side cut may be measured together on the plan. From Line 1 through Line 4 to Line 5 is $\frac{1}{2}''$ plus $\frac{13}{16}''$ for a total run of $1\frac{5}{16}''$. This is an important point and will be used in the layout of the VR for an intersecting roof of unequal pitch.

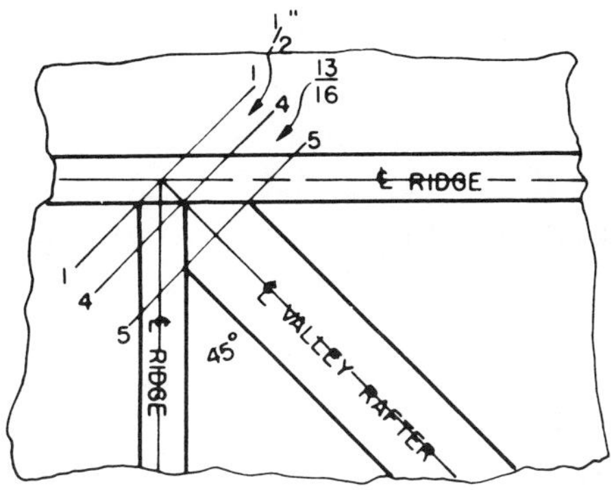

FIGURE 14.7

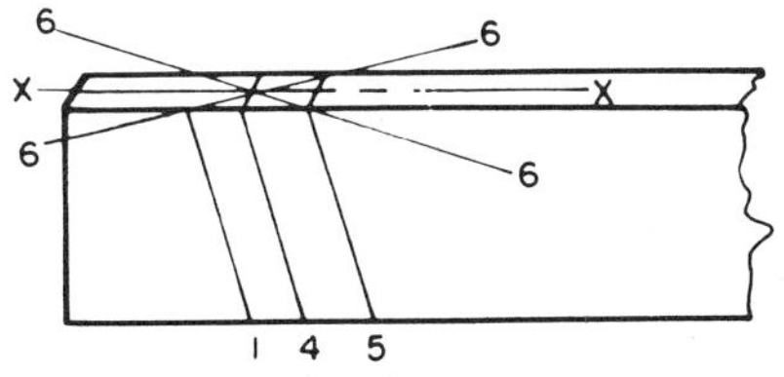

FIGURE 14.8

8. Birdsmouth side cuts

a. Figure 14.9 shows that portion of the VR where it intersects the major and minor roof wall plates. Plumb side cuts may be made on the birdsmouth so that the rafter will fit snug into the wall plate intersection.

b. Carefully note that the run of the side cuts of the birdsmouth is measured on the plan from Line 2 towards the tail end of the rafter at Line 7. The run is equal to half the thickness of the stock, in this case $\frac{13}{16}''$.

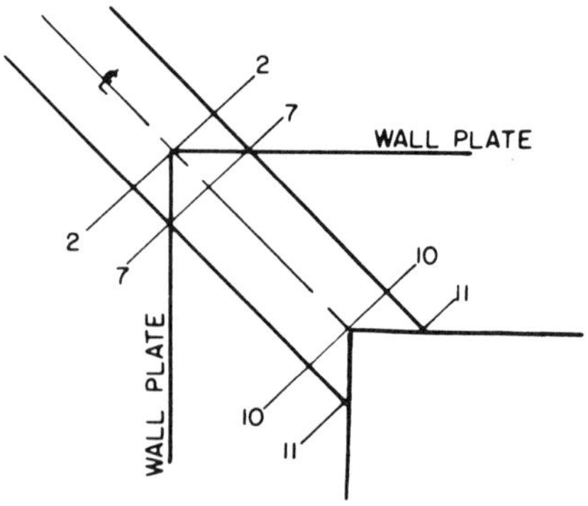

FIGURE 14.9

c. Measure back $\frac{13}{16}$″ (on the stock) at right angles from Line 2 and draw plumb Line 7. Square this line over to the opposite face of the stock.

d. Draw a ℄ on the underside of the rafter as at *x–x* (Figure 14–10). Join the outside edges of the squared-under Lines 8 and 9 to the center of the squared-under Line 2. Cut down to Line 3 on Lines 7 and 8 and then on Lines 7 and 9. Complete the operation by cutting on Line 3. See Figure 14.10.

e. A very common building practice is to make a square plumb cut on Line 7 only. In this case only the outside edges of the plumb cut will fit against the wall plates.

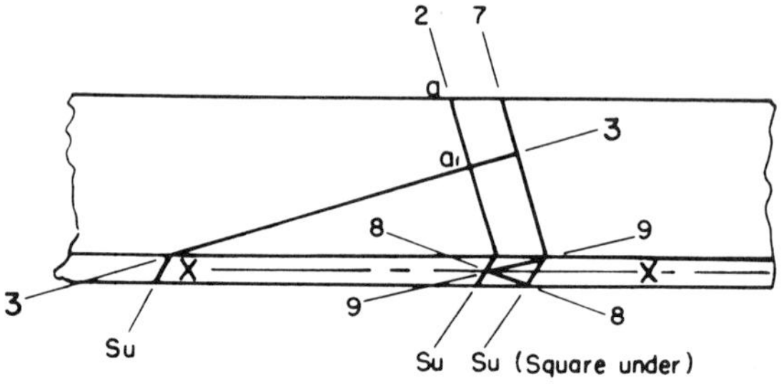

FIGURE 14.10

9. Line length of the tail of the valley rafter

a. The LL of the tail of the VR is laid off from plumb Line 2. The calculations for the LL should be made on the face of the rafter a little to the right of this line.

b. Read on the rafter tables of the framing square (on page 42) under 6″ on the second line:

Length of Hip or Valley per Foot Run . . . 18.00

(It will be recalled that this roof is specified to have a span of 28′–0″, the pitch $\frac{1}{4}$, and the overhang 32″, or 2′–8″.)

c. Multiply 18″ by the number of feet of run in the CR tail. The LL of the VR tail is therefore 18″ × $2\frac{2}{3}$, which equals 4′–0″.

d. Measure 4′–0″ along the back of the VR from Line 2, and draw the plumb Line 10. See Figure 14.11.

e. Check the LL of the tail with the framing square step-off method. Review Chapter 11.

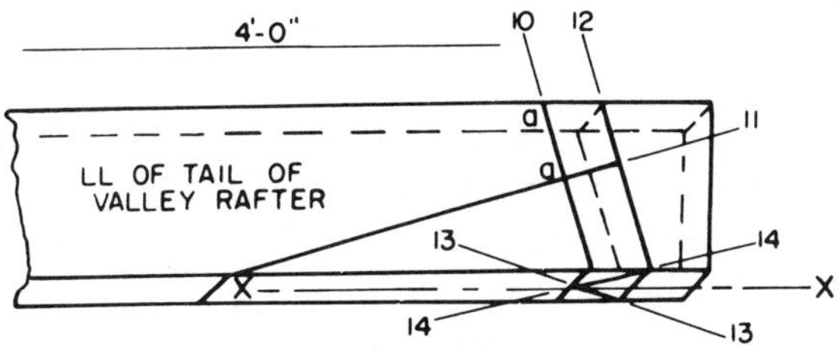

FIGURE 14.11

10. Tail end side cuts of the valley rafter

a. The fishtail end cuts of the VR are similar to the side cuts at the ridge.

b. The run of the side cuts of the VR tail is measured between Lines 10 and 12 on the working drawing. It is $\frac{13}{16}$″. The run is the same as the run between Lines 4 and 5 at the ridge. See Figure 14.7.

c. Measure the run of the side cut ($\frac{13}{16}$″) at right angles from Line 10 towards the end of the stock, and draw Line 12. See Figure 14.11.

d. On the bottom edge of the stock draw a ℄ as at *x–x*, and join the outer edges of Lines 13 and 14 to ℄ on the squared-over Line 11.

e. Draw a level Line 13 to intersect Line 10 at the same relative position as the birdsmouth seat cut Line 3 intersects Line 2. Cut on Line 11 and then on Lines 12 and 13 and Lines 12 and 14.

f. For the side cut lines, read the framing square rafter tables (page 42) under 6″ on the fifth line:

Side Cut of Hip or Valley Use $11\frac{5}{16}$″

You may set the stair gauges to $11\frac{5}{16}$″ and 12″. Apply the square to the back of the stock, and draw the side cut lines on the 12″ side. See Figure 14.11.

11. Valley jack rafter

a. The valley jack rafter (VJR) plumb cuts at the ridge and the valley are the same as the CR of the roof of which they form a part. The side cut angle is the same as that of the JR.

b. The 16″ OC spacings of the major and minor roofs should be laid off from the intersection of the ℄ of the ridges. See Figure 14.12. The run of the VJR on a regular roof is equal to its distance from the intersection of the ridges' ℄s. The run of the shortest VJR is therefore 16″.

c. The LL of the CR per foot run for a $\frac{1}{4}$ pitch roof is 13.42″ (see the rafter tables). Multiply 13.42 by the number of feet of run of the shortest VJR. This is $1\frac{1}{3}$′, and $13.42 \times 1\frac{1}{3} = 17\frac{7}{8}$″. The common difference in lengths of succeeding VJRs is $17\frac{7}{8}$″.

d. Make a full-scale working drawing of the 45° intersection of the VJR with the valley rafter. See Figure 14.12.

e. Lay off from the original plumb cut Line 1 of a CR the LL of the shortest VJR as at Line 5, Figure 14.13.

f. Measure the run between Lines 5 and 6 on the plan. In this case it is $1\frac{1}{8}$″. Lay off at right angles forward from Line 5, and draw Line 6. Draw a ℄ on the back of the stock as at *x–x*. On the back of the stock draw Line 7 through the intersection of the JR ℄ and Line 5 to Line 6. This is the side cut.

g. Check the side cut angle either with the No. 3 pitch board or with the framing square. On the fifth line of the rafter tables, page 42, under the 6″ mark read:

Side Cut of Jacks Use . . . $10\frac{3}{4}$

Set the stair gauges at $10\frac{3}{4}$″ and 12″, respectively, and cut on 12″.

h. Lay off the VJRs in pairs with a common difference in lengths of $17\frac{7}{8}$″. Study Figure 14.13.

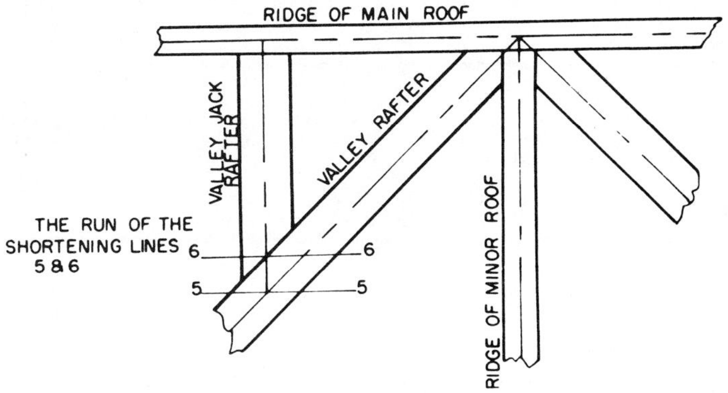

FIGURE 14.12. Intersection of VJR with VR.

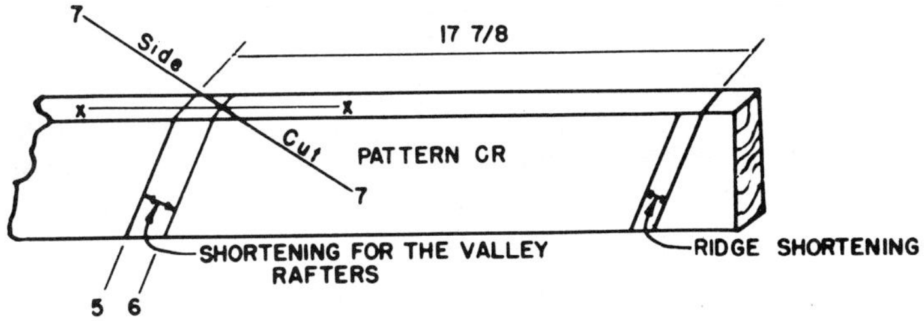

FIGURE 14.13

15

Jack Rafters

Some roofs are architecturally designed to give a distinctive appearance to a home. In such cases the roofs are often cut up into a series of hips, valleys, and gables. Figure 15.1 shows a plan of a roof with hips, valleys, and one gable.

1. Specifications

Plan of the major roof 28′–0″ × 48′–0″. Plan of the minor roof 8′–0″ × 10′–0″. Pitch $\frac{1}{4}$. Ridge $\frac{3}{4}$ × $5\frac{1}{2}$″ (finished size). Overhang 2′–0″. Rafters 24″ OC. CRs 2 × 4. Hips 2 × 8.

2. Supporting valley rafter

The Supporting Valley Rafter (SVR) is so called because it forms a valley rafter and supports a valley rafter. Its LL is equal to that of a HR for a roof of the same pitch and span. The LL should be determined by either the graphic or the framing square method. It is 18′–0″.

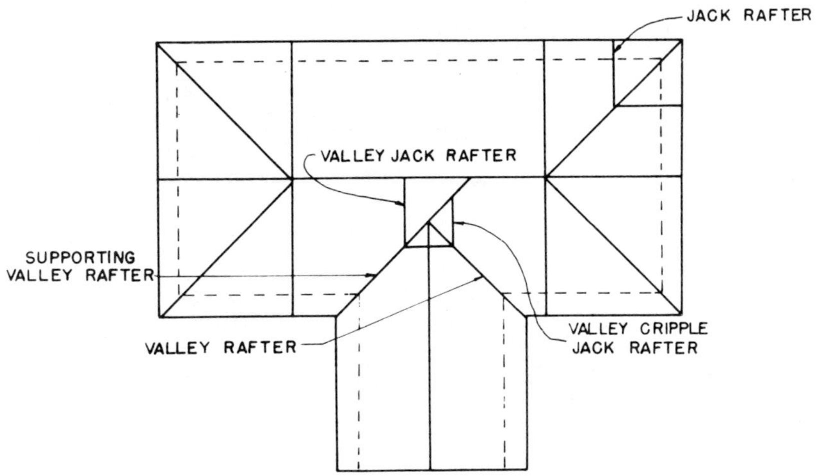

FIGURE 15.1. Plan of roof with hips, valleys, and one gable.

3. Ridge side cut

a. Figure 15.2 is a composite drawing showing a plan view of portions of the SVR at its intersection with the ridge, the VR, the wall plates, and the lines of the CR tail ends. The carpenter's working drawing should show all the roof members of this drawing to full-scale thickness. The SVR intersects the ridge at a 45° angle on the plan.

b. Figure 15.3 shows the SVR side cut. From plumb Line 1 on the SVR stock, lay off at right angles the shortening for the ridge. This is measured between Lines 1 and 4 on the plan. See Figure 15.2. It is half the diagonal thickness of the ridge board. For this roof it is $\frac{1}{2}''$ for all practical purposes. Draw the shortening plumb Line 4. Lay off at right angles from Line 4 half the thickness of the SVR stock. In this case it is $\frac{13}{16}''$. From the top edge of the stock at Line 5, draw the side cut line to intersect with the ℄ on Line 4. Draw the side cut Line 6. Cut on Lines 6 and 5.

4. The birdsmouth

a. The SVR is neither dropped nor raised. *It is backed on one side only above the point of its intersection with the VR.*

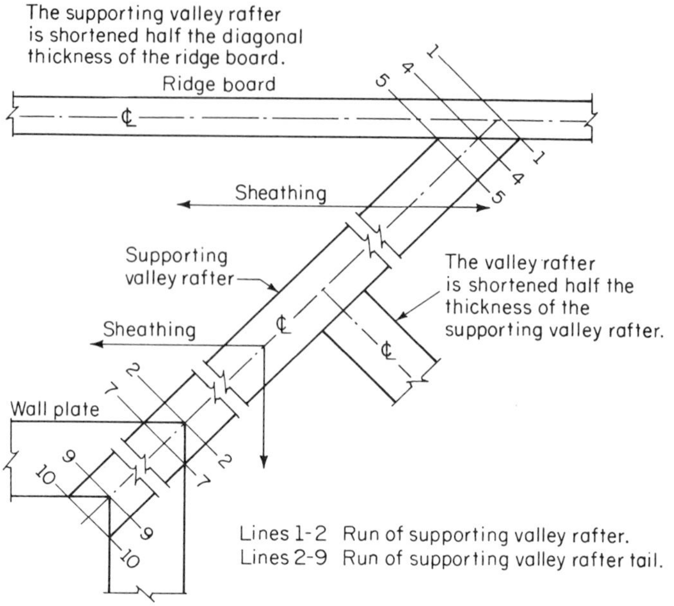

FIGURE 15.2. Plan of portions of the SVR.

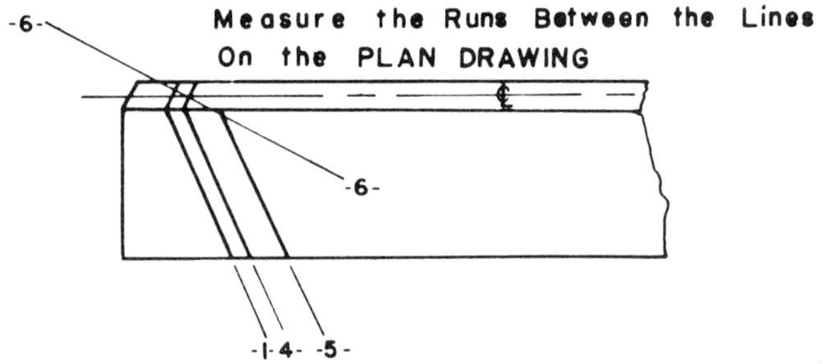

FIGURE 15.3

Measure down, on Line 2 of the CR, the amount of rafter over the plate; then measure a like amount down on Line 2 of the SVR. From this point draw the level cut of the birdsmouth as at Line 3, Figure 15.4.

b. From LL Line 2, lay off at right angles (towards the tail end of the rafter) the run measurement between Lines 2 and 7 on the plan as at Figure 15.2. Square Lines 2 and 7 under the bottom edge of the rafter and up the opposite face. See Figure 15.4. Draw a ℄ *x–x* on the bottom edge of the stock. Join the outside edges of Line 7 to the squared-under Line 2 at the ℄. Cut on Lines 8 and 7 to the seat cut line. Cut on the level Line 3 to remove the birdsmouth.

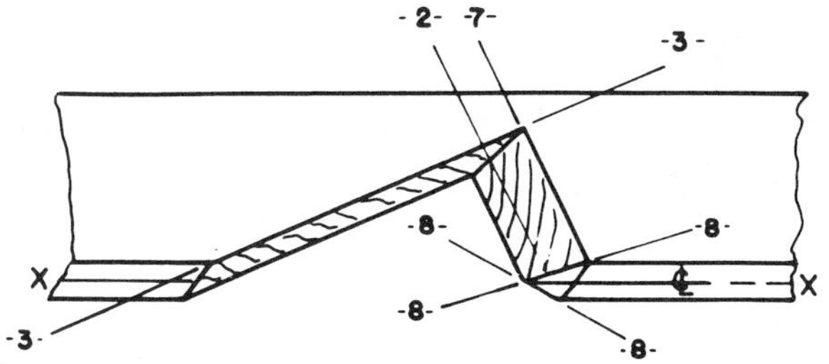

FIGURE 15.4. The supporting valley rafter birdsmouth.

5. Tail cut

a. The tail LL of the SVR is laid off from Line 2 in exactly the same manner as if it were an HR. See Figure 15.5. The overhang is 2′–0″, and the pitch is $\frac{1}{4}$. The LL is therefore 3′–0″. From Line 2 lay off the LL, and draw the tail plumb Line 9.

b. From Line 9 lay off at right angles (towards the end of the stock) the run measurement between Lines 9 and 10 on Figure 15.2. Draw plumb Line 10. Square Lines 9 and 10 under the bottom and up the opposite face of the rafter. Draw a ℄ *x–x* on the bottom edge, and join the outside edges of Line 10 to the ℄ at the squared-under Line 9.

c. The level line of the SVR tail cut is the same vertical measurement down on Line 9 as the seat cut line is down on Line 2. Draw the level Line 11. Cut on Line 11, and then cut on Line 10 and 12 from both sides.

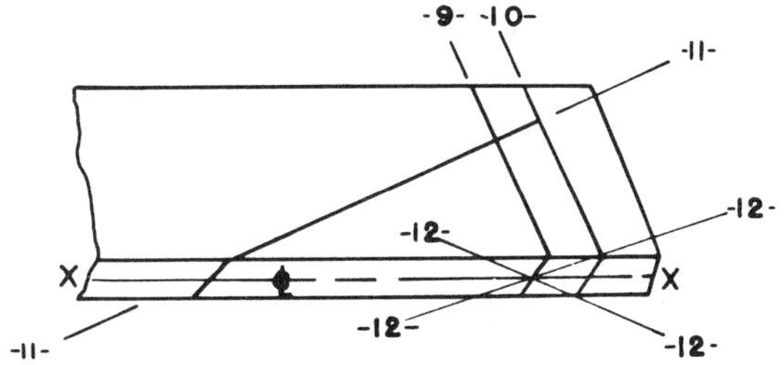

FIGURE 15.5

6. Valley rafter

a. The valley rafter birdsmouth and tail cuts are the same as those of the SVR. The LL of the VR is laid out by the same methods as would be used for an HR. It has the same pitch as the SVR, and the run is that of the minor roof.

b. The VR intersects the SVR at a 90° angle. It will require shortening half the thickness of the SVR stock. It will have a square cut. See Figure 15.2.

7. Jack rafter and the valley-cripple jack rafter

a. Study the different types of JRs at Figure 15.1. The JRs run from the wall plate to the ridge. The valley-cripple jack rafters run from the valley rafter to the supporting valley rafter.

b. From the intersection of the supporting valley rafter and the valley rafter ℄s, lay off (for this roof) at 24″ OC the shortest valley-cripple jack rafter. The shortest rafter will have a *run* equal to twice the distance it is from the point of intersection of the two valley rafters. See Figure 15.6.

c. The common difference in the lengths of the valley-cripple jack rafters is twice the OC spacings.

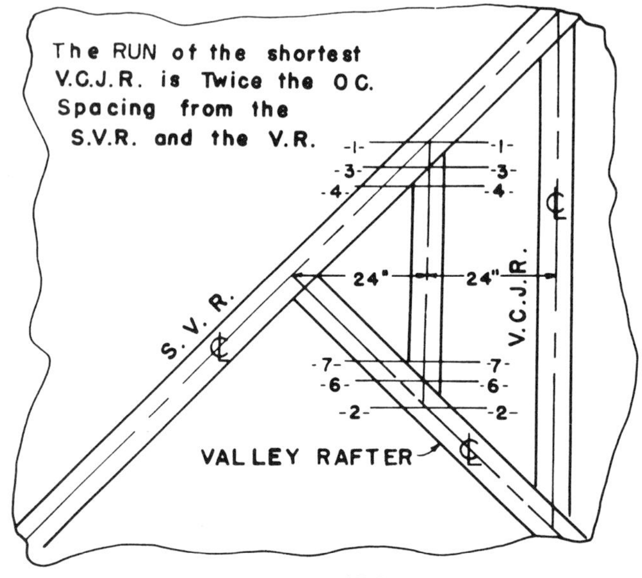

FIGURE 15.6

d. The valley-cripple jack rafter will require shortening at both ends. See Figure 15.7.

e. Lay off the LL as in Figures 15.6 and 15.7. Allow for the shortening, which is half the diagonal thickness of the rafter against which it fits, in this case $1\frac{1}{8}''$ to Line 3. Lay off half the thickness of the valley-cripple-jack-rafter stock as at Line 4. Draw the side cut Line 5.

f. The rafter is shortened at both ends. The side cuts are both made either with the No. 3 pitch board, or with the framing square set for the side cut of the hip or jack rafter for the pitch of the roof to be cut. For this roof cut on Lines 4 and 5 then on Lines 7 and 8. See Figure 15.7.

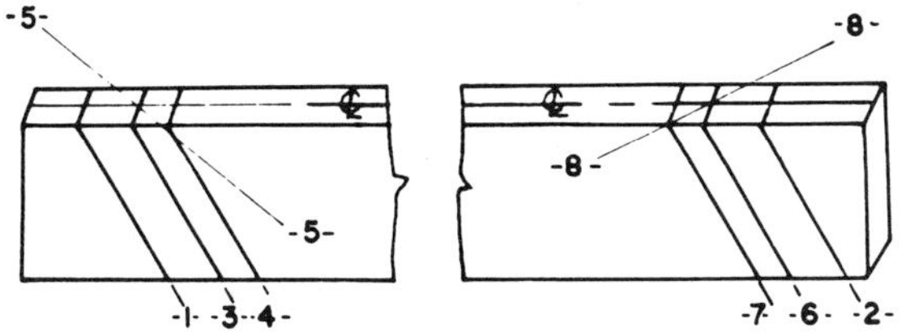

FIGURE 15.7

8. Valley jack rafter

This member (VJR) fits against the ridge and the valley rafter. See Figure 15.1. It must be shortened for both the ridge and the valley rafter.

16

Intersecting Roof of Unequal Pitch: Level Cornice, Equal Cornice Run

To obtain a pleasing effect, architects and designers of buildings often give a roof more than one slope. This chapter deals with such a roof, which is shown in plan and elevation in Plate IX on page 145. Major and minor roofs have identical rises but different slopes. The cornices have the same run, and they are level.

1. Specifications

MAJOR ROOF. Outside plate measurements 28′–0″ × 46′–0″. Pitch $\frac{1}{4}$. Ridge 1 × 6 (nominal). Cornice run 32″. CRs 16″ OC. Hips 2 × 8. Valley rafters 2 × 8. Rafter tails to have plumb and level cuts to receive a fascia board and soffit.

MINOR ROOF. Span 12′–0″. Length of the projection 10′–0″. Pitch to be established. Ridge 1 × 6 (nominal). Rafter tail to have plumb and level cuts to receive a fascia board and soffit. Cornice to be at the same level as the major roof and also with a run of 32″.

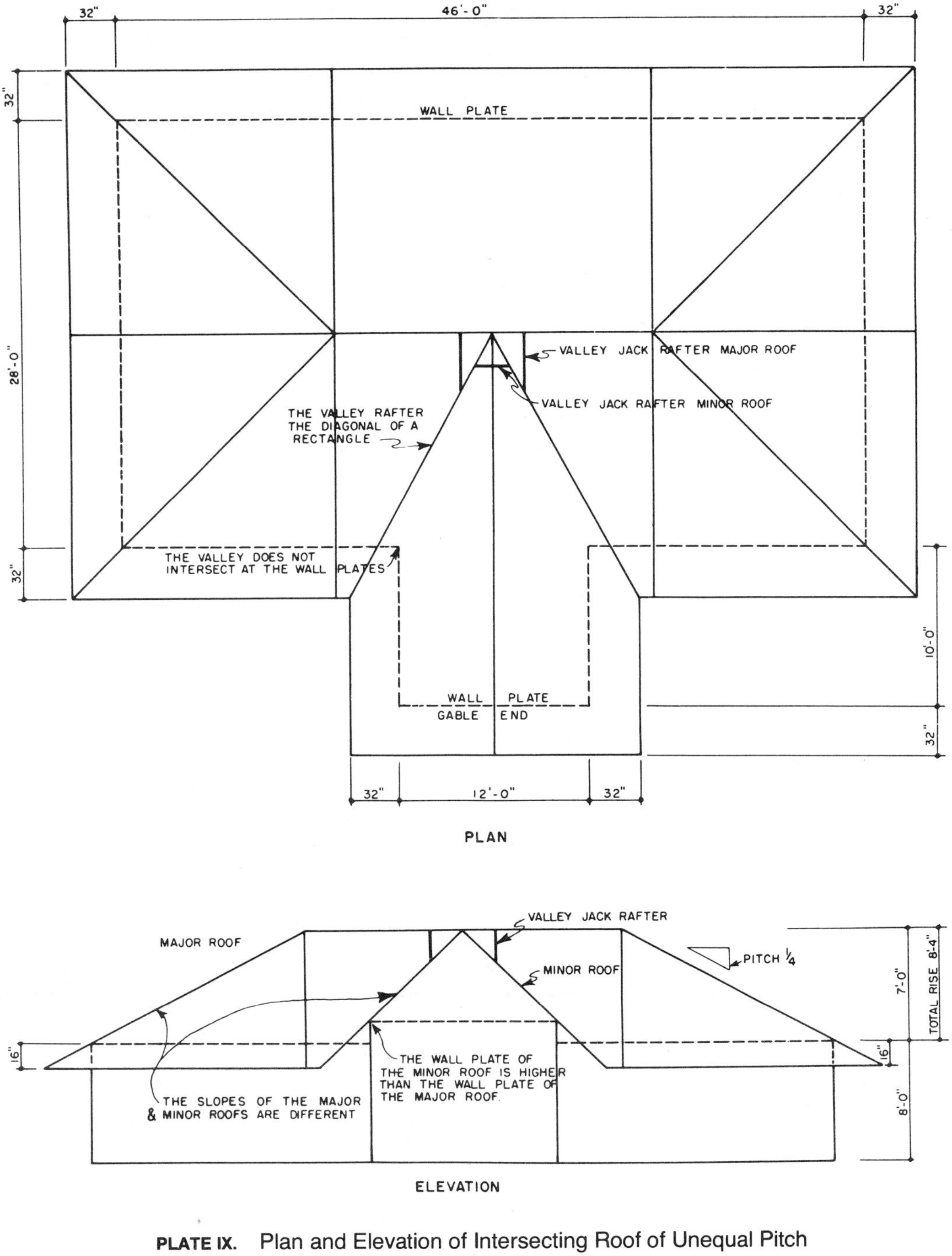

PLATE IX. Plan and Elevation of Intersecting Roof of Unequal Pitch

2. Carpenter's working drawing

a. In building practice a single-line partial plan of the roof is drawn to the scale of 1″ to 1′–0″, as shown in Figure 16.1.

b. Each roof member is later drawn to full scale thickness on the single line $\frac{1}{12}$ scale drawing. *This is very important and must be remembered.*

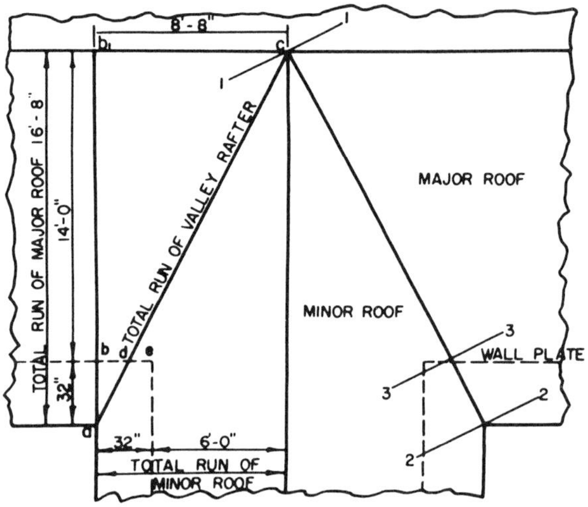

FIGURE 16.1

3. Total run of the valley rafter

a. Unlike the VR for a roof of equal pitch, the VR for an intersecting roof of unequal pitch is the diagonal of a rectangle. The runs of the HRs and VRs are not parallel. They are also of different lengths.

b. The VR of an unequal-pitch roof intersects both roofs at the ridge, and also at the **tail end** of the rafter. It does not intersect the major and minor roof wall plates. See Figure 16.1, Line 1–2.

c. The total LL is from the plumb line at the ridge (excluding the shortening) to the intersection of the overhang at the cornice.

4. Run of the tail of the valley rafter

The birdsmouth is measured back from the tail plumb line.

5. Determining the total run of the valley rafter

a. The total run of the VR is the diagonal of a rectangle. One side is equal to the run of the major roof including the cornice run. The other side is equal to the run of the minor roof including the cornice run. See Figure 16.1.

b. There are three ways to determine the total run of the VR:

by mathematics
by measuring the total run on the scaled drawing
by measuring across the correct place on the framing square.

(See Figure 16.2.)

6. Mathematical total run of the valley rafter

Step 1: The total run of the CR of the major roof is equal to half the span plus the cornice run, that is, 14′–0″ plus 32″. The total run is 16′–8″.

Step 2: The total run of the CR of the minor roof is equal to half the span plus the cornice run, that is, 6′–0″ plus 32″. The total run is 8′–8″.

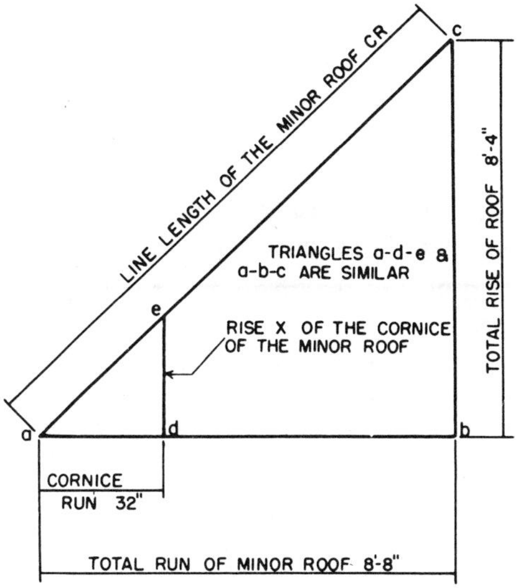

FIGURE 16.2

Step 3: The total run of the VR is the diagonal of a rectangle, the sides of which are 16′–8″ and 8′–8″, respectively. This equals, by calculation, 18′–9$\frac{3}{8}$″ to the nearest $\frac{1}{8}$″.

7. Measured run of the valley rafter

a. Provided that an accurate scaled drawing has been made, the run of the VR may be measured. The total run is from Line 1 at the intersection of the ridges' ℄s to Line 2 at the intersection of the cornice runs. See Figure 16.4.

b. The total run may also be measured by laying off to scale (along the machined edges) of any piece of composition board the major and minor total runs and the total rise, and then measuring the diagonal. The larger the scale used, the more exact the reading.

c. The total run of the VR may be made by measuring across the framing square. Using the $\frac{1}{12}$ scale, set one stair gauge to the total run of the major roof; set the other gauge to the total run of the minor roof. Measure the diagonal with a $\frac{1}{12}$ scale rule.

d. The mathematically-calculated total run should always be checked by measuring from the scaled drawing.

8. Location of the valley rafter birdsmouth on the major roof wall plate

a. The ℄ of the birdsmouth of the VR intersects the major roof wall plate a little distance from the minor roof wall plate. This may be determined by mathematics, or it may be measured on the scaled drawing. See Figure 16.1.

b. The mathematical calculation may be made as follows:

Step 1: The total run of the major roof *a–bi* is 16′–8″.
The total run of the minor roof *bi–c* is 8′–8″.
The total run of the VR *a–c* is 18′–9$\frac{3}{8}$″ to the nearest $\frac{1}{8}$″.
See Step 3, Para. 6.

Step 2: The run of the cornice *a–b* is 32″.
The distance *b–d* is to be calculated.

Step 3: The triangle in Step 1 *(a–bi–c)* is a triangle similar to the one in Step 2 *(a–b–d)*.

Step 4: It follows that *a–b* is to *b–d* as is *a–bi* to *bi–c*. Substituting the measurements:

32″ is to *b–d* as is 16′–8″ to 8′–8″

$$\frac{32}{bd} = \frac{16'\text{–}8''}{8'\text{–}8''} = \frac{200}{104}$$

Step 5: Expressed in inches, 32 is to *b–d* as 200 is to 104 inches.

32	200)3328(16	128	$16\frac{5}{8}''$ to the
104	200	200	nearest $\frac{1}{8}''$
128	1328		
320	1200		
3328	128		

The distance *b–d* is $16\frac{5}{8}''$ to the nearest $\frac{1}{8}''$.

Step 6: The position of the ℄ of the VR birdsmouth from the minor roof wall plate is *b–e* minus *b–d*, where *b–e* is the run of the cornice and *b–d* is calculated to be $16\frac{5}{8}''$.

Step 7: Expressed as a formula, the position of the ℄ of the VR from the minor roof wall plate is:

$$\text{Cornice run} - \frac{\text{minor roof total run}}{\text{major roof total run}} \times \text{cornice run}$$

$$32 - \frac{8'\text{–}8''}{16'\text{–}8''} \times 32 = 32 - \frac{104}{200} \times 32$$

$$32 - \frac{3328}{200} = 32 - 16\tfrac{5}{8}'' = 15\tfrac{3}{8}''$$

Step 8: Check the mathematical calculation with the scaled drawing.

Step 9: On the job, measure $15\frac{3}{8}''$ from the intersection of the two wall plates along the major roof. This is the position of the ℄ of the VR birdsmouth.

9. Difference in heights of wall plates: mathematical method

a. Carefully examine the elevation at Plate IX, page 145. The heights of the wall plates are different. This is because both the cornices have the same level and run; but the slopes of the roofs are different.

b. The difference in the heights of the wall plates may be calculated mathematically, or the difference may be measured directly from a full-scale drawing of both cornices.

c. Specifications:

Major roof span 28′–0″, run of cornice 32″, pitch $\frac{1}{4}$.
Minor roof span 12′–0″, run of cornice 32″, pitch unknown.

Step 1: Consider the total rise of the major roof.

Step 2: The rise of the major roof is $\frac{1}{4}$ of 28′–0″, which is 7′–0″.

Step 3: **The rise of the major roof cornice** is $\frac{1}{4}$ of twice the cornice run. Thus $\frac{1}{4}$ of (32″ plus 32″) is 16″.

Step 4: The total rise of the roof is 7′–0″ plus 16″, which is 8′–4″.

Step 5: The total rise of the minor roof is the same as the total rise of the major roof; namely, 8′–4″. See Plate IX, page 145.

Step 6: The total run of the major roof is equal to the run of the major roof CR plus the run of the cornice. This is 14′–0″ plus 32″, which is 16′–8″.

Step 7: The total run of the minor roof is equal to the run of the minor roof CR plus the run of the cornice. This is 6′–0″ plus 32″, which is 8′–8″. See Figure 16.1.

Step 8: A triangle formed by the total rise, the total run, and the total length of the CR of the minor roof is a similar triangle formed by the rise of the minor roof cornice and the run and the LL of the overhang.

Step 9a: The minor roof CR has a total rise of 8′–4″ or 100″. It has a total run of 8′–8″ or 104″.

Step 9b: The minor roof cornice has a rise unknown and a run of 32″. See Figure 16.2.

Step 10: 100″ is to *x*″ as 104″ is to 32″

$$\frac{100 \times 32''}{104} \text{ is } 30\tfrac{3}{4}'' \text{ approx.}$$

Step 11: The difference in the heights of the major and minor roof wall plates is equal to the rise of the minor roof cornice minus the rise of the major roof cornice. This is $30\frac{3}{4}$ minus 16″ which equals $14\frac{3}{4}$″.

Step 12: When the roof is being constructed, the minor roof wall plate should be built up $14\frac{3}{4}$″ higher than the major roof wall plate.

10. Difference in heights of wall plates: full scale graphic

a. A quick, simple, and practical way of finding the differences in the heights of the wall plates is to draw both the cornice outlines to full scale.

b. Select or make a straight edge on a piece of composition board, and lay off the cornice as at *a–b* Figure 16.3(a).

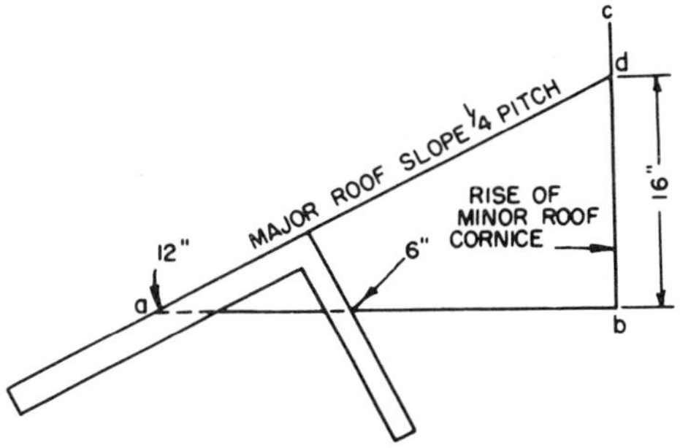

FIGURE 16.3(a)

c. Draw a perpendicular of indefinite height from *b* to *c.* The line *b–c* represents the wall or building line. Using the framing square and the stair gauges, place the 12″ mark at **a** and the 6″ mark on the line *a–b.* Draw and extend the $\frac{1}{4}$ pitch slope of the roof to intersect *b–c* at *d.* The rise of the major roof cornice is measured at *b–d.* It is 16″.

d. Reset the stair gauges on the framing square to the scale of the total run and the total rise of the minor roof. The scaled total run is $8\frac{8}{12}$″, and the scaled total rise is $8\frac{4}{12}$″.

e. Lay the framing square on the drawing with the $8\frac{8}{12}$″ mark at *a* and the $8\frac{4}{12}$″ mark on the line *a–b* as in Figure 16.3(b).

f. Note carefully that the run of the framing square is on the line of slope in each case.

g. The rise of the minor roof cornice is measured from *b* to *e.* It is $30\frac{3}{4}$″. The rise of the major roof cornice is measured from *b* to *d.* It is 16″. The difference in the rises is $14\frac{3}{4}$″. When constructing the roof, the minor roof wall plate should be built up $14\frac{3}{4}$″ higher than the major roof wall plate. See Plate IX, page 145.

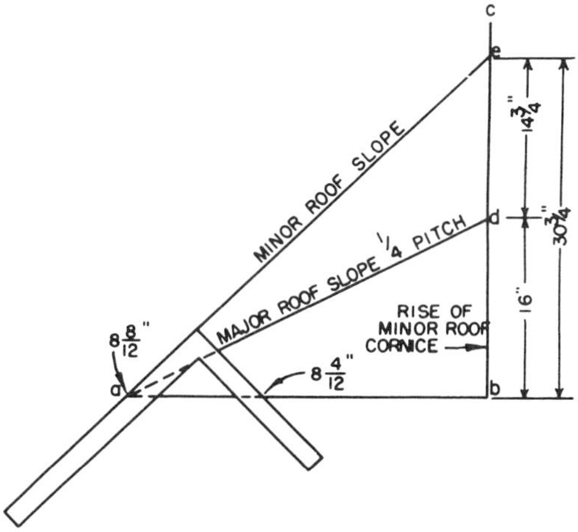

FIGURE 16.3(b)

11. Layout of the valley rafter

a. Unless a VR is made in the form of a half truss, it is the weakest part of any roof. In this section it will be assumed that 2 × 8 stock is to be used.

b. A very careful study should be made of Figure 16.4 with this part of the text.

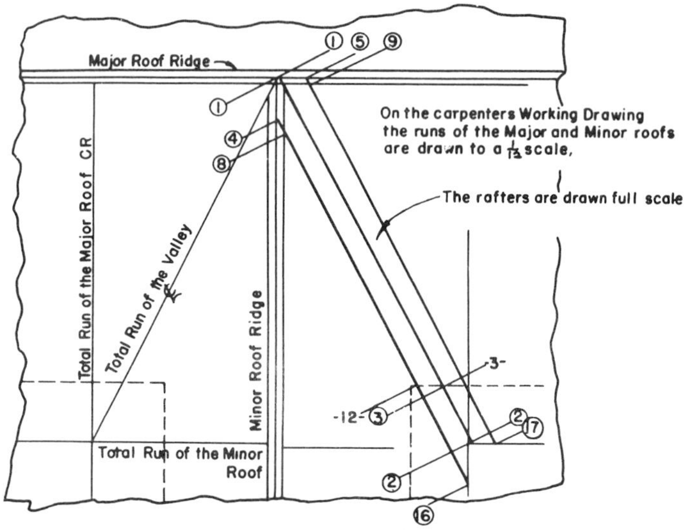

FIGURE 16.4

12. Plumb line of the valley rafter

a. Use the $\frac{1}{12}$ scale on the framing square. Set one stair gauge to to the total rise $8\frac{4}{12}''$ on the square. Set the other stair gauge to the total run of the VR $18\frac{4}{12}''$ on the opposite side of the square.

b. Near the left-hand end of the rafter stock with the crown side away from the carpenter, draw plumb Line 1. See Figure 16.5.

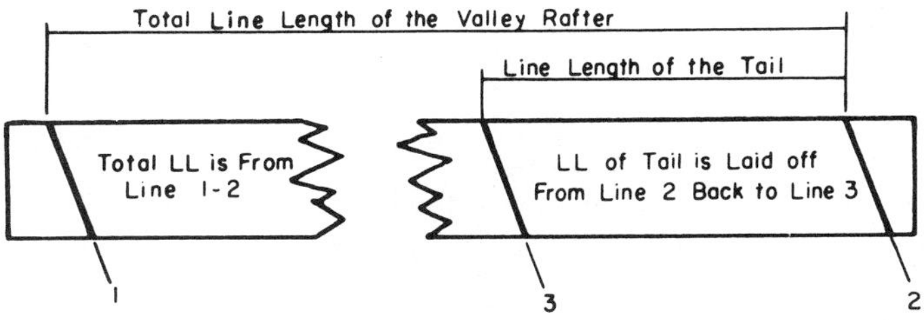

FIGURE 16.5

13. Total line length of the valley rafter

a. The total LL of the VR is the hypotenuse of a right-angle triangle. The base is the total run of the rafter. The height is the total rise of the roof.

b. The total run is $18'\text{–}9\frac{3}{8}''$. The total rise is $8'\text{–}4''$. The total LL is $\sqrt{(225\frac{3}{8})^2 + 100.^2}$ This is $20'\text{–}6\frac{5}{8}''$ to the nearest $\frac{1}{8}''$.

c. Measure from plumb Line 1 on the back of the rafter $20'\text{–}6\frac{5}{8}''$. This is the total LL from the intersection of the ridges' ℄ s to the extreme end of the rafter tail. Draw the plumb Line 2.

d. The total LL may be measured on the development drawing. See the graphic presentation of all the rafter developments in Plate XI. See page 155.

14. Mathematical location of the plumb line of the birdsmouth on the valley rafter

a. The location of the ℄ of the plumb cut of the birdsmouth on the rafter stock is determined by measuring from Line 2 of the LL of the tail of the overhang and drawing the plumb Line 3.

b. The LL of the tail may be calculated, or it may be measured on the scale drawing. See Figure 16.6.

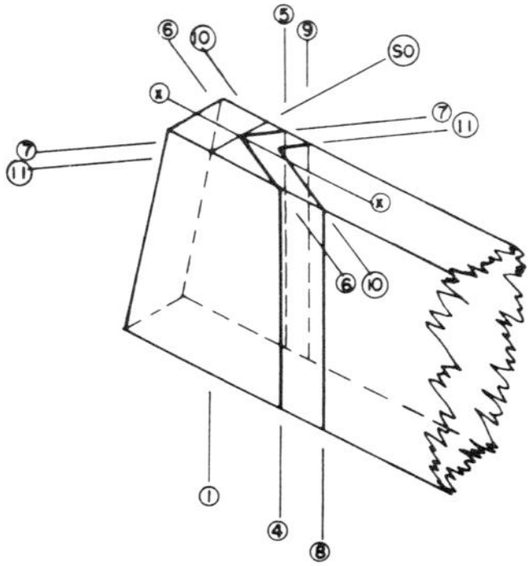

LINE LENGTHS

Line	1 - 2	Total Line Length of the VR
Line	2 - 3	Line Length of the Tail

THE RIDGE SIDE CUTS: *Right Side*

Lines	1 - 4	The run between the ℄ of the ridge and the outside edge of rafter stock measured on the ℄ of the VR on a full-scale plan view (minor roof side)
	1 - 5	The run measured as above but on the major roof side
	6 - 6	The developed angle parallel to the side cut on the minor roof side
	7 - 7	The developed angle parallel to the side cut on the major roof side
	1 - 8	The Total Run of the side cut measured on plan on the minor roof side
	10 - 10	Side cut. Cut on lines 8 & 10 for the minor roof side
	11 - 11	Side cut. Cut on lines 9 & 11 for the major roof side

Note that lines 10 & 11 meet out of center

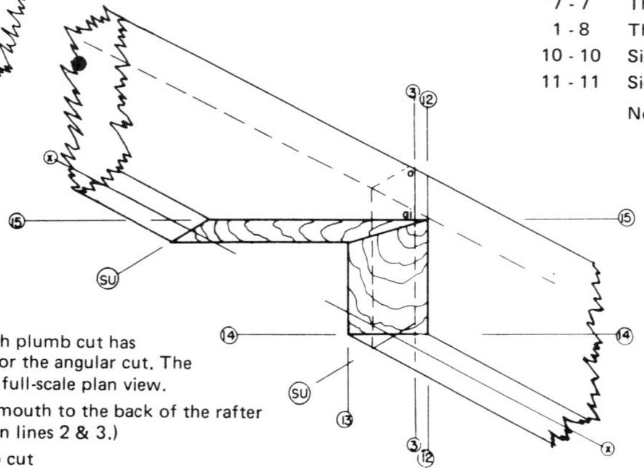

THE BIRDSMOUTH: *Left Side*

Lines	3 - 12	The run distance the birdsmouth plumb cut has moved on the minor roof side for the angular cut. The measurement is taken from the full-scale plan view.
	a - ai	Vertical measure over the birdsmouth to the back of the rafter as on the CR layout (Measure on lines 2 & 3.)
	14 - 14	Angle of the birdsmouth plumb cut
	15 - 15	Seat cut (The VR is neither dropped nor backed.)
	14 - 12	Cutting lines for the birdsmouth plumb cut.

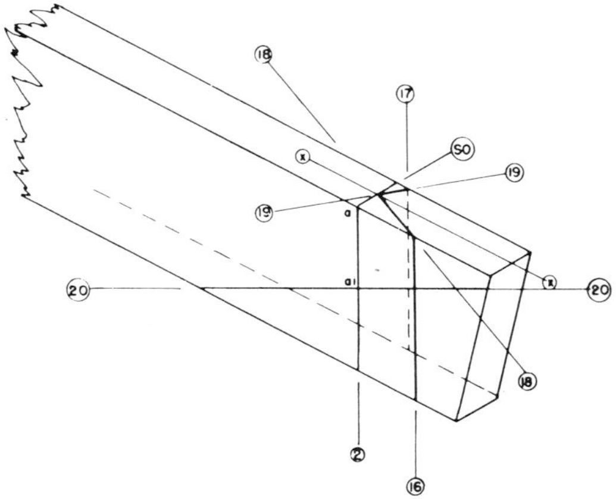

THE TAIL: *Right Side*

Lines	2 - 16	Run of the long Side cut line (as at 1 - 4)
	2 - 17	Run of the short Side cut line (as at 1 - 5)
	18 - 18	Long Side cut line (as at 6 - 6)
	19 - 19	Short Side cut line (as at 7 - 7)
	20 - 20	Tail Level cut
	16 - 18	Cutting lines for the long Side tail cut
	17 - 19	Cutting lines for the short Side tail cut

Note that the intersection of lines 18 & 19 meet on the ℄ of the back of the stock.

PLATE X. Layout of the Intersecting Roof of Unequal-Pitch Valley Rafter

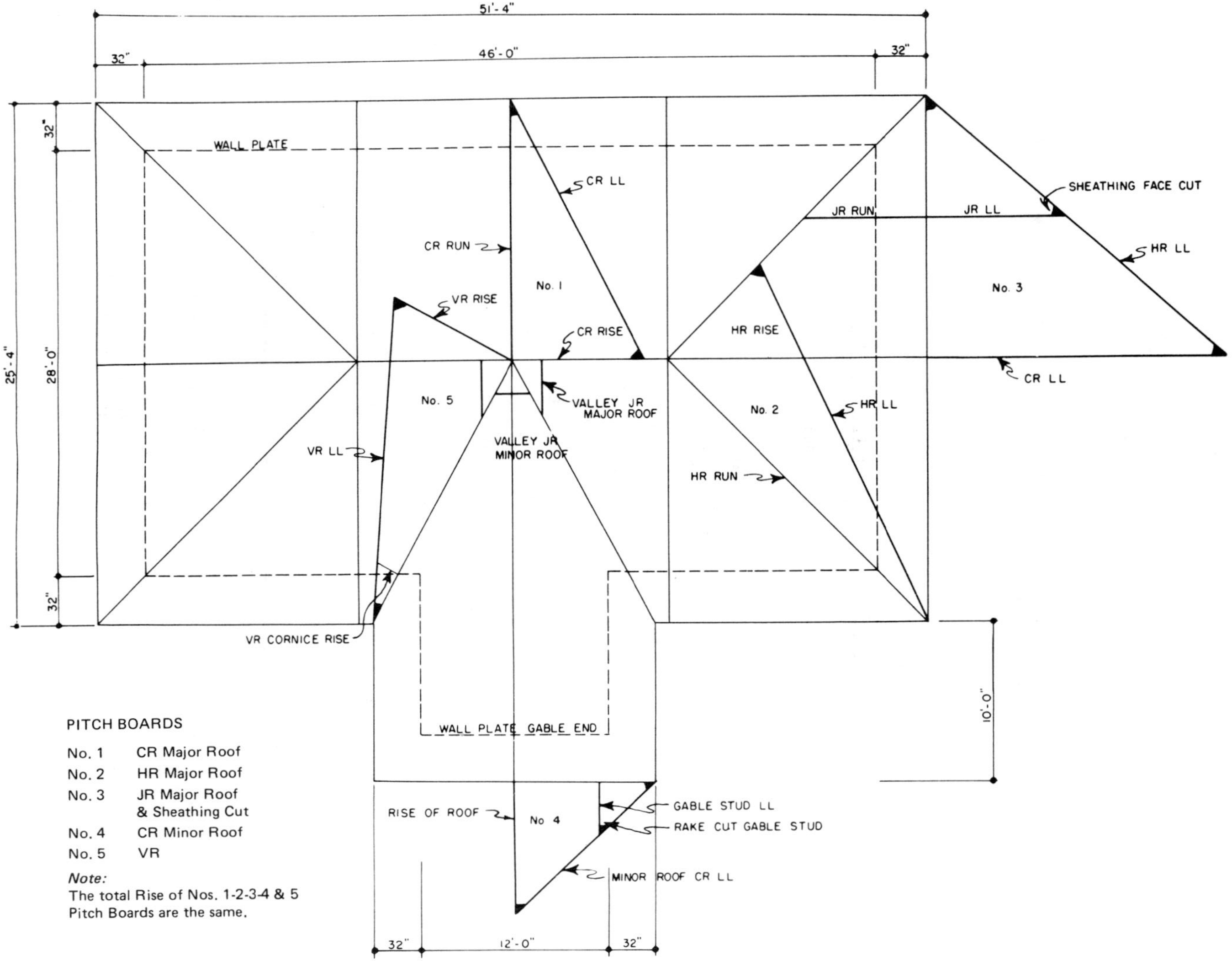

PITCH BOARDS

No. 1 CR Major Roof
No. 2 HR Major Roof
No. 3 JR Major Roof & Sheathing Cut
No. 4 CR Minor Roof
No. 5 VR

Note:
The total Rise of Nos. 1-2-3-4 & 5 Pitch Boards are the same.

PLATE XI. Developed Drawing for Roof of Unequal Pitch

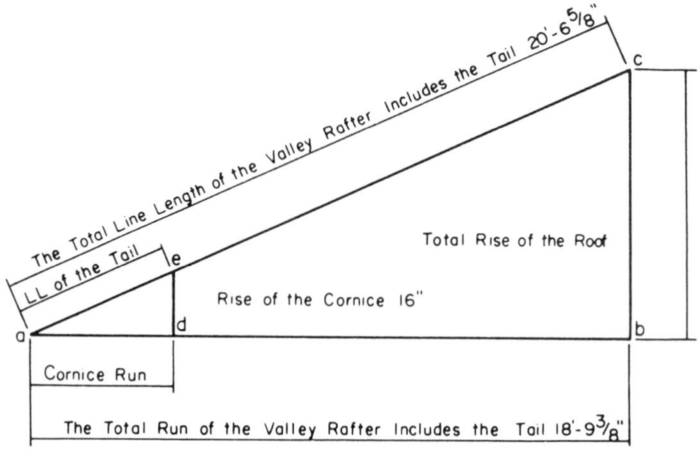

FIGURE 16.6

c. The mathematical calculation may be made by comparing the sides of two similar triangles.

Step 1: The total rise of the roof is 8′–4″, or 100″.
The total run of the VR is 18′–9$\frac{3}{8}$″, or 225$\frac{3}{8}$″.
The total LL of the VR is 20′–6$\frac{5}{8}$″, or 246$\frac{5}{8}$″.

Step 2: The rise of the cornice is 16″.
The LL of the tail of the VR is to be calculated.

Step 3: The total rise of the roof is to the rise of the cornice as is the LL of the VR to the LL of the tail of the rafter.

Step 4: Expressed in inches:

100:16 as 246$\frac{5}{8}$ is to the LL of the tail

$$\frac{100}{16} = \frac{246\frac{5}{8}}{\text{LL of tail}}$$

$$\text{LL of tail} = \frac{16}{100} \times 246\tfrac{5}{8}$$

246.62	100)3945.92(39.4592 = 39$\frac{1}{2}$″ approx.
16	
1479.72	
2466.2	The LL of the tail of the VR is
3945.92	39$\frac{1}{2}$″ to the nearest $\frac{1}{8}$″.

Step 5: Measure the LL of the tail from Line 2 on the rafter. Draw the ℄ of the plumb cut of the birdsmouth Line 3. See Figure 16.5 page 153. Line 3 must now be squared-over to the opposite face of the rafter.

15. Location of the plumb line of the birdsmouth on the valley rafter by the graphic method.

a. Draw a right-angle triangle to scale as in Figure 16.6 with:
the base *a–b* equal to the total run of the VR,
the height *b–c* equal to the total rise of the roof, and
the hypotenuse *a–c* equal to the LL of the VR.

b. Draw a 16″ perpendicular line *d–e* from the base to intersect the hypotenuse. Measure *a–e* which is the LL of the rafter tail.

c. Make a very careful study of Plate XI, page 155, where the graphic development of all the critical members of the roof is shown.

16. Plumb lines at the ridge

a. It must be very thoroughly understood that the carpenter's working drawing of the outlines of the wall plates and the runs of the rafters is drawn to the $\frac{1}{12}$ scale; that is, 1″ equals 1′–0″.

b. **It must be equally understood that the partial plans of any rafters are drawn to full scale.** This is done so that the runs of the cuts can be measured exactly in full scale on the drawing.

c. Since the VR run is the diagonal of a rectangle, the plumb lines on the minor roof side of the rafter will have longer runs than those on the opposite side. Study Figure 16.7, where the thickness of the rafter stock is drawn in.

d. The run between Lines 1 and 4 is measured on the plan. This measurement is laid off at right angles from plumb Line 1 on the VR stock, and plumb Line 4 is drawn. See Figure 16.8.

e. Line 5 (shown in dashes) is similarly laid off on the opposite side of the rafter. First measure between Lines 1 and 5 on plan.

f. Draw a ℄ on the back of the rafter stock. Draw the bevel Lines 6 and 7 from the edges of the rafter at Lines 4 and 5 respectively to join the ℄ of Line 1 on the back of the rafter.

g. It will be noted that only the ℄ (single line) of the ridge is shown at this stage of the development drawing. Study Plate IX, page 155, and Figure 16.4.

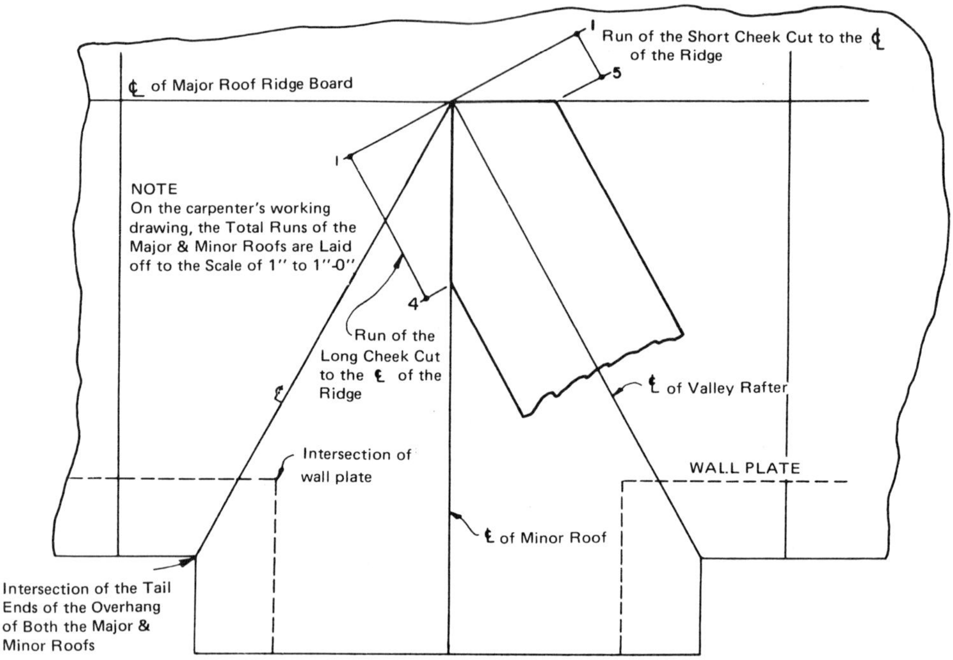

FIGURE 16.7

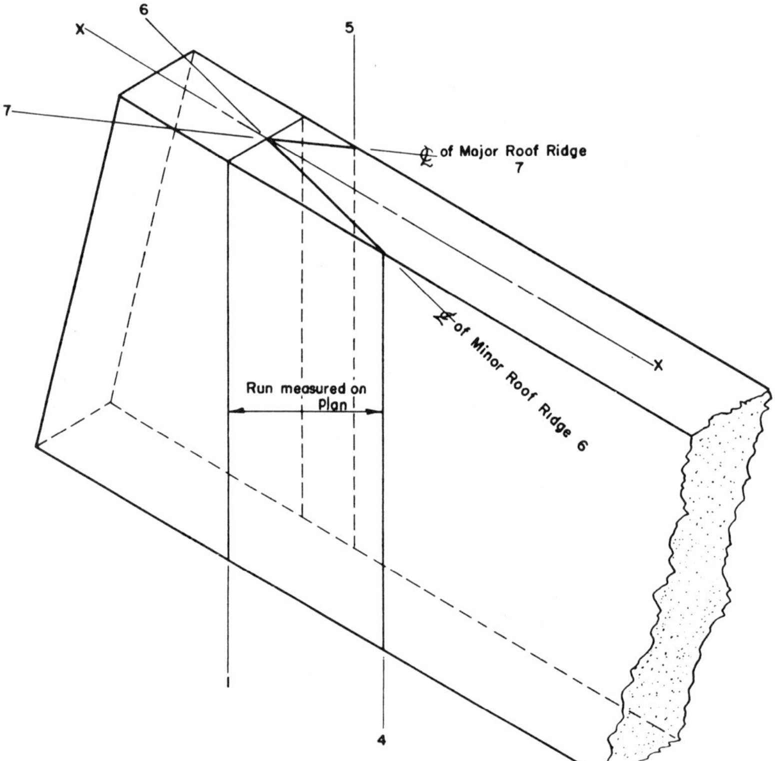

FIGURE 16.8

17. Shortening for the ridges

a. Draw the full-size thickness of the ridge and the full-size plan view of the upper portion of the VR as in Figure 16.9.

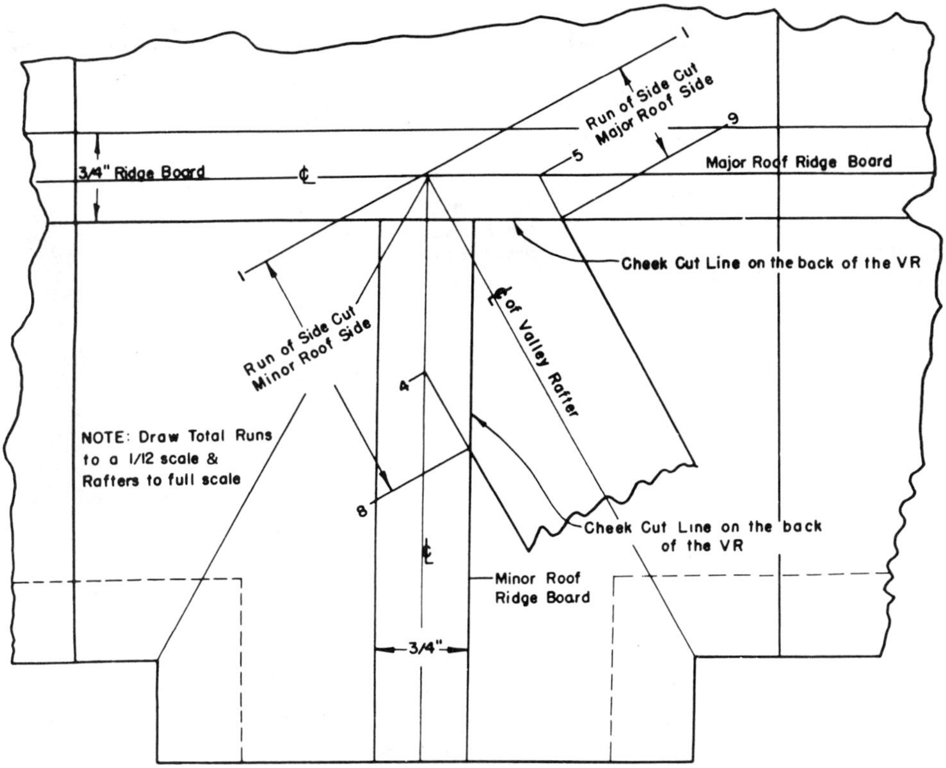

FIGURE 16.9

b. Measure the run of the side cut lines between Lines 1 and 8 on the plan. From Line 1 on the rafter stock at right angles, lay off this measurement, and draw the plumb side cut Line 8.

c. Measure the run of the side cut lines between Lines 1 and 9 on the plan. From Line 1 on the rafter stock at right angles (on the opposite side of the rafter), draw the plumb side cut Line 9; this line is shown in dashes in Figure 16.10, page 160.

d. Set a carpenter's bevel from the face of the rafter stock across the back of the rafter to the bevel Line 6. Slip the bevel to the plumb Line 8. Draw the side cut Line 10 parallel to Line 6. In a similar manner, after adjusting the bevel to Line 7, draw Line 11.

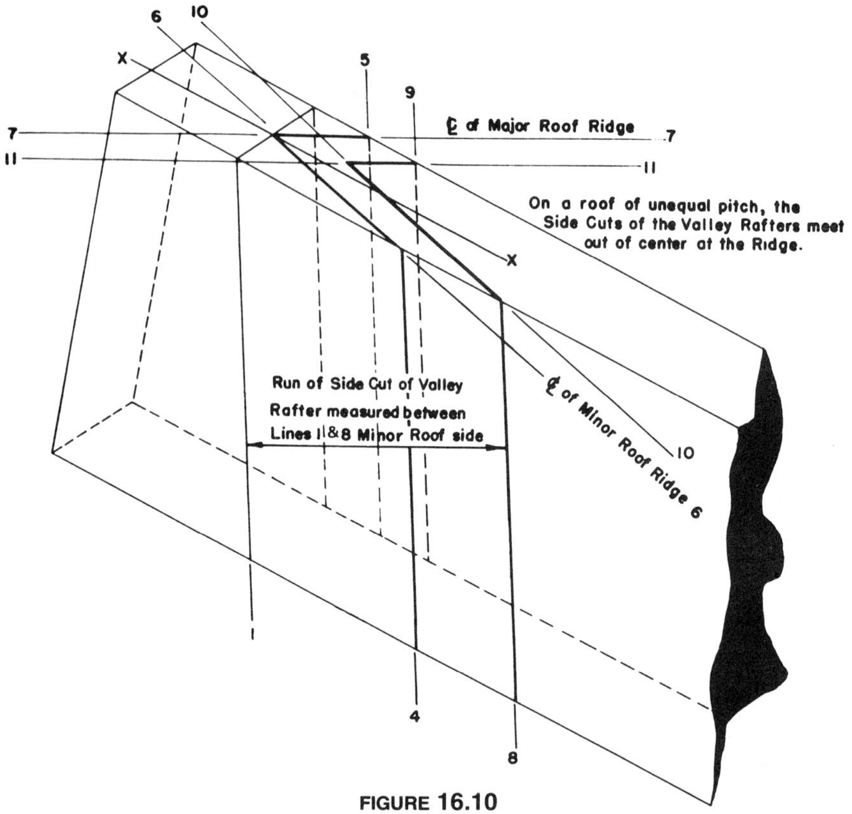

FIGURE 16.10

e. It should be most carefully noted that the side cut lines, after allowance is made for the shortening for the ridge, meet out of center. This is because the VR is the diagonal of a rectangle. Cut on Lines 8 and 10 and on Lines 9 and 11.

18. Valley rafter seat cut

a. Figure 16.11(a) shows the major roof pattern CR. On Line 2 at *a–ai* is shown the perpendicular height of the CR from the seat cut to the back of the rafter.

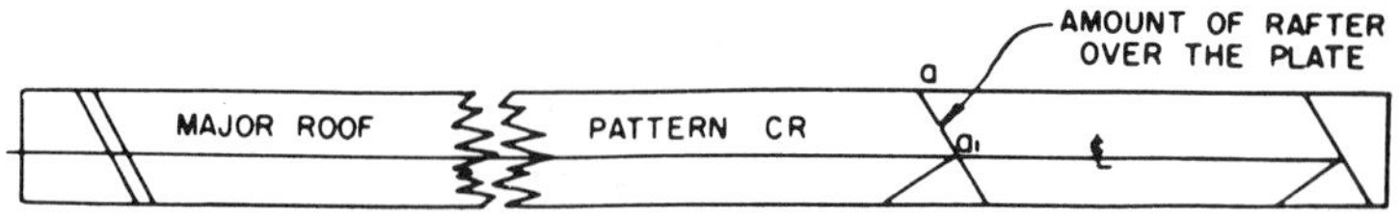

FIGURE 16.11(a)

b. Figure 16.11(b) shows the pattern VR. On Line 3 at *a–ai* is shown the amount of the rafter above the plate. This is measured on the CR and made the same on the VR. From the point *a–ai* on Line 3 draw at right angles from the seat cutting line, Line 15. Square over Line 3 to the opposite face and also draw the seat cutting line, the measurement *a–ai*, on the back of the stock. Since a 2 × 8 rafter is to be used, a very large birdsmouth will be removed.

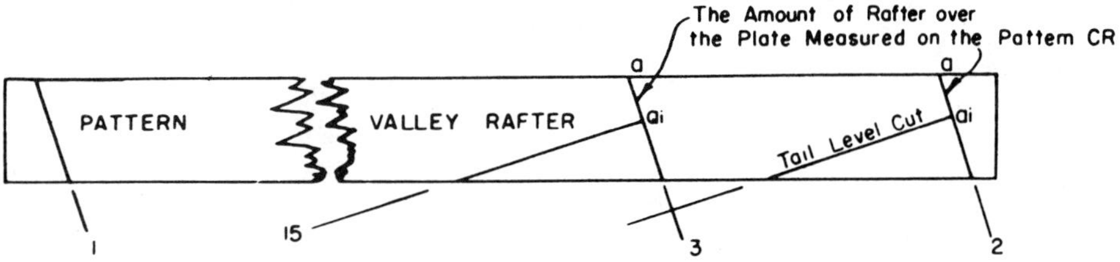

FIGURE 16.11(b)

c. The VR is neither dropped, dubbed, nor backed. The center of the back of the rafter is in the same plane as the backs of the CRs of both the major and minor roofs.

19. Valley rafter birdsmouth angled plumb cut

a. Figure 16.12 shows that portion of the plan of the VR where it crosses the major roof wall plate. Since the VR crosses the wall plate at an angle, it requires the plumb cut of the birdsmouth to be angled.

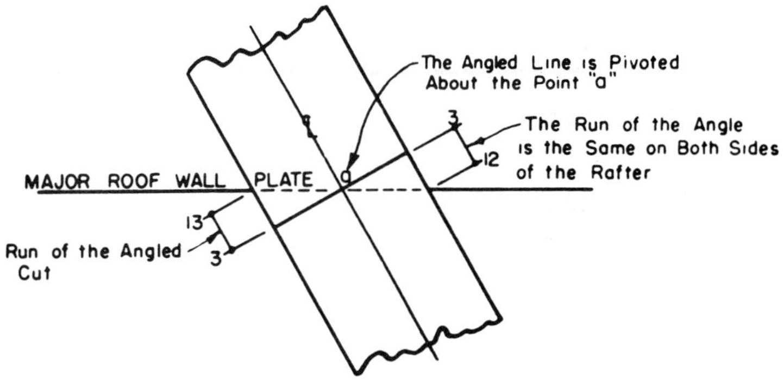

FIGURE 16.12

b. Measure on the plan the run between Lines 3 and 12 and between Lines 3 and 13; they are the same.

c. For the right-hand VR, lay off from Line 3 at right angles towards the tail the run measured on the plan between Lines 3 and 12; draw the plumb cutting Line 12 as in Figure 16.13.

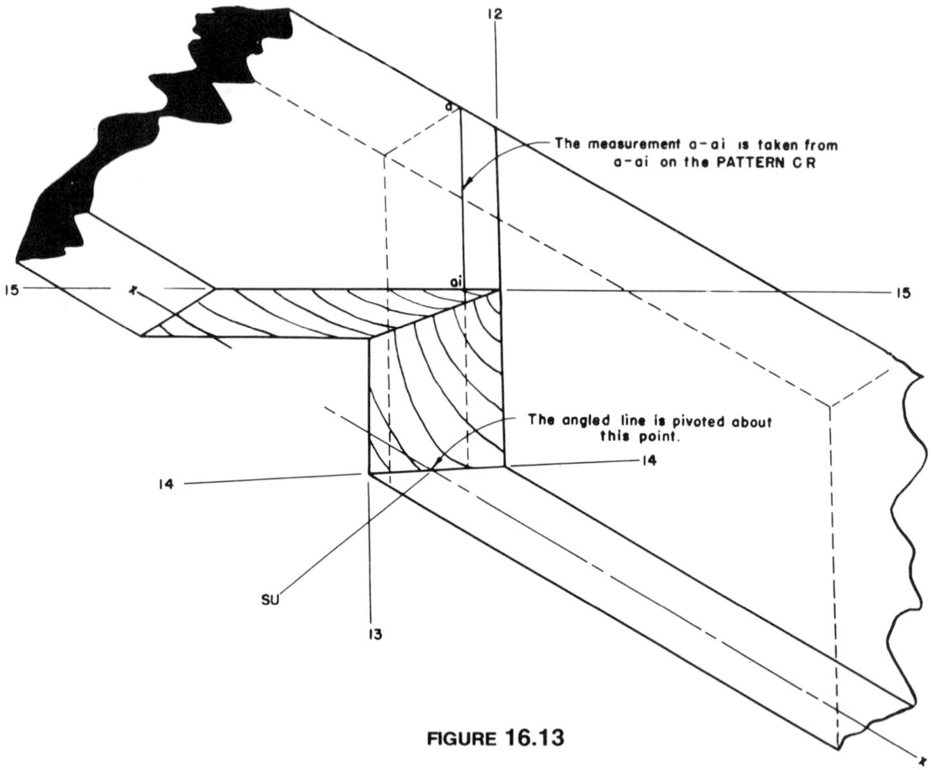

FIGURE 16.13

d. On the opposite side of the rafter, lay off from Line 3 at right angles towards the ridge the run measured between Lines 3 and 13. Draw the plumb cutting Line 13.

e. On the underneath side of the rafter, join cutting Lines 12 and 13. This is the bevel cutting Line 14.

f. Cut down on the bevel Line 14 to the seat cutting Line 15.

20. Tail side cuts

a. The total LL of the VR is from Line 1 to Line 2 on the rafter layout, that is, from the ℄ of the ridge to the tail end. The run of the side cuts of the tail is additional to the total LL. (See Figure 16.14.)

b. Measure on the plan the run between Lines 2 and 16. Lay this run off at right angles from Line 2, and draw the plumb Line 16. (See Figure 16.14.)

c. Measure on the plan the run between Lines 2 and 17. On the opposite side of the rafter, lay this run off at right angles to the squared-over Line 2 and draw the plumb Line 17.

d. Draw a ℄ on the back of the rafter, and lay off the side cutting lines as in Figure 16.14, Lines 18 and 19. The carpenter's cutting lines are Lines 16 and 18 and Lines 17 and 19.

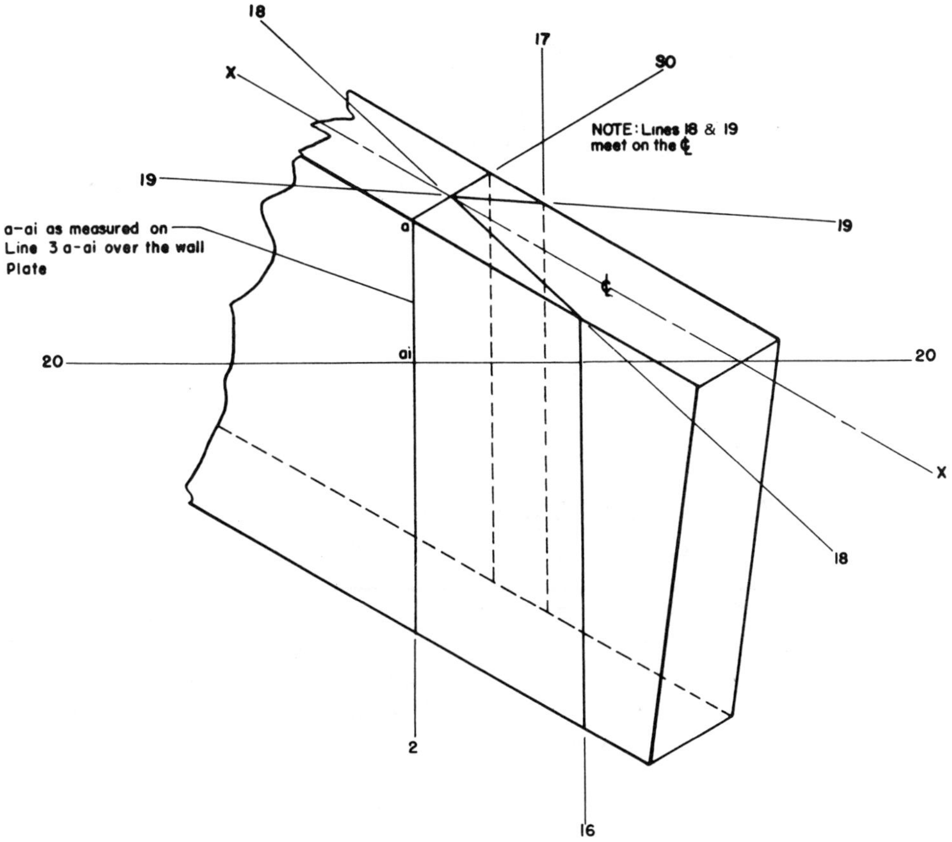

FIGURE 16.14

21. Tail level cut

From the back of the rafter, measure down and spot on Line 2 the same distance as from the back of the rafter to the birdsmouth on Line 3. From the measured spot draw the level cut on the tail, Line 20, at right angles to Line 2. Cut on Line 20.

22. Complete developed drawing

a. Plate XI shows a complete developed drawing. The LL of the rafters shown may be determined by one or more of the following methods:

   1. developing the lengths on a 1″ to 1′–0″ scaled-drawing and measuring them
   2. mathematical calculations
   3. reading and interpreting the rafter tables
   4. adapting the framing square from the graphic method

b. The lengths and cuts of the CR and the gable studs of the minor roof should be determined by the preceding methods (1), (2), and (4).

c. To make a complete summary of the layout of the VR, Plates IX, X, and XI, pages 145, 154, and 155, should be very closely studied together.

23. Sheathing, purlin, windblock, and valley jack rafter cuts

a. An intersecting roof of unequal pitch with a level cornice will require developed cuts for all the minor roof members intersecting at the VR. The major roof will have a regular pitch, and all its roof members excepting those that intersect at the VR will have regular cuts.

b. There are no figures given on the framing square for the layout of any of the members of the minor roof. The cuts are developed by the graphic method. The framing square may then be readily adapted by placing the stair gauges at predetermined settings.

24. Specifications

All dimensions and specifications are as detailed in this chapter in article 1 page 144.

25. Carpenter's working drawing

Figure 16.15 is a pictorial drawing of the roof, and Figures 16.16 and 16.17 show the carpenter's working drawings. These drawings must be carefully studied; they should be drawn to a $\frac{1}{12}$ scale.

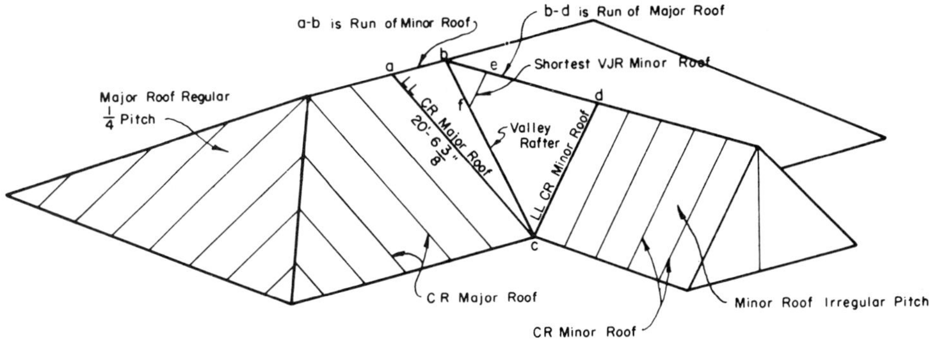

FIGURE 16.15

26. Sheathing face cuts: graphic method

a. Examine Figure 16.15. The two right-angle triangles *a–b–c* and *b–c–d* have one side equal; that is, the hypotenuse (the VR).

b. The major roof right-angle triangle has *a–b* equal to the total run of the minor roof (8′–8″); *a–c* equals the total LL CR of the major roof (20′–6$\frac{3}{8}$″); and *b–c* equals the total LL of the VR.

c. The minor roof right-angle triangle has *b–d* equal to the run of the major roof (16′–8″); *d–c* equals the total LL of the minor roof CR (12′–0$\frac{1}{4}$″); and *b–c* equals the total LL of the VR. The total LL of the minor roof CR should be checked mathematically.

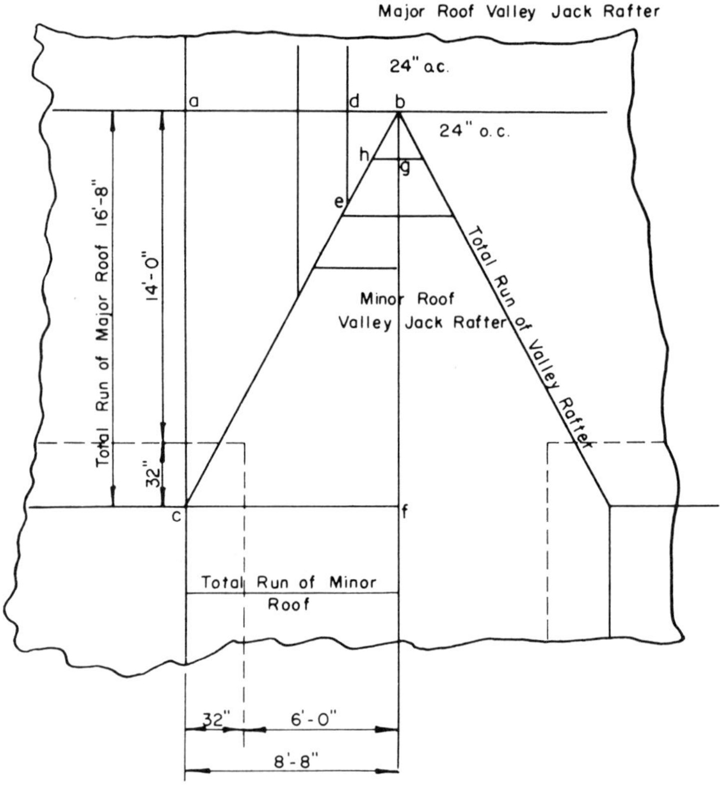

FIGURE 16.16

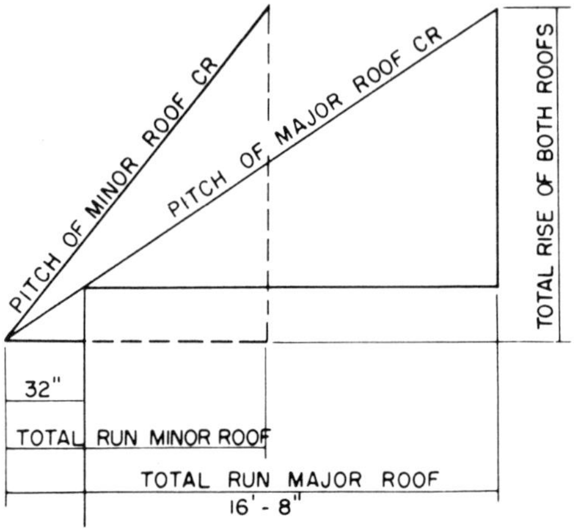

FIGURE 16.17. Comparative pitches of major and minor roofs.

27. Major roof sheathing cut at valley

a. Make a pitch board for the major roof sheathing cut to a $\frac{1}{12}$ scale. Figure 16.18 shows the pitch board applied to a piece of roof sheathing board. Study Figures 16.18 and 16.19 together.

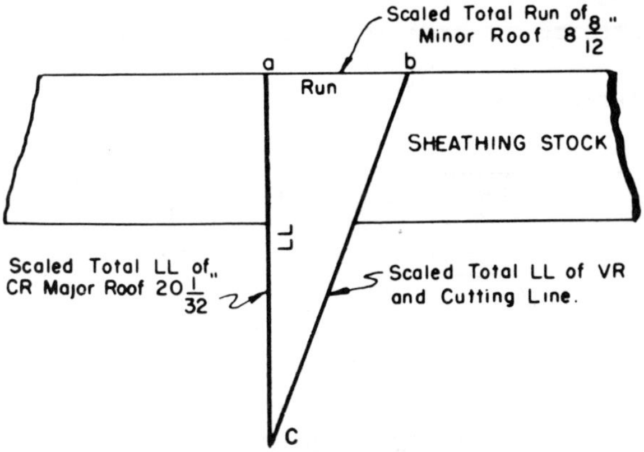

FIGURE 16.18

b. Figure 16.19 shows the framing square adapted to the graphic method. It has one stair gauge set at $8\frac{8}{12}$" for the run of the minor roof; the other is set at $20\frac{1}{32}$" for the total LL of the CR of the major roof. Both the pitch board and the framing square settings are for the face cut of the sheathing of the major roof.

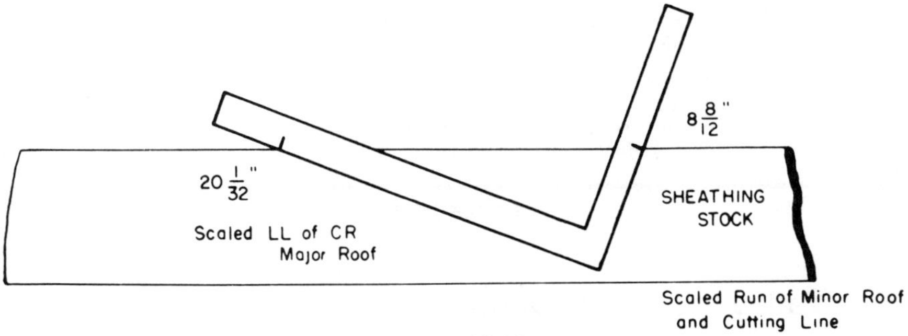

FIGURE 16.19

28. Minor roof sheathing face cut at valley

a. Make a pitch board for the minor roof sheathing face cut to a $\frac{1}{12}$ scale. Examine Figure 16.15. *b–d* equals the total run of the major roof (16′–8″), *d–c* equals the total LL of the minor roof CR (12′–0$\frac{1}{4}$″), and *b–c* equals the total LL of the VR. Figure 16.20 shows the pitch board applied to a piece of roof sheathing board.

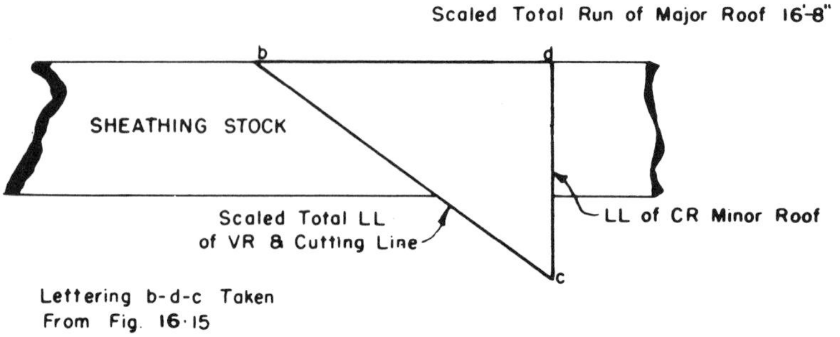

FIGURE 16.20

b. Figure 16.21 shows the framing square adapted to the graphic method. It has one stair gauge set at 16$\frac{2}{3}$″ for the run of the major roof; the other is set at 12$\frac{3}{32}$″ for the total LL of the minor roof.

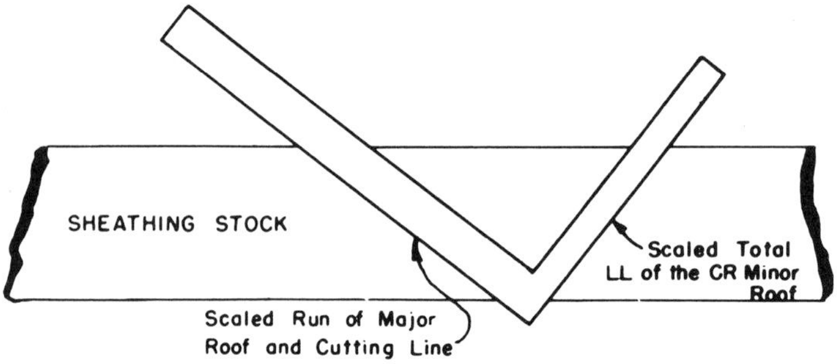

FIGURE 16.21

29. Valley jack rafter plumb and side cut: major roof

a. The plumb cut of the VJR of the major roof is the same as that of the major roof CR.

b. The side cut of the VJR of the major roof at its intersection with the VR is made with the same pitch board or framing square setting as is used for the major roof sheathing face cut.

30. Valley jack rafter plumb and side cut: minor roof

a. The plumb cut of the VJR of the minor roof is the same as that of the minor roof CR.

b. The side cut of the VJR is made with the same pitch board or framing square setting as is used for the major roof sheathing face cut.

31. Line length of unequal-pitch roof valley jack rafters

The VJR on center spacings for the major and minor roofs should be laid off from the intersection of the ℄ of the ridges (see Figure 16.16 page 166). The rafters for the different roofs will have different plumb and level cuts and they will not intersect at the VR. The run of the shortest VJR may be measured from the plan. This measurement will be the common difference in length of all the VJRs for the side on which it is measured.

32. Calculated run of the shortest major roof valley jack rafter

In Figure 16.16 the two right-angle triangles (on the major roof side) *a-b-c* and *d-b-e* are similar: *a-b* is to *d-b* as *a-c* is to *d-e*. Using the measurements, 8′–8″ is to 2′–0″ as 16′–8″ is to *x*. Or, stated in inches, 104 is to 24 as 200 is to *x* inches. This is approximately $46\frac{1}{8}''$, which is the common difference in the lengths of all the VJRs for the major roof side.

33. Layout of the shortest valley jack rafter for major roof

a. The pitch is $\frac{1}{4}$, and the run is $46\frac{1}{8}''$, or 3.93′. The LL per foot run for a $\frac{1}{4}$ pitch roof is 13.32. Therefore, $13.32'' \times 3.93 = 4'\text{–}4\frac{3}{4}''$ approximately. This is the common difference in the lengths of the VJRs on the major roof side.

b. Figure 16.22(a) shows the major roof sheathing face-cut-pitch-board applied to the back of a piece of rafter stock for the side cut of the VJR. Figure 16.22(b) shows the framing square setting and placing for the same cut.

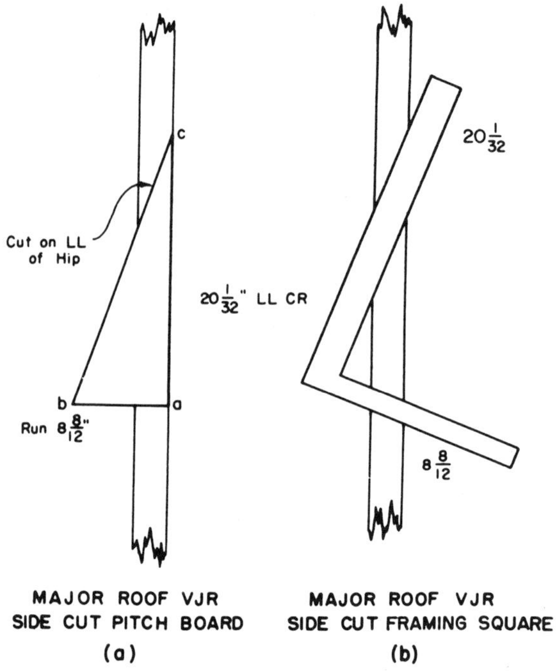

FIGURE 16.22

c. The plumb cut at the ridge and valley is the same as the plumb cut of the CR for the side of the roof of which they both form part.

d. Lay off the shortest VJR for the major roof, as in Figure 16.23.

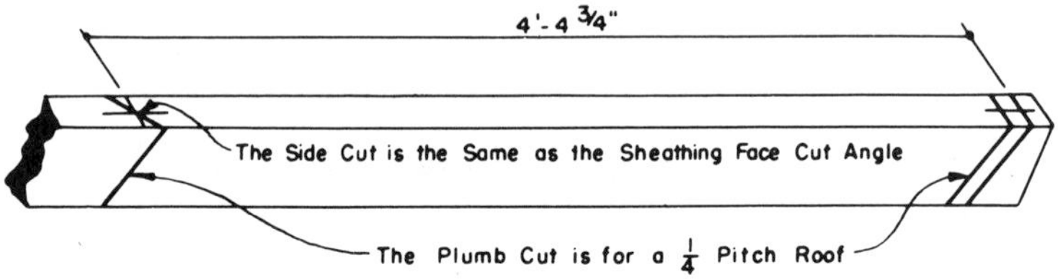

FIGURE 16.23

34. Line length of the shortest valley jack rafter for minor roof

The pitch of the minor roof is irregular, and no direct readings may be made from the rafter tables. In Figure 16.15, page 155, the minor roof right-angle triangles *b–d–c* and *b–e–f* are similar where:

b–d is the total run of the major roof	16′–8″	200″
b–e is the OC measurements of the VJRs	2′–0″	24″
d–c is the total LL of the minor roof CR	12′–0$\frac{1}{4}$″	144.27″
e–f is the LL of the VJR (unknown)		
b–d is to *b–e* as *d–c* is to *e–f*		

200:24 as is 144.27:*x* inches. The unknown is 17.31″, and this is the LL of the shortest VJR on the minor roof side. It is also the common difference in the lengths of all the VJRs on the minor roof.

35. Layout of the shortest valley jack rafter for minor roof

a. Figure 16.24(a) shows the minor roof sheathing face-cut-pitch-board applied to the back of the rafter stock for the side cut of the VJR. Figure 16.24(b) shows the framing square setting for the same cut.

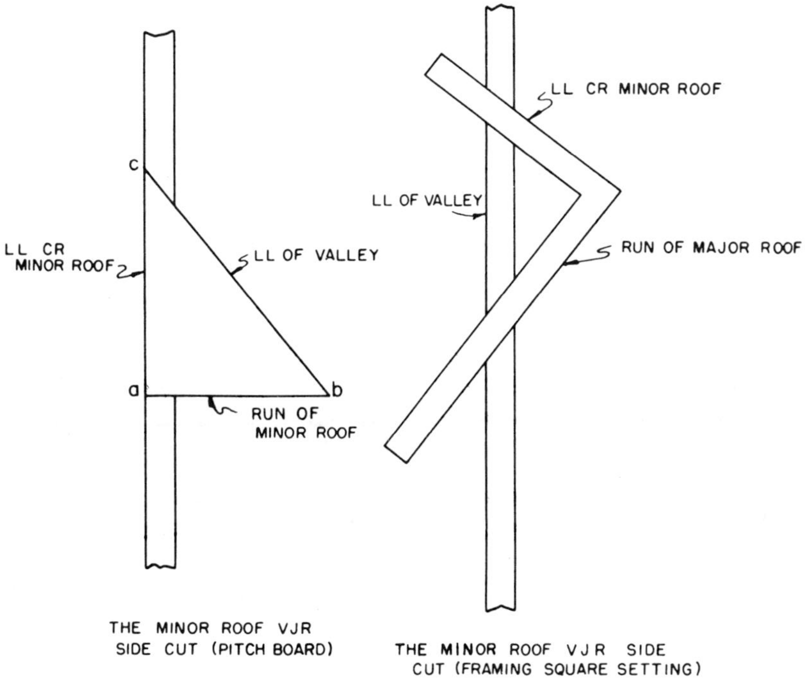

FIGURE 16.24

b. The plumb cut at the ridge and valley is the same as the plumb cut of the CR for the side of the roof of which they both form part.

c. Lay off the shortest VJR for the minor roof as in Figure 16.25.

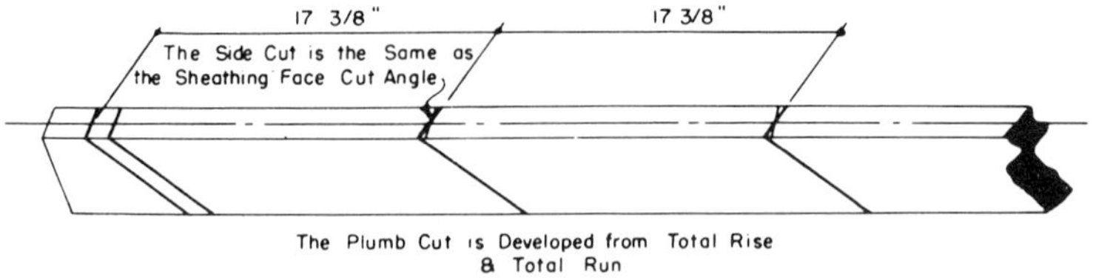

FIGURE 16.25

36. Purlin cut

a. Figure 16.26 shows the sections and the plan views of the major and minor roof purlins. Observe that, since the roofs have different pitches, the plan views of the two purlins are different.

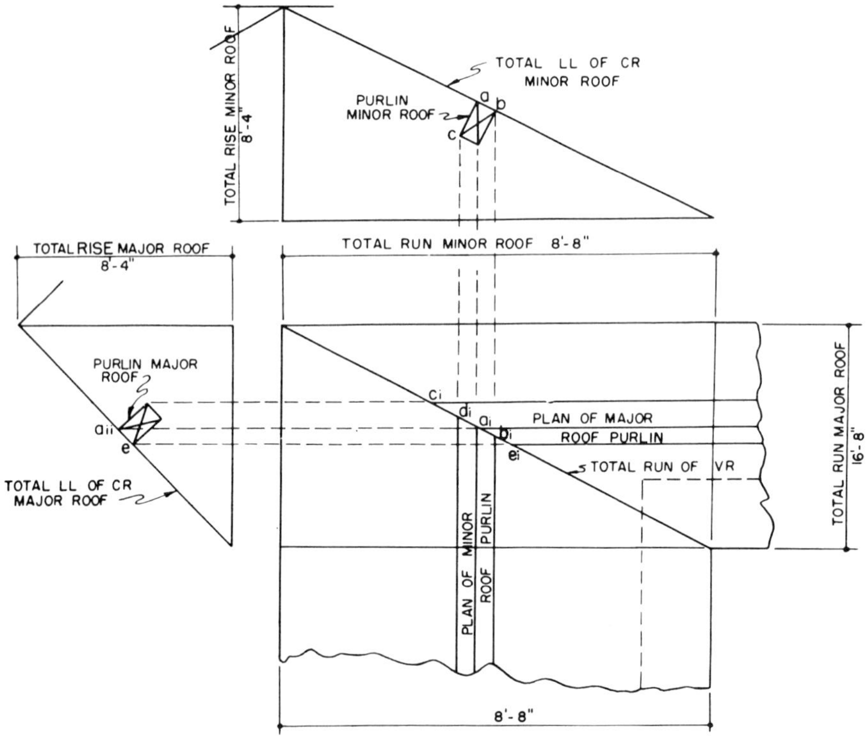

FIGURE 16.26

The carpenter, when building, would thoroughly secure these members by nailing through the VR into the purlin from both sides.

b. The side cut of the purlin has been dealt with in sections 28 and 29.

37. Purlin face cut: minor roof

a. Because the VR is the diagonal of a rectangle, both the purlins intersecting at this point must be developed by drawings. Study Figure 16.27.

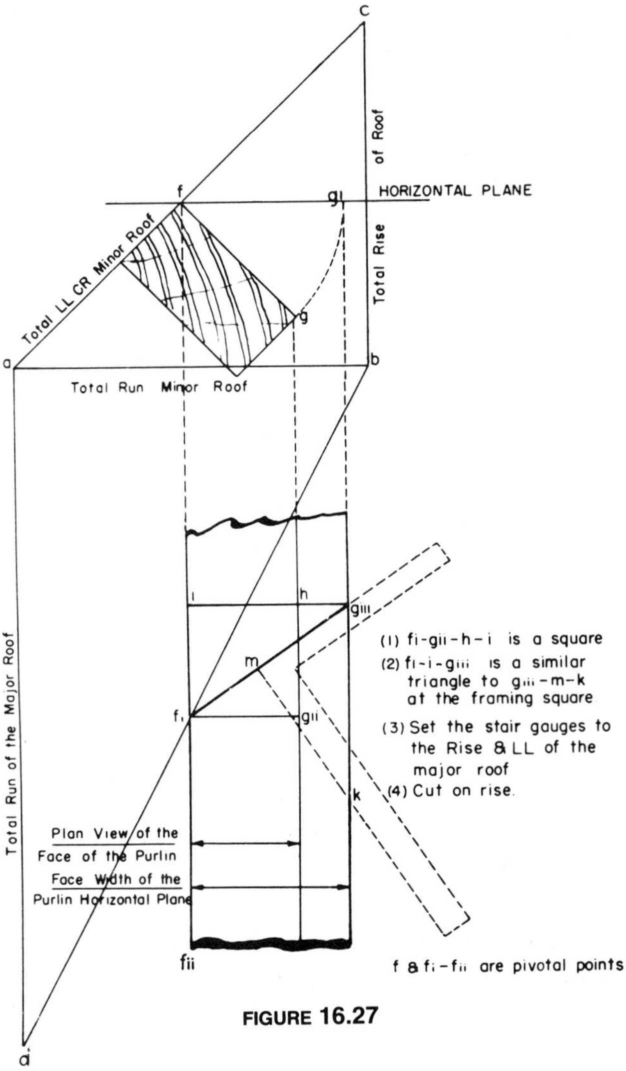

FIGURE 16.27

b. On a piece of building paper draw Figure 16.27 to scale:

   a–b–c to determine the pitch
   a–d–b to determine the angle of the run of the VR
   f–g which is the face of the purlin in section
   f–gi which is the face of the purlin revolved in a horizontal plane
   f to *fi–fii*, which are pivotal points
   fi–gii, the plan view of the face of the purlin
   fi–gii–h–i, a square (on a regular intersecting roof the miter of the purlin on the plan would be *fi–h–i*)
   i–giii, the face of the purlin on the horizontal plane equal to *f–gi*
   fi–giii, the developed true miter of the face of the purlin

c. The total run angle of the VR is the miter angle of the purlin on the plan shown at *b–d.*

d. The triangle *fi–giii–i* is a similar triangle to the one shown on the outline of the framing square at *giii–m–k.*

e. Set the stair gauges on the framing square at *giii* for the rise of the roof and for the LL of the CR at *k.* Lay off the cutting line for the true face miter angle on the rise *m–giii.*

38. Purlin face cut: major roof

Study Figure 16.28, and compare this drawing with Figure 16.27; the method of layout is the same. The cutting line for the true face miter angle for any purlin intersecting at a VR may be made by setting one stair gauge on the framing square to the rise of the roof, and the other to the LL of the CR for the roof of which the purlin forms part. Draw the face angle, and cut on the rise.

39. Purlin face cut for any roof

Rule: *The face angle of the purlin for any roof may be made by setting (to scale) one stair gauge on the framing square to the total rise of the roof and the other to the total LL of the CR for the roof of which the purlin forms part. Cut on rise.*

Note: *A complete study should now be made of the hopper section dealing with the unequal-pitch hoppers.*

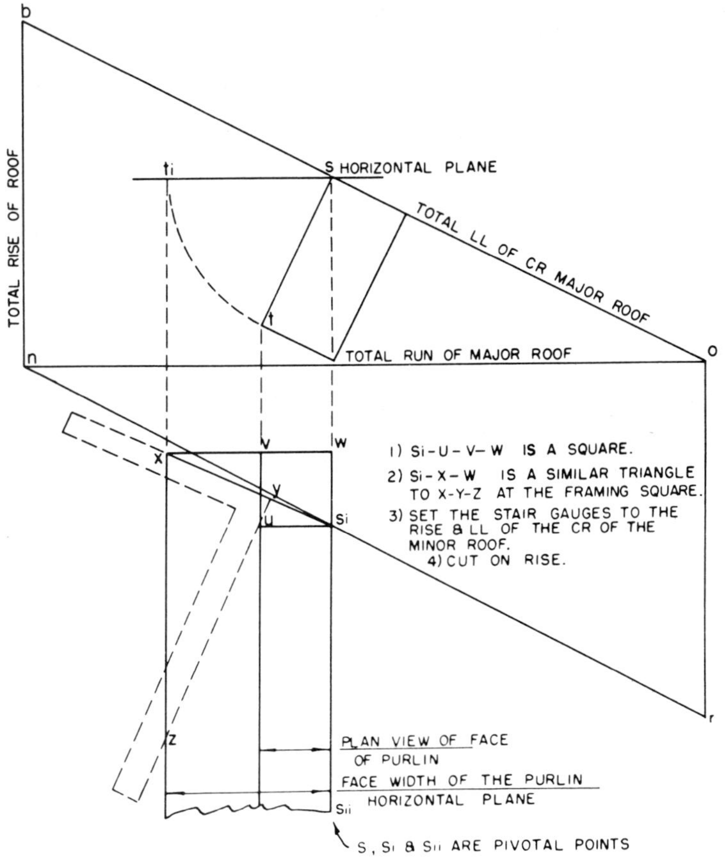

FIGURE 16.28

40. Summary

Theory of residential roof framing has been presented in the first sixteen chapters.

Applications of roof framing theory are offered in the next seven chapters. Framing measurements are from both the imperial and metric measurement systems.

The following is a summary with respect to terminology, as used in this book. Reference may be made to this summary as succeeding chapters are considered.

1. *Pitch* is used with imperial measurements. See chapter 2.

2. *Slope* is used with metric measurements. See chapter 6.

3. A *pitch board* usually has imperial measurements. See chapter 4.

Example:

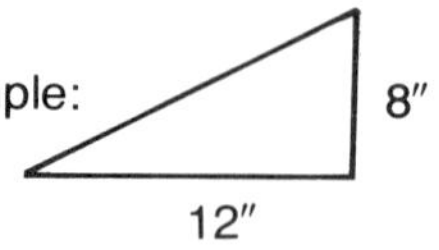

$\frac{1}{3}$ Pitch

4. A *slope board* usually has metric measurements.

Example:

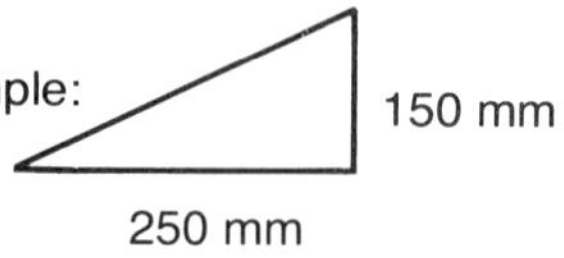

5. A *slope triangle* usually has metric measurements.

Example:

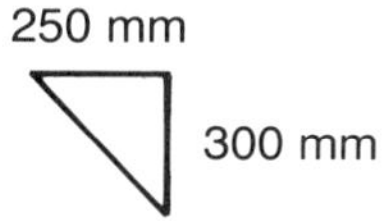

6. The definition of the word *pitch* determines the dimensions of the pitch board. Pitch is $\frac{\text{Rise}}{\text{Span}}$ but *Rise* and *Run* measurements only are used.

Example: For $\frac{1}{3}$ pitch, the pitch board may be dimensioned

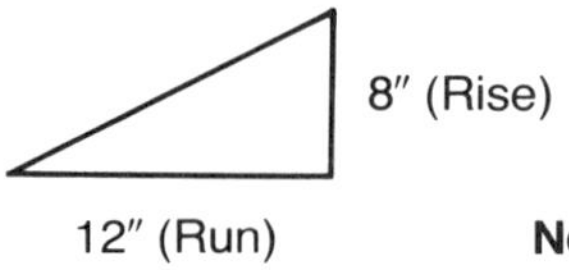

Note: $\frac{1}{3} = \frac{8\text{ (Rise)}}{24\text{ (Span)}}$

7. The definition of the word *slope* determines the dimensions of the slope board. Slope is $\frac{\text{Rise}}{\text{Run}}$, therefore rise and run measurements are used. A $\frac{3}{5}$ (0.6) slope might be shown as:

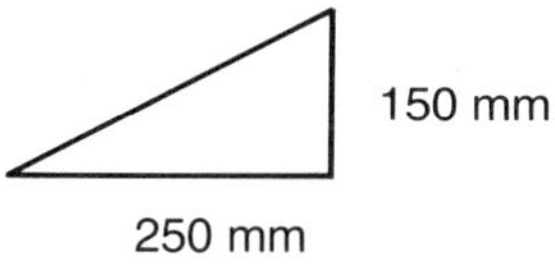

that is, $\frac{150\text{ mm}}{250\text{ mm}} = \frac{3}{5}$

(See Table 16.1 on page 171) **Note:** A $\frac{3}{5}$ slope is different from a $\frac{3}{5}$ pitch.

TABLE 16-1

Equivalent Pitch and Slope Measurements

Pitch = $\frac{Rise}{Span}$				Slope = $\frac{Rise}{Run}$			
Common Fractions	*Cuts*	*Decimal Fractions*	*Pitch Board* x / 12	*Decimal Fractions*	*250 mm* x (slope triangles)	*1000 mm* x (slope triangles)	*Slope Board* x / 250 mm
Full	24 in 12	1.0	24	2.0	500	2000	500
		0.9		1.8	450	1800	450
		0.8		1.6	400	1600	400
$\frac{3}{4}$		0.75	18	1.5	375	1500	375
		0.7		1.4	350	1400	350
$\frac{2}{3}$		0.67		1.33	333	1332	333
$\frac{5}{8}$		0.625	15	1.25	312.5	1250	312.5
		0.6		1.2	300	1200	300
$\frac{1}{2}$	12 in 12	0.5	12	1.0	250	1000	250
		0.4		0.8	200	800	200
$\frac{1}{3}$		0.33+	8	0.67	167.5	667	167.5
		0.3		0.6	150	600	150
$\frac{1}{4}$		0.25	6	0.5	125	500	125
	5 in 12	0.21	5	0.42	105	420	105
		0.2		0.4	100	400	100
		0.1		0.2	50	200	50

8. A slope triangle, usually not drawn to scale, is dimensioned, as is a slope board, in terms of $\frac{\text{Rise}}{\text{Run}}$.

Example:

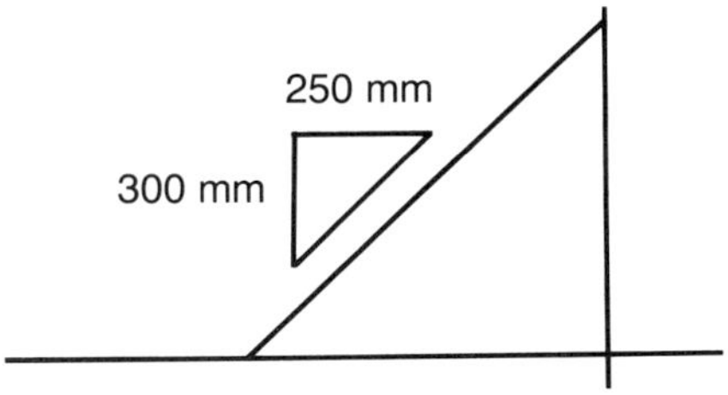

9. The words "equal" and "unequal" are applied to both terms *pitch* and *slope.*

Example:

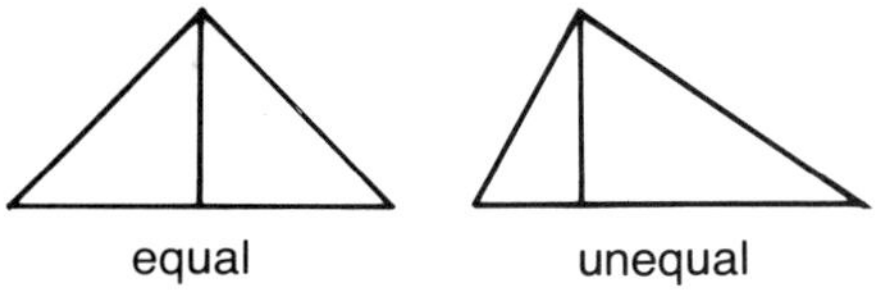

10. Angular measurements may be given in degrees.

Example:

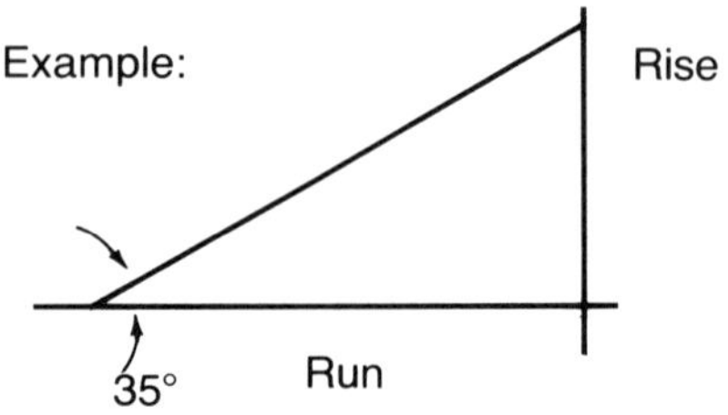

Applications of most of these terms are given in Table 16.1, "Equivalent Pitch and Slope Measurements," which is essentially a reference data sheet for the following chapters.

17

Shop Assignments: Rafter Layout Problems

The solving of the following eighteen problems will help to relate roof framing theory to typical on-the-job rafter layout situations.

Four layout tests follow the problems. These layout tests will help to establish the standards of measurements. Read the specifications for each building and roof; then lay out pattern rafters as required. Sketches should be made before laying out the rafter on full-size stock and before answering questions. A sketch when carefully made shows that one understands the problem. The dimensioning of sketches should be equally well done. Answers involving lumber dimensions and quantities should be written as retail lumbermen would write them.

Rafter layout tools include a roofing square, a pair of stair clips, a twenty-five foot (10 m) tape, a sharp black pencil, and a colored pencil for the marking of cutting lines. A sixteen-foot 2 × 4, or metric equivalent, is required for most problems. It should be placed across two sawhorses or benches with the crown side away. To measure LLs when working alone, drive a small nail into the stock at the first plumb line, and hook one end of tape over it.

Two sets of problems are presented: one of imperial measurements and one of metric measurements. One set or both sets may be attempted.

Problem 1–1

Building: 20′–0″ × 32′–0″. COTTAGE

Roof: Gable: $\frac{1}{3}$ pitch: 2 × 6 ridge board; 12″ eave; rafters: 24″ OC; plumb cut rafter tails.

Problem: Lay out the common rafter (CR). Use the graphic method.

Question: What rafter stock would you order to make this roof?

Sketch the plan view of the roof in the space below.

Note: Allow for 12″ overhang of the CR where the specifications read *12″ eave.*

Problem 1–2

Building: 6100 mm × 9760 mm. COTTAGE

Roof: Gable: 250 mm 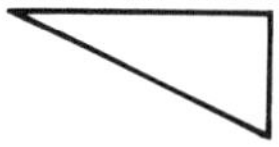 200 mm ; 19 mm × 140 mm ridge board; 300 mm eave; rafters 600 mm OC; plumb cut rafter tails.

Problem: Lay out the common rafter (CR). Use the graphic method.

Question: What rafter stock would you order to make this roof? (Sketch the plan view of the roof.)

Note: Allow for 300 mm overhang of the CR where specifications read 300 mm eave.

Problem 2–1

Building: 12′–0″ × 22′–0″. GARAGE

Roof: Gable: $\frac{1}{6}$ pitch; $\frac{3}{4}$″ ridge board; 14″ eave; rafters: 16″ OC; plumb cut rafter tail.

Problem: Lay out the CR. Use the graphic method. Check with a framing square.

Questions:
1. What rafter stock would you order?
2. Allowing for a 12″ gable projection, how many 4′–0″ × 8′–0″ panels of plywood are required for the roof?

Problem 2–2

Building: 3660 mm × 6710 mm. GARAGE

Roof: Gable: Slope 250 mm  100 mm ; 19 mm × 140 mm ridge board; 350 mm eave; rafters 400 mm OC; plumb cut rafter tails.

Problem: Lay out the CR. Use the graphic method. Check with a roofing square.

Questions:

1. What rafter stock would you order?
2. How many panels of 9 mm × 1200 mm × 2400 mm are required to sheath the roof?
    a. Allow for a 300 mm gable projection.
    b. Allow 3% extra for waste.

Problem 3–1

Building: 16′–0″ × 40′–0″. CHICKEN HOUSE

Roof: Gable: $\frac{1}{2}$ pitch; $1\frac{1}{8}''$ ridge board; $2\frac{3}{4}''$ seat bearing; 9″ eave; plumb and level cut rafter tail with $2\frac{3}{4}''$ plumb cut; rafters 24″ OC.

Problem: Lay out the CR.

Question: What rafter stock is needed?

Sketch the plan view of this roof.

Problem 3–2

Building: 4880 mm × 12200 mm. CHICKEN HOUSE

Roof: Gable: Slope 1 m 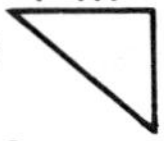 1 m ; 38 mm ridge board; 70 mm seat bearing; 230 mm eave; plumb and level cut rafter tail with 70 mm plumb cut; rafters 600 mm OC.

Problem: Using a metric framing square lay out the CR.

Question: What rafter stock is needed?

Sketch the plan view of this roof.

Problem 4–1

Building: 13′–6″ × 22′–0″.

Roof: Gable: cut 5 in 12; $\frac{3}{4}$″ ridge board; 3″ seat bearing; 10″ eave; square cut rafter tail; rafters 16″ OC.

Problem: Lay out the CR. Use a framing square.

Question: What will the rafters cost?

Sketch the section of the roof.

Problem 4–2

Building: 4118 mm × 6710 mm.

Roof: Gable: cut 250 mm in 600 mm; 19 mm ridge board; 76 mm seat bearing; 250 mm eave; square cut rafter tail; rafters 400 mm OC.

Problem: Lay out the CR. Use a rafter square.

Question: What will the rafters cost?

Sketch the section of the roof.

Problem 5–1

Building: 16′–0″ × 25′–0″. SHELTER

Roof: Gable with two pitches: $\frac{3}{4}$″ ridge is 4′–0″ from front; pitch of front $\frac{1}{2}$; eave at front 8″; eave at back 12″; rafters 24″ OC; plumb and level cut tail with equal cuts.

Problem: Lay out both rafters.

Questions:
1. What is the rear pitch?
2. What rafter stock order is required for this roof?

Sketch the section view of the roof in the space below.

Problem 5–2

Building: 4880 mm × 7625 mm. SHELTER

Roof: Gable with two slopes; 19 mm ridge is 1220 mm from front;
slope of front

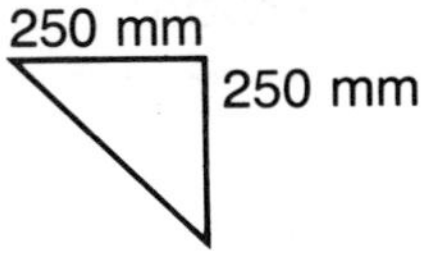

; eave at front 200 mm; eave at back 300 mm; rafters 600 OC; plumb and level cut tail with equal cuts.

Problem: Lay out both rafters.

Questions:

1. What is the slope of the front, expressed as a fraction?
2. What is the rear slope?
3. What rafter stock order is required for this roof?
4. What is the meaning of the word *slope?* Of the word *pitch?*

Problem 6–1

Building: 13′–0″ × 36′–0″. MACHINE SHED

Roof: Shed type; $\frac{1}{3}$ pitch; front and rear eaves 10″; plumb cut rafter tail at top and bottom; 3″ seat bearings; 24″ OC; 10″ overhang of gable ends.

Problem: Lay out the CR (omit four steps).

Question: How many bundles of cedar shingles are required to shingle the roof? Four bundles of shingles cover a *square* if laid 5″ to the weather.

Problem 6–2

Building: 3965 mm × 10980 mm. MACHINE SHED

Roof: Shed type: 0.67 slope; front and rear eaves 250 mm; plumb cut rafter tails at the top and the bottom; 76 mm seat bearings; 600 mm OC; 250 mm overhang of gable ends.

Problem:	Lay out the CR (omit four steps).
Question:	If shingles are laid 125 mm to the weather, how many bundles of cedar shingles are required to shingle the roof?

Problem 7–1

Building:	28′–0″ × 36′–0″. HOUSE WITH ENTRY (gable 4′–0″ × 8′–0″)
Roof:	Gable; main roof $\frac{1}{3}$ pitch; entry roof $\frac{2}{3}$ pitch; ridges $\frac{3}{4}$″; eave of main roof 14″; rafters 16″ OC.

Problem 7–2

Building:	8540 mm × 10980 mm. HOUSE WITH ENTRY (gable 1220 mm × 2440 mm)
Roof:	Gable; main roof 0.67 slope; entry 1.33 slope; ridges 19 mm; eave of main roof 350 mm; rafters 400 mm OC.

Tasks for Problems 7–1 and 7–2:

1. Lay out the CR of the main roof.
2. Lay out the CR of the entry roof. (See Figure 17.1.)
3. Make the eave of the entry level with the main eave.

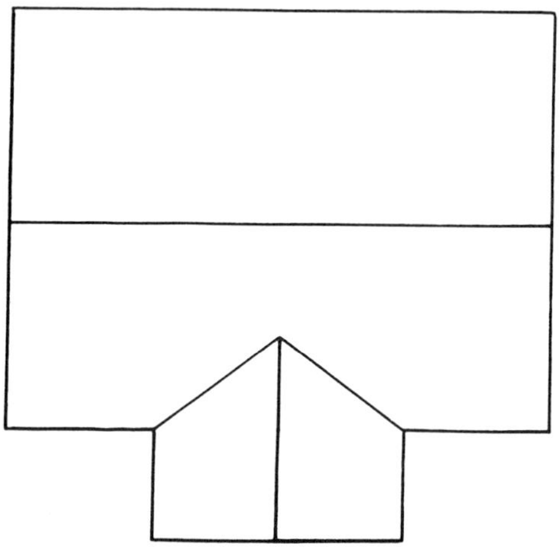

FIGURE 17.1 Plan of house with entry.

Problem 8–1

Building: 20′–0″ × 60′–0″. HOG HOUSE

Roof: Gable of unequal pitch (see Figure 17.2); front pitch is $\frac{1}{2}$; eave 10″ front and back; $\frac{3}{4}$″ ridge; rafters, 24″ OC; plumb cut rafter tails; 10″ gable projection.

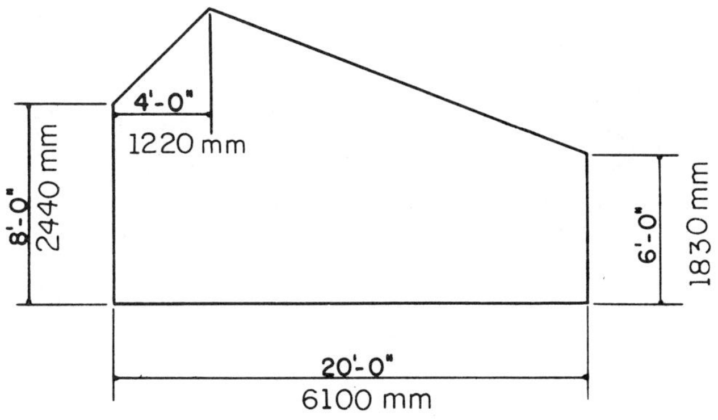

FIGURE 17.2

Problem 8–2

Building: 6100 mm × 18300 mm. HOG HOUSE

Roof: Gable of unequal slopes (see Figure 17.2); front slope is 1.0; eave 250 mm front and back; 19 mm ridge; rafters 600 mm OC; plumb cut rafter tails; 250 mm gable projection.

Tasks for Problem 8–1 and for Problem 8–2:

1. Lay out both rafters.
2. Calculate the roof area.

Problem 9–1

Building: 30′–0″ × 60′–0″. BARN

Roof: Gambrel (see Figure 17.3); no eave on rafter *A;* no ridge board; miter between top end of *A* and bottom end of *B*.

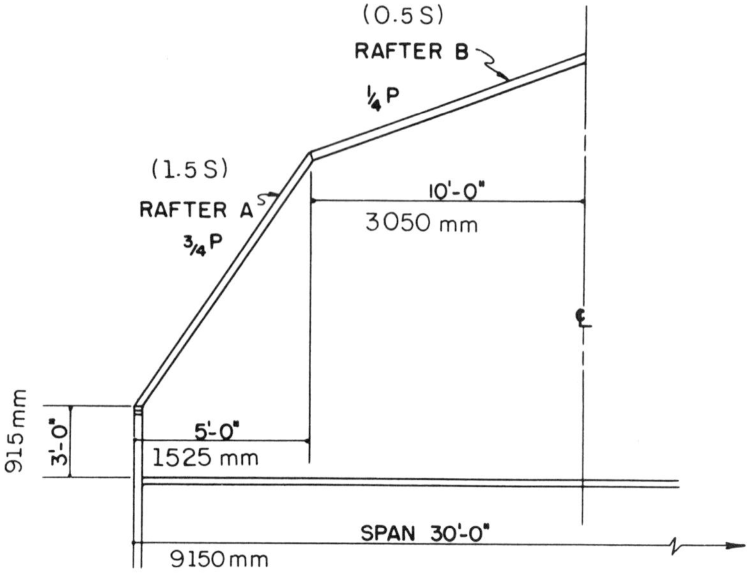

FIGURE 17.3

Problem 9–2

Building: 9150 mm × 18300 mm. BARN

Roof: Gambrel (see Figure 17.3); no eave on rafter A; no ridge board; miter between top end of A and bottom end of B.

Tasks for Problem 9–1 and for Problem 9–2:

1. Lay out both rafters.
2. Calculate the measurement of the loft floor to the ridge.

Problem 10–1

Building: 16′–0″ × 24′–0″. STORE HOUSE

Roof: Gable: rise 3′–0″; 2 × 6 ridge board; 12″ rafter tail; rafters on 16″ centers; 2 × 4 gable studs; 1 × 6 collar tie with upper edge 2′–0″ from top of plate.

Problem 10–2

Building: 4880 mm × 7320 mm. STORE HOUSE

Roof: Gable; rise 915 mm; 38 mm × 140 mm ridge board; 305 mm rafter tail; rafters on 400 mm centers; 38 mm × 89 mm gable studs; 19 mm × 140 mm collar ties with upper edge 600 mm from top of plate.

Tasks for Problem 10–1 and for Problem 10–2:

1. Lay out the CR.
2. Lay out the longest gable stud.
3. Lay out the collar tie.

Problem 11–1

Building: 16′–0″ × 20′–0″. CABIN WITH REGULAR HIP ROOF

Roof: Cottage: $\frac{1}{4}$ pitch; double cheek cut hips; 12″ eave; plumb cut tail; 16″ OC; hips dropped: $1\frac{1}{8}$″ ridge.

Problem 11–2

Building: 4880 mm × 6100 mm. CABIN WITH REGULAR HIP ROOF

Roof: Cottage: slope 0.5; double cheek cut hips; 305 mm eave; plumb cut tail; 400 mm OC; hips dropped; 28 mm ridge.

Tasks for Problem 11–1 and for Problem 11–2:

1. Lay out the HR.
2. Calculate the stock required for the hip rafters.
3. Sketch the roof framing plan view.

Note: This cottage roof is a regular hip roof.

Problem 12–1

Building: 12′–0″ × 30′–0″. PUMPING STATION

Roof: Cottage: $\frac{1}{3}$ pitch; plumb and level cut tail; rafters 16″ OC; hips dropped; double cheek cut hip; 14″ eave; $\frac{3}{4}$″ ridge; 2″ plumb line above CR seat.

Problem 12–2

Building: 3660 mm × 9150 mm. PUMPING STATION

Roof: Cottage: slope 0.67; plumb and level cut tail; rafters 400 mm OC; hips dropped; double cheek cut hip; 355 mm eave; 19 mm ridge; 50 mm plumb line above CR seat.

Tasks for Problem 12–1 and Problem 12–2:

1. Lay out the hip rafter (HR).
2. Calculate the stock required for the hips.
3. Calculate the length of the ridge board.
4. Sketch the roof-framing plan view.

Problem 13–1

Building: 12′–0″ × 20′–0″. GARAGE

Roof: Cottage: cut 5 in 12; single cheek cut hips; 12″ eave; ridge $\frac{3}{4}$″; plumb cut tail; rafters 24″ OC.

Problem 13–2

Building: 3660 mm × 6100 mm. GARAGE

Roof: Cottage: slope 0.42; single cheek cut hips; 305 mm eave; ridge 19 mm; plumb cut tail; rafters 600 mm OC.

Tasks for Problem 13–1 and for Problem 13–2:

1. Lay out the HR.
2. Calculate the length of the ridge board.
3. Calculate the LL of the first JRs.
4. Sketch the HR to ridge framing detail.

Problem 14–1

Building: 8′–0″ × 24′–0″. VERANDA

Roof: Half hip against main building; $\frac{1}{3}$ pitch; single cheek cut hip; eave 10″; rafters 24″ OC; $\frac{3}{4}$″ ridge nailed to side of building; plumb cut tails.

Problem 14–2

Building: 2440 mm × 7320 mm. VERANDA

Roof: Half hip against main building; 0.67 slope; single cheek cut hip; eave 250 mm; rafters 600 mm OC; 19 mm ridge nailed to side of building; plumb cut tails.

Tasks for Problem 14–1 and for Problem 14–2:

1. Lay out the HR.
2. Calculate the LL of the CR.
3. Calculate the LL of the 1st JR; the 2nd JR; and the 3rd JR.
4. Sketch the HR to ridge framing detail.

Problem 15–1

Building: 21′–0″ × 21′–0″. DOUBLE GARAGE

Roof: Regular hip roof; $\frac{1}{4}$ pitch; 12″ eave with plumb and level cut tail ($2\frac{1}{2}$″ plumb); roof framed with four HRs, four CRs and with JRs at 16″ OC.

Problem 15–2

Building: 6405 mm × 6405 mm. DOUBLE GARAGE

Roof: Hip type; slope 250 mm 125 mm ; 305 mm eave with plumb and level cut tail (65 mm plumb); roof framed with four HRs, four CRs, and with JRs at 400 mm OC.

Tasks for Problem 15–1 and for Problem 15–2:

1. Lay out a HR. Omit four steps when stepping off the LL with the framing square.
2. Lay out a CR.
3. Lay out a set of JRs on the CR.
4. Calculate the area of the roof.

Problem 16–1

Building: 14′–8″ × 20′–0″. COTTAGE

Roof: Hip type; $\frac{1}{2}$ pitch; double cut hips at ridge; $\frac{3}{4}''$ ridge board; plumb cut rafter tail; eave 10″; rafters 24″ OC.

Problem 16–2

Building: 4475 mm × 6100 mm. COTTAGE

Roof: Hip type; slope 1000 mm 1000 mm ; double cut hips at ridge; 19 mm ridge board; plumb cut rafter tail; eave 250 mm; rafters 600 mm OC.

Tasks for Problem 16–1 and for Problem 16–2:

1. Lay out one HR.
2. Lay out one CR.
3. Lay out first, second, and third JRs.
4. Determine:

a. the number of CRs in the roof.
b. the number of HRs in the roof.
c. the number of JRs in the roof.

Problem 17–1

Building: A SMALL HOUSE WITH INTERSECTING GABLE ROOFS OF UNEQUAL PITCHES

Major Roof: 16′–0″ × 30′–0″; $\frac{1}{4}$ pitch; $\frac{3}{4}$″ ridge; plumb cut rafter tails; cornice run 12″; rafters 2 × 4s at 24″ OC.

Minor Roof: 12′–0″ × 4′–0″; $\frac{3}{4}$″ ridge; cornice run 12″ with the fascia at the same level as the fascia of the major roof; 2 × 4 rafters at 24″ OC. See Figure 17.4.

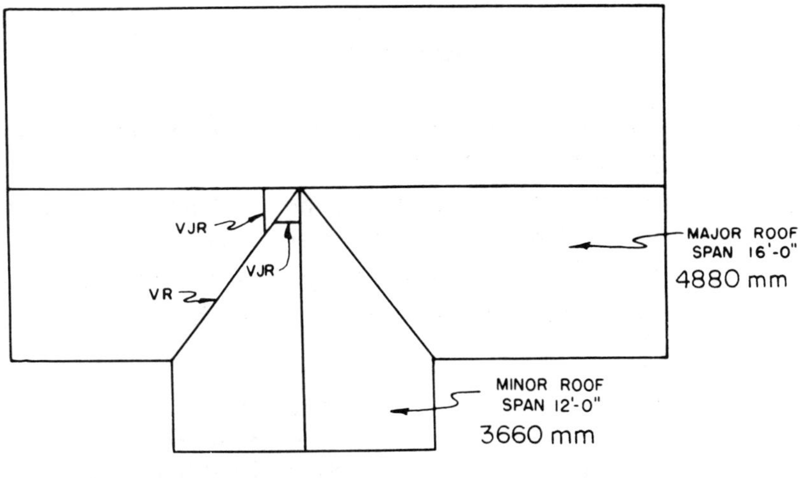

FIGURE 17.4

Problem 17–2

Building: A SMALL HOUSE WITH INTERSECTING GABLE ROOFS OF UNEQUAL SLOPES

Major Roof: 4880 mm × 9150 mm; slope 500 / 250 ; 19 mm ridge;

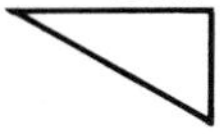

plumb cut rafter tails; cornice run 305 mm; rafters 38 mm × 89 mm at 600 mm OC.

Minor Roof: 3660 mm × 1220 mm; 19 mm ridge; cornice run 305 mm with the fascia at the same level as the fascia of the major roof; 38 mm × 89 mm rafters at 600 mm OC. See Figure 17.4.

Tasks for Problem 17–1 and for Problem 17–2:

1. Lay out a minor roof CR.
2. Lay out a VR.
3. Lay out two VJRs placed 24″ (600 mm) from the center lines of the ridge boards.

Note: When the specifications read *cornice run 12″ (305 mm),* allow 12″ (305 mm) overhang of the CR.

Problem 18–1

Building: A PUMP-HOUSE WITH INTERSECTING GABLE ROOFS OF UNEQUAL PITCHES

Major Roof: 12′–0″ × 22′–0″; ½ pitch; ¾″ ridge; plumb cut rafter tails; 12″ cornice run; 2 × 4 rafters at 24″ OC.

Minor Roof: 12′–0″ × 4′–0″; ¾″ ridge: 12″ cornice run with the fascia in line with the fascia of the main roof; 2 × 4 rafters at 24″ OC.

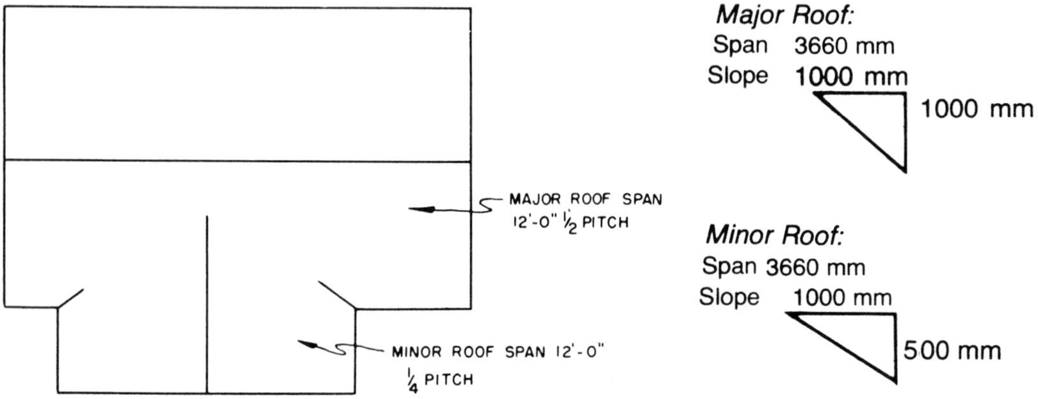

FIGURE 17.5

Problem 18–2

Building: A PUMP-HOUSE WITH INTERSECTING GABLE ROOFS OF UNEQUAL SLOPES

Major Roof: 3660 mm × 6710 mm; slope 1000 mm

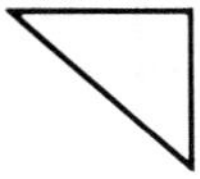

19 mm ridge; plumb cut rafter tails; 305 mm cornice run; 38 mm × 89 mm rafters at 600 mm OC.

Minor Roof: 3660 mm × 1220 mm; 19 mm ridge; 305 mm cornice run with the fascia in line with the fascia of the main roof; 38 mm × 89 mm rafters at 600 mm OC.

Test 1–1

A garage gable roof is to be built to the following specifications:

Outside plate measurements	12′–0″ × 22′–0″
Pitch	$\frac{1}{3}$
Ridge	1″ nominal
Eave	16″
Rafters	2 × 4s at 16″ OC
Rafter tail plumb cut only	
Collar tie	1 × 6 with upper edge 2′–0″ from the top of the plate

Test 1–2

A garage gable roof is to be built to the following specifications:

Outside plate measurements	3660 mm × 6710 mm
Slope	1000 mm 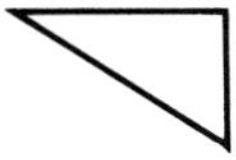 667 mm
Ridge	19 mm
Eave	400 mm
Rafters	38 mm × 89 mm at 400 mm OC
Rafter tail plumb cut only	
Collar tie	19 mm × 140 mm with upper edge 600 mm from the top of the plate

Common Rafter, Gable Stud, and Collar Tie Scoring Sheet

Pitch _____ Span _____ Ridge _____ Overhang _____ Rafters _____ OC

Slope: 

	Possible Score		*Score*	*Calculations*
Pattern Common Rafter				
1. Crown side	1			
2. Plumb line at ridge	4			
3. Calculation for LL	7			
4. Measured LL. Tolerance $\pm\frac{1}{8}''$ ($\pm$2 mm)	5			
5. Birdsmouth plumb and level lines	3			
6. Shortening	4			
7. Presentation and calculation for rafter tail LL. Tolerance $=\frac{1}{8}''$ ($\pm$2 mm)	7			
8. Tail-end layout	4			
		35		
Pattern Gable Stud				
1. LL. Tolerance $=\frac{1}{8}''$ ($\pm$2 mm)	10			
2. Shortening	5			
3. Roof rake cut. Line layout	5			
4. Layout of notching for CR	5			
		25		
Pattern Collar Tie				
1. Length—measured along the upper edge	10			
2. Roof rake cut. Line layout	5			
		15		
		75		

Percentage score (Factor 1.33) ________________________

Date: ____________________ Layout time: ____________________

INSTRUCTIONS

For Test 1–1 and for Test 1–2 (Use metric measurements for Test 1–2):

A. On one side and one edge of a 2 × 4 (38 mm × 89 mm), lay out a pattern CR.

B. On the other side and edge, lay out a second-longest gable stud.

C. On a 1 × 6 (19 mm × 140 mm), lay out a pattern collar tie.

D. Use black pencil lines for the layouts. Use a colored pencil to show where all cuts are to be made.

E. The "Common Rafter, Gable Stud, and Collar Tie Scoring Sheet" may be used as a guide. The checking points, measurement tolerances, and possible scores are given. Although all calculations should be worked directly on the rafter stock, it would be worthwhile for reference purposes to record the correct calculations on the scoring sheet. A space is there for this purpose.

Test 2–1

A garage hip roof is to be built to the following specifications:

Outside plate measurements	12′–0″ × 22′–0″
Pitch	$\frac{1}{4}$
Ridge	1″ nominal
Eave	16″
Rafters	2 × 4s at 24″ OC
Rafter tails plumb cut	
The HR is to be dropped and the plate dubbed.	

Test 2–2

A garage hip roof is to be built to the following specifications:

Outside plate measurements	3660 mm × 6710 mm
Slope	250

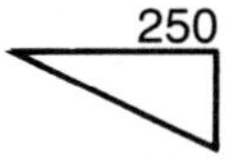

125

Ridge	19 mm

Eave	400 mm
Rafters	38 mm × 89 mm at 600 mm OC
Rafter tails plumb cut	
The HR is to be dropped and the plate dubbed.	

INSTRUCTIONS

For Test 2–1 and for Test 2–2 (Use metric measurements for Test 2–2):

A. On one side and one edge of a 2 × 4 (38 mm × 89 mm), lay out a pattern CR.

B. On the same side and edge as the CR layout, and using the same birdsmouth and rafter tail, lay out the complete lefthand set of JRs.

C. On the other side and edge lay out a pattern HR.

D. Use black pencil lines for the layouts. Use a colored pencil to show where all cuts are to be made.

E. The scores should be recorded on the "Common Rafter, Hip Rafter, and Jack Rafter Scoring Sheet."

Tests 3A and 3B (imperial measurements) and 3A.1 and 3B.1 (metric measurements)

An equal-pitch projection is to be built to a hip roof as shown in Figure 17.6.

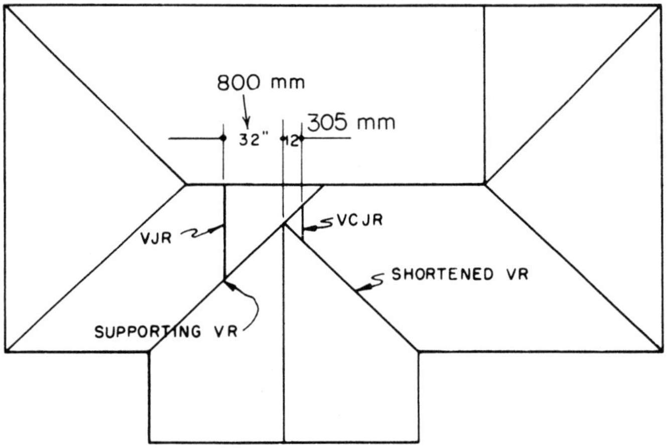

FIGURE 17.6

Common Rafter, Hip Rafter, and Jack Rafter Scoring Sheet

Pitch _____ Span _____ Ridge _____ Overhang _____ Rafters _____ OC

Slope: 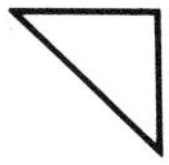

		Possible Score	*Score*	*Calculations*
	Pattern Common Rafter			
1.	Crown side	1		
2.	Plumb lines at ridge	4		
3.	Calculation for LL	7		
4.	Measured LL. Tolerance $\pm\frac{1}{8}''$ (±2 mm)	5		
5.	Birdsmouth plumb and level lines	3		
6.	Shortening	4		
7.	Presentation and calculation for rafter tail LL. Tolerance $\pm\frac{1}{8}''$ (±2 mm)	7		
8.	Tail-end layout	4		
			35	
	Pattern Hip Rafter			
1.	Plumb line at the ridge	2		
2.	Calculation for LL	5		
3.	Measured LL. Tolerance $\pm\frac{1}{8}''$ (±2 mm)	2		
4.	Original level line at the birdsmouth	4		
5.	Shortening at the ridge	4		
6.	Side cut lines	4		
7.	Dubbing	4		
8.	Dropping	5		
9.	Presentation and calculation for rafter tail LL. Tolerance $\pm\frac{1}{8}''$ (±2 mm)	5		
10.	Measured LL of rafter tail	6		
11.	Tail-end layout	4		
			45	
	Jack Rafters Layout			
1.	LL of longest JR. Tolerance $\pm\frac{1}{8}''$ (±2 mm)	5		
2.	Shortening	5		
3.	Side cut line	5		
4.	Measurement of common difference for remaining JRs	5		
			20	
			100	

Date: ____________________ Layout time: ____________________

SPECIFICATIONS

For Tests 3A and 3B (imperial measurements)

Major Roof:

Outside plate measurements	16′–0″ × 30′–0″
Pitch	$\frac{1}{4}$
Ridge	2 × 6 nominal size
Eave (cornice run)	24″
Rafters	2 × 6s at 16″ OC
Backs of rafters $3\frac{1}{2}$″ above the plate	
Rafter tails plumb cut	

Minor Roof:

Span	12′–0″
Length	4′–0″
Ridge	2 × 6 nominal size
Eave (cornice run)	24″
Rafters	2 × 6s at 16″ OC
Backs of rafters $3\frac{1}{2}$″ above the plate	
Rafter tails plumb cut	

SPECIFICATIONS

For Tests 3A.1 and 3B.1 (metric measurements)

Major Roof:

Outside plate measurements	4880 mm × 9150 mm
Slope	250 mm

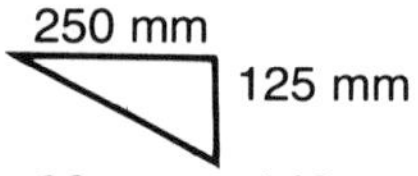

Ridge	38 mm × 140 mm
Eave (cornice run)	600 mm
Rafters	38 mm × 140 mm at 400 mm OC
Backs of rafters 90 mm (perpendicular measurement) above the plate	
Rafter tails plumb cut	

Minor Roof:

Span	3660 mm
Length	1220 mm
Ridge	38 mm × 140 mm
Eave (cornice run)	600 mm
Rafters	38 mm × 140 mm at 400 mm OC

Backs of rafters 90 mm
(perpendicular measurement)
above the plate
Rafter tails plumb cut

INSTRUCTIONS

For Test 3A (imperial) and Test 3A.1 (metric)

A. On one side and one edge of a 2 × 6 (38 mm × 140 mm), lay out a pattern CR for the major roof.

B. On the same side and edge, lay out the valley jack rafter for the major roof. Use the CR ridge layout.

C. On the other side and edge, lay out the supporting valley rafter.

D. Use black pencil lines for the layouts. Use colored pencil lines to show where all cuts are to be made.

E. The scores should be recorded on the "Common Rafter, Valley Jack Rafter, and Supporting Valley Rafter Scoring Sheet"

INSTRUCTIONS

For Test 3B (imperial) and Test 3B.1 (metric)

A. On one side and one edge of a 2 × 6 (38 mm × 140 mm), lay out a pattern CR for the minor roof.

B. On an unmarked space on the same side and edge, lay out the valley-cripple-jack rafter.

C. On the other side and edge, lay out the shortened valley rafter.

D. Use black pencil lines for the layouts. Use colored pencil lines to show where all cuts are to be made.

E. The scores should be recorded on the "Common Rafter, Valley-Cripple-Jack Rafter, and Shortened Valley Rafter Scoring Sheet."

Common Rafter, Valley Jack Rafter, and Supporting Valley Rafter Scoring Sheet

Pitch _____ Span _____ Ridge _____ Overhang _____ Rafters _____ OC

Slope: 

	Possible Score	*Score*	*Calculations*
Pattern Common Rafter			
1. Crown side	1		
2. Plumb line at ridge	4		
3. Calculation for LL	7		
4. Measured LL. Tolerance $\pm\frac{1}{8}''$ ($\pm$2 mm)	5		
5. Birdsmouth plumb and level lines	3		
6. Shortening	4		
7. Presentation and calculation for rafter tail LL. Tolerance $\pm\frac{1}{8}''$ ($\pm$2 mm)	4		
8. Tail-end layout	2		
		30	
Valley Jack Rafter			
1. Plumb line at the ridge	3		
2. Calculation of LL	7		
3. Measured LL. Tolerance $\pm\frac{1}{8}''$ ($\pm$2 mm)	3		
4. Shortening at the ridge	4		
5. Shortening at the supporting valley rafter end	4		
6. Side cut layout at the supporting valley rafter end	4		
		25	
Supporting Valley Rafter			
1. Crown side	1		
2. Plumb line at ridge	2		
3. Calculation for LL	7		
4. Measured LL. Tolerance $\pm\frac{1}{8}''$ ($\pm$2 mm)	4		
5. Back of rafter over the plate	4		
6. Birdsmouth	3		
7. Shortening at the ridge	5		
8. Side cut layout at the ridge	4		
9. Presentation and calculation for rafter tail LL. Tolerance $\pm\frac{1}{8}''$ ($\pm$2 mm)	7		
10. Measured LL of rafter tail	4		
11. Tail side cut layout	4		
		45	
		100	

Date: ____________________ Layout time: ____________________

Common Rafter, Valley-Cripple-Jack Rafter, and Shortened Valley Rafter Scoring Sheet

Pitch _____ Span _____ Ridge _____ Overhang _____ Rafters _____ OC

Slope: 

	Possible Score	*Score*	*Calculation*
Pattern Common Rafter			
1. Crown side	1		
2. Plumb line at ridge	4		
3. Calculation for LL	7		
4. Measured LL. Tolerance $\pm\frac{1}{8}''$ ($\pm$2 mm)	5		
5. Birdsmouth plumb and level lines	3		
6. Shortening	4		
7. Presentation and calculation for rafter tail LL. Tolerance $\pm\frac{1}{8}''$ ($\pm$2 mm)	4		
8. Tail-end layout	2		
		30	
Valley-Cripple-Jack Rafter			
1. Plumb line	1		
2. Calculation for LL	7		
3. Measured LL. Tolerance $\pm\frac{1}{8}''$ ($\pm$2 mm)	1		
4. Shortening at the supporting valley rafter	4		
5. Side cut at the supporting valley rafter	4		
6. Shortening at the shortened valley rafter	4		
7. Side cut layout at the shortened valley rafter	4		
		25	
Shortened Valley Rafter			
1. Crown side	1		
2. Plumb line at the supporting valley	2		
3. Calculation for LL	10		
4. Measured LL. Tolerance $\pm\frac{1}{8}''$ ($\pm$2 mm)	4		
5. Back of rafter over plate	5		
6. Birdsmouth	2		
7. Shortening at the valley rafter	6		
8. Presentation and calculation for rafter tail LL. Tolerance $\pm\frac{1}{8}''$ ($\pm$2 mm)	7		
9. Measured tail length	4		
10. Tail-end layout	4		
		45	
		100	

Date: ____________________ Layout time: ____________________

Test 4–1 (imperial measurements) and Test 4–2 (metric measurements)

An unequal-pitch projection is to be built to a gable roof as shown in Figure 17.7.

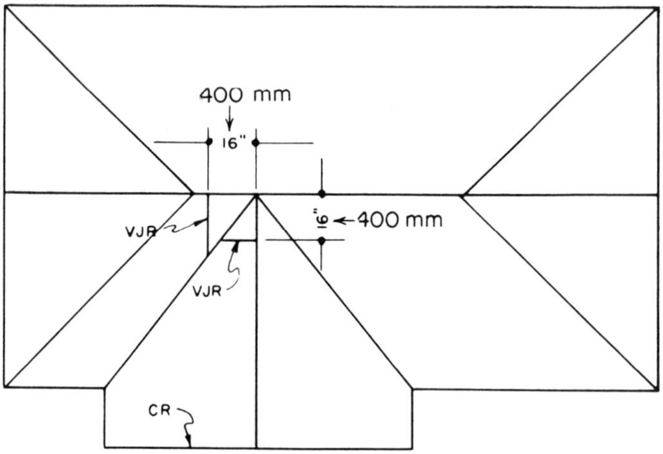

FIGURE 17.7

SPECIFICATIONS

For Test 4–1 (imperial measurements)

Major Roof:

Outside plate measurements	16′–0″ × 30′–0″
Pitch	$\frac{1}{4}$
Ridge	1 × 6 nominal size
Eave	18″
Rafters	2 × 6s at 16″ OC with backs $3\frac{1}{2}$″ above the plate.

Rafter tails are plumb and level cut with a 3″ plumb cut

Minor Roof:

Span	12′–0″
Length	4′–0″

Eave to be at the same level as the eave of the major roof with a run of 18″

Rafters	2 × 6s at 16″ OC
Rafter tails are to be plumb and level cut	

SPECIFICATIONS

For Test 4–2 (metric measurements)

Major Roof:

Outside plate measurements	4880 mm × 9150 mm
Slope	250 mm 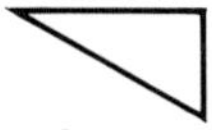 125 mm
Ridge	19 mm × 140 mm
Eave	450 mm
Rafters	38 mm × 140 mm at 400 mm OC with backs 90 mm above the plate.
Rafter tails are plumb and level cut with a 75 mm plumb cut.	

Minor Roof:

Span	3660 mm
Length	1220 mm
Eave to be at the same level as the eave of the major roof with a run of 450 mm.	
Rafters	38 mm × 140 mm at 400 mm OC
Rafter tails are to be plumb and level cut.	

INSTRUCTIONS

For Tests 4–1 and 4–2:

A. Determine the difference in plate levels.

B. Draw a plan view showing the ridge intersection, the VR, and the two VJRs.

C. On one side and one edge of a 2 × 6 (38 mm × 140 mm), lay out the VR.

D. On the other side and edge, lay out the major roof VJR.

E. Also on this side and edge, lay out the minor roof VJR.

F. The scores should be recorded on the "Scoring Sheet for the Roof of Unequal Pitches."

G. Use black pencil lines for the layouts. Use colored lines to show where all cuts are to be made. Use building paper for the drawings required for A and B.

Scoring Sheet for the Roof of Unequal Pitches

	Possible Score		Score	Calculations
Difference in plate levels		10		
Plan View Drawing				
1. Plan size	6			
2. Eaves size	4			
3. Valley rafter plan	8			
4. Valley pitch board	6			
5. Major roof VJR	3			
6. Minor roof VJR	3			
		30		
Valley Rafter				
1. Plumb lines	4			
2. Total LL	8			
3. Tail LL	4			
4. Shortening at ridge	3			
5. Side cuts at ridge	3			
6. Birdsmouth	6			
7. Fish tail	2			
		30		
Major Roof VJR				
1. Plumb lines	3			
2. LL	6			
3. Shortening at ridge	2			
4. Shortening at VR	2			
5. Side cut at VR	2			
		15		
Minor Roof VJR				
1. Plumb lines	3			
2. LL	6			
3. Shortening at ridge	2			
4. Shortening at VR	2			
5. Side cut at VR	2			
		15		
		100		

Date: -------------------- Layout time: --------------------

The foregoing problems and tests were examples of typical rafter layouts.

The following tasks in Chapter 18 require layouts, cutting, and assembly of roof members.

18

Shop Assignments: Roof Framing Jobs

The four small-roof jobs described in this chapter are designed to encourage accurate measuring and careful construction when using full-size stock. Specifications are given in both imperial and metric measurements.

Job 1: Build an equal-pitch hip roof.

Job 2: Build an equal-pitch hip roof with an equal-pitch intersecting roof. The span of the minor roof is the same as that of the major roof.

Job 3: Build an equal-pitch hip roof with an equal-pitch intersecting roof. The span of the minor roof is less than that of the major roof.

Job 4: Build an equal-pitch hip roof with an unequal-pitch intersecting roof. The span of the minor roof is less than the span of the major roof. The total rise for each is the same. Therefore the pitch of the minor roof is not the same as the pitch of the major roof.

Note: These jobs are outlined on the following pages.

Rafters for the roof-framing jobs should be cut from full-size 2 × 4 (38 mm × 89 mm) stock and assembled on a reusable plate.

A suggested method of plate construction follows. Note that only two separate plate assemblies are required for the four framing jobs.

For Job 1 use the 2′–6′′ × 6′–0′′ (800 mm × 1800 mm) plate assembly. See Figure 18.1.

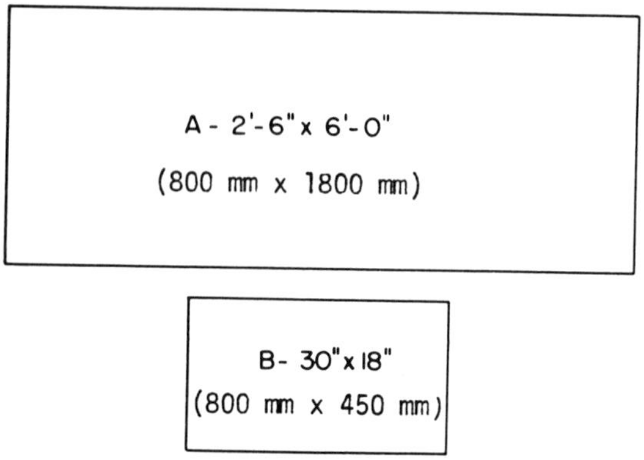

FIGURE 18.1. Plate assemblies for Jobs 1 and 2.

For Job 2 use the 2′–6″ × 6′–0″ (800 mm × 1800 mm), with a 30″ × 18″ (800 mm × 450 mm) plate assembly attached. See Figure 18.1.

For Job 3 use the 2′–6″ × 6′–0″ (800 mm × 1800 mm) plate assembly with an 18″ × 30″ (450 mm × 800 mm) plate assembly attached. See Figure 18.2.

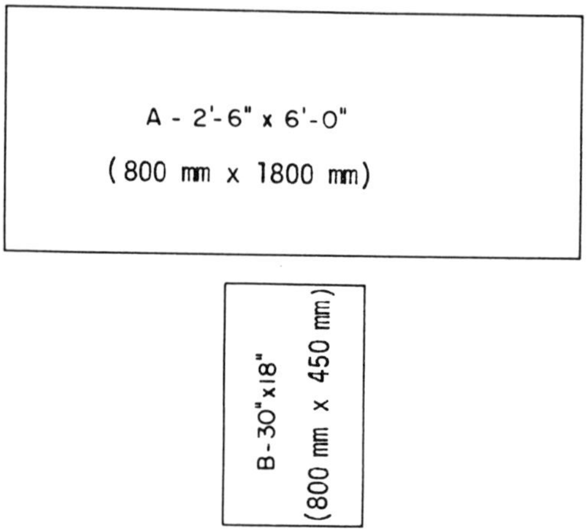

FIGURE 18.2. Plate assemblies for Jobs 3 and 4.

For Job 4 use the same assembly as for Job 3.

1. The 1 × 2 (19 mm × 38 mm) strips are nailed to the 2 × 4 (38 mm × 89 mm) base and are replaceable (Figure 18.3). The plate layout is done on the upper surface.
2. Assembly *B* may be fastened to *A* by means of bolts or nails.

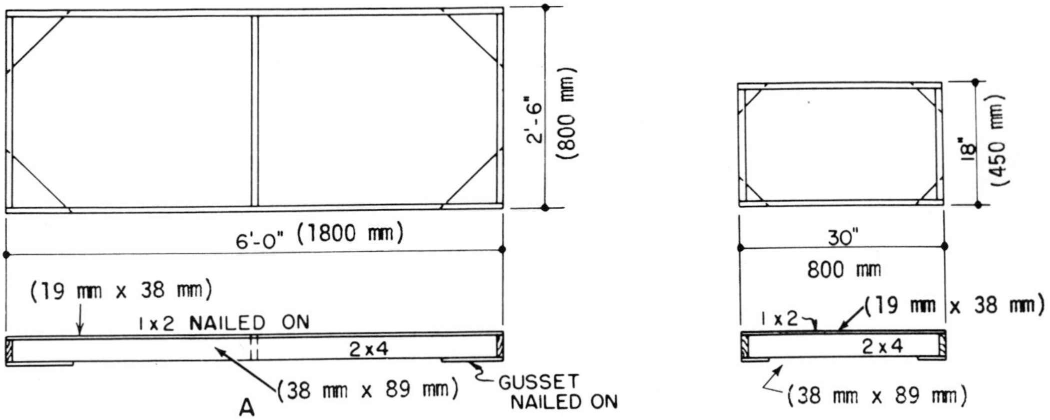

FIGURE 18.3

Roof Framing Job 1

Plate: 2′–6″ × 6′–0″ (800 mm × 1800 mm).

Roof: Equal-pitch hip type: 1″ × 6″ (19 mm × 140 mm) ridge board; double-cheek cut hips; 4″ (100 mm) eave; plumb cut rafter tails; CRs 2″ (50 mm) above the plate (Figure 18.4).

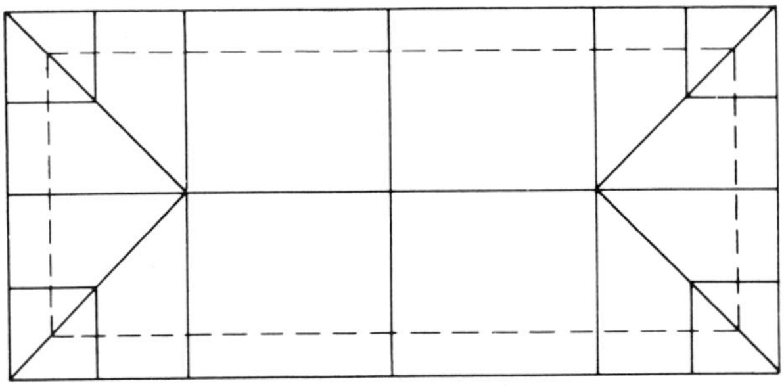

FIGURE 18.4. Roof plan for Job 1.

Pitch _____ (After the plan view drawing has been completed, ask your instructor what the pitch[1] of the roof is to be.)

PROCEDURE:

1. Check the plate for size and squareness. Dub one corner.
2. On building paper lay out a full-size plan of one end of the roof. Include one end of the ridge, three CRs, two HRs, and four JRs at 7½″ (200 mm) OC. Show the measuring lines in color, and then number them in order of layout on the rafter stock.
3. Lay out and cut the ridge board, 8 CRs, and all the HRs and JRs.
4. Set up the plate on sawhorses.
5. Frame the roof.

Note: Use full-size 2 × 4 (38 mm × 89 mm) stock for rafters. Make complete line layouts on the rafter stock for all pattern rafters. Number all the plumb lines on the pattern rafters to correspond with the numbers on the building paper layout.

When you have completed the roof framing job check it with an eight foot (3 m) straightedge. Be sure that the backs of the rafters are in line and that the rafter tails are in line. Devise a method of checking the pitch of the roof after its construction by using a carpenter's level and a framing (roofing) square.

ADDITIONAL JOBS:

6. Lay out and cut a pair of sheathing boards to be fitted over the hip rafter. Check the miter and face angles. Nail the boards in place.
7. Lay out and cut a pair of windblocks to be fitted between one HR and two adjacent JRs. Nail them in place.
8. Lay out and cut a pair of purlin segments to be fitted under and against a hip rafter. Nail them in place.
9. On the building paper layout (see 2 above), add one more JR at an angle other than 90 degrees to the plate. Using information given on this plan view, lay out and cut the rafter. Nail it in place, and then check its height above the plate. When this

[1]For a metric roof use the term *slope* rather than *pitch*.

has been done, you are able to understand most roof framing problems—including that of the unequal-pitch-roof VR layout.

10. Lay out and cut an elliptical hole in a piece of roof sheathing to take a 2″ (50 mm) round vertical pipe.

Roof Framing Job 2

Plate: A basic plate 2′–6″ × 6′–0″ (800 mm × 1800 mm) with an auxiliary plate 18″ × 30″ (450 mm × 800 mm) bolted to it.

Roof: A hip roof, as in Job 1, with an equal-pitch[2] intersecting roof of span 30″ (800 mm) (Figure 18.5).

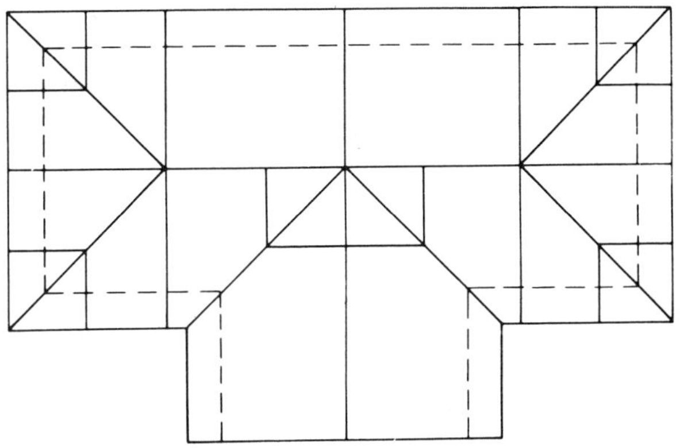

FIGURE 18.5. Roof plan for Job 2.

PROCEDURE:

1. Build a hip roof on the basic plate as in Job 1.
2. Check the measurements of the added plate.
3. On the building paper lay out a full size plan view of the intersecting roof. Include two end CRs, the ridge board, two valley rafters (VRs), and four valley jack rafters (VJRs) at 7½″ (200mm). Show the measuring lines in color, and then number in order of layout on the rafter stock.
4. Lay out and cut all the roof members.
5. Frame the intersecting roof.

[2]For a metric roof use *equal-slope*.

Roof Framing Job 3

Plate: A basic plate 2′–6″ × 6′–0″ (800 mm × 1800 mm) with the same auxiliary plate 18″ × 30″ (450 mm × 800 mm) turned 90 degrees and bolted to it.

Roof: A hip roof, as in Job 1, with an equal-pitch[3] intersecting roof of span 18″ (450 mm) (Figure 18.6).

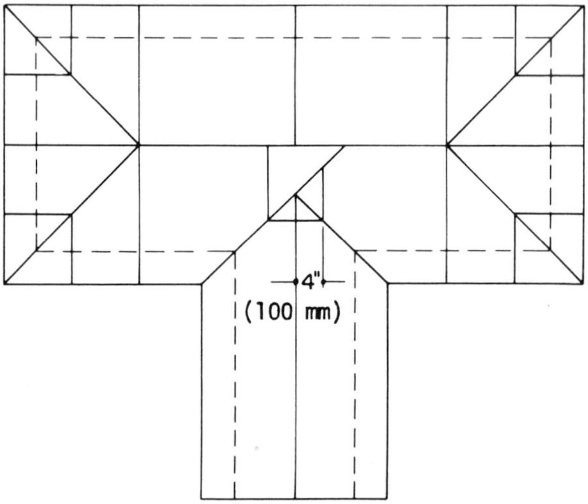

FIGURE 18.6. Roof plan for Job 3.

PROCEDURE:

1. Build a hip roof on the basic plate as in Job 1.
2. Check the measurements of the added plate.
3. On building paper lay out a full-size plan view of the intersecting roof. Include two end CRs, the ridge board, two VRs, and four VJRs. Show the measuring lines in color. Number them in the order of layout on the rafter stock.
4. Lay out and cut all the roof members.
5. Frame the intersecting roof. Back the supporting VR.

[3]For a metric roof use *equal-slope*.

Roof Framing Job 4

Plate: See Job 3.

Roof: A hip roof as in Job 1 with an unequal-pitch[4] intersecting roof rising to the same height. The cornice of the minor roof is to be in line with the cornice of the major roof (Figure 18.7).

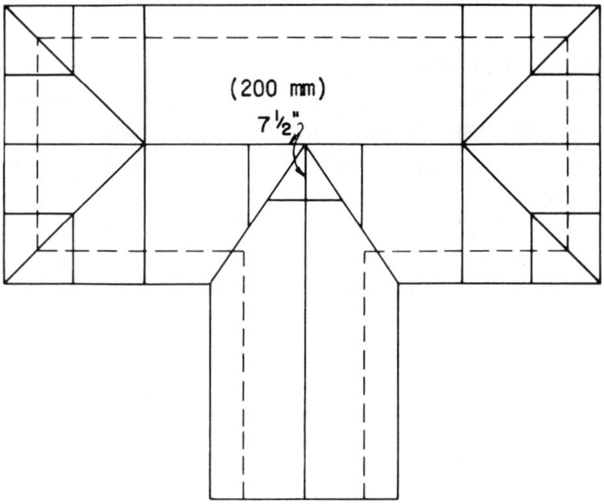

FIGURE 18.7. Roof plan for Job 4.

PROCEDURE:

1. Remove the equal-pitch intersecting roof of Job 3.
2. Re-check the measurements of the added plate. The span of the minor roof should be 18″ (450 mm).
3. On building paper lay out a full-size plan view of the minor roof and segments of the adjacent major roof cornice and ridge board. Include two CRs and VRs, and four JRs. *Note that the VR will not pass directly over the point of intersection of the plates.*

   Lay out the rafter center lines. Measure the rafter stock to be used. Complete the plan view of the rafters. Show the cutting lines in color. Number the measuring lines in the order of layout on the rafter stock.

[4]For a metric roof use *unequal-slope*.

4. Complete a comparative pitch drawing to determine the change in height of the plate (or depth of the birdsmouth) required for the CRs.
5. Lay out and cut the ridge board and the two CRs. Nail them in place. Check the cornice heights.
6. Lay out and cut the VRs. Nail them in place.
7. Lay out and cut the VJRs. Nail them in place. Lay a carpenter's rule down the back of a VJR. Note where the rule strikes the VR.

Note: The drawing for the VR of this roof was full size. The run of the rafter and the thickness of the rafter are shown on the plan view. The measurements are transferred carefully to the rafter stock. When the problem arises of laying out the VR for an unequal pitch roof of a house, it is not practical to draw a full-size plan view of its RUN. **It is convenient to draw the two roof spans and the VR run one twelfth full size, imperial measurement, or one tenth full size, metric measurement. The VR thickness and the space between measuring lines remain full size.**

Shop Drawing Assignments

Complete the missing views

To be able to visualize the whole building is often the carpenter's problem, particularly if he is given only some of the views. The partly finished three-view drawings shown on Plate XII are to be completed. A plan view, a front elevation, and a side elevation are required for each drawing (except number five which requires the addition of a right end elevation).

Date ______________

Score ______________

Complete the plan views

In Figures 18.8 and 18.9 on page 214 are shown the wall plates of two equal-pitch roofs in plan view. Complete the single line drawings to show the rafters and ridge boards in plan view.

Date ______________

Score ______________

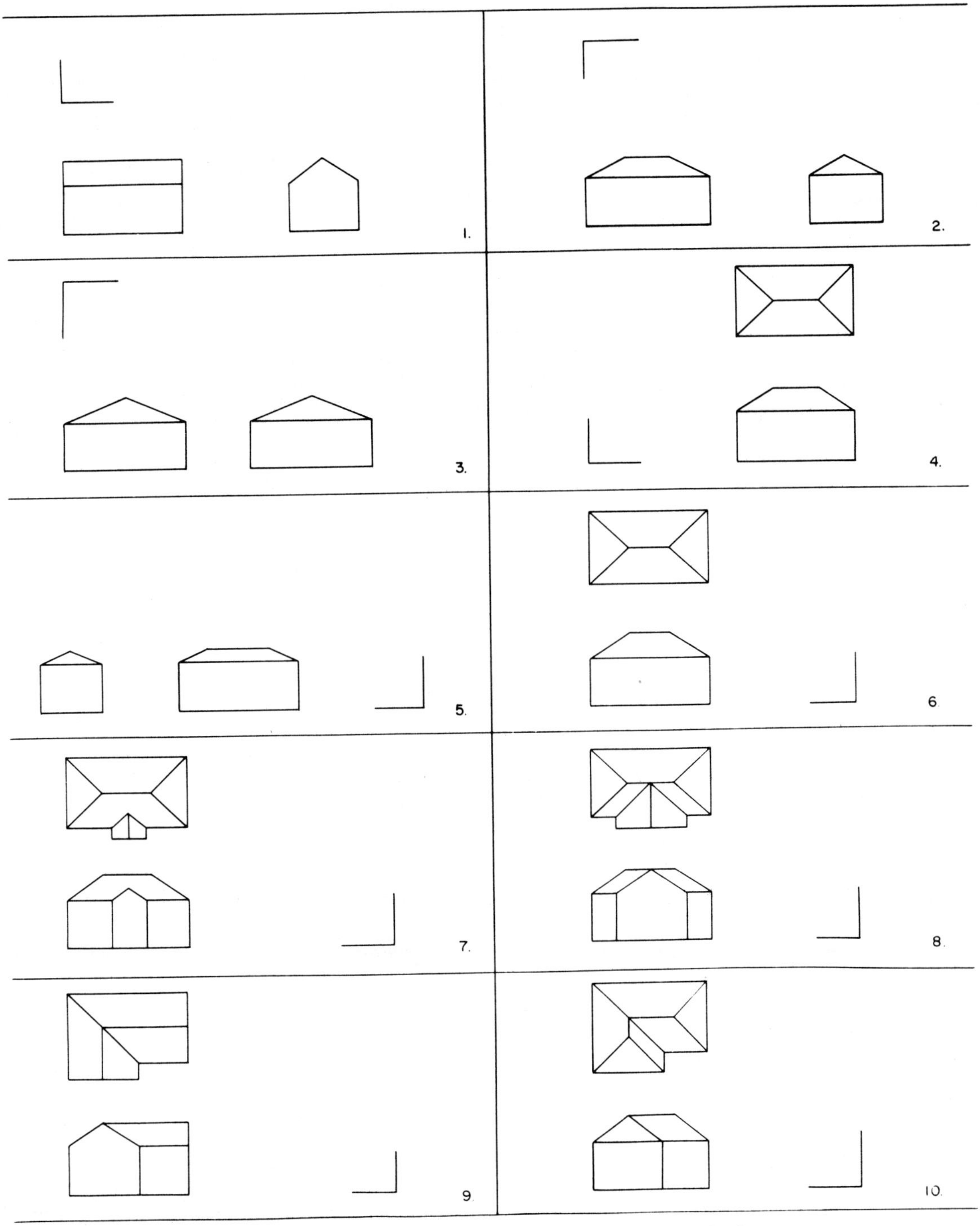

PLATE XII. Elevations and Plan Views: A Shop Drawing Assignment.

FIGURE 18.8.

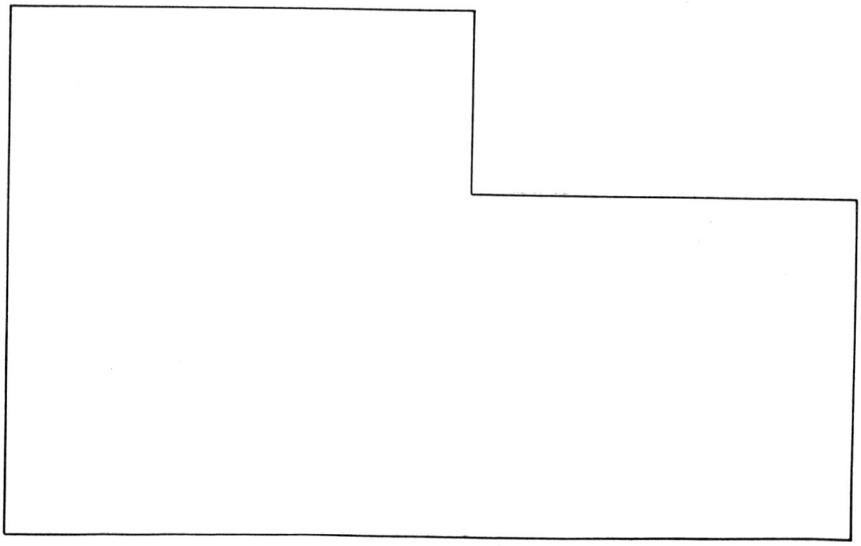

FIGURE 18.9

Equal- and unequal-pitch (slope) roofs have been framed using CRs, HRs, JRs, VRs, VJRs, sheathing, windblocks, and purlins.

To advance the understanding of roof framing theory, pitch board/ slope board development is discussed in the next chapter.

19

Shop Drawing: Pitch Boards/Slope Boards and the Framing Square: I

The development model and the shop drawing assignment indicate how measurements of run, rise, and line-length may be used to determine sheathing cuts and side cuts of JRs.

1. Development of pitch boards[1]/slope boards

The development of pitch boards for the rafters and sheathing of the equal-pitch hip roof is a step towards understanding general roofing theory. To complete the pitch boards for a specific problem, the carpenter requires a knowledge of roofing terms and mathematics and the ability to construct a three-dimensional project from given information.

The method of pitch board development for a specific roof is shown below. The drawings are small—too small for taking useful measurements. The roof framer would be well advised to develop a drawing on building paper to the scale given in the text. Many drawings for the problem are shown. This is done to show the step-by-step development. The roof framer should develop one drawing only, by using one set of roof specifications, imperial or metric.

IMPORTANT [1]Where metric measurements are used, substitute the term *slope board* for the term *pitch board.*

ROOF SPECIFICATIONS:

	Imperial	Metric
Span	12′–0″	5000 mm
Length	20′–0″	7500 mm
Pitch	$\frac{1}{3}$	$\frac{2}{5}$ (0.4)
Rafters	24″ OC	500 mm OC
Rafter overhang	Nil	Nil
Scale of drawing	1″ = 1′–0″	1:10

MATERIALS AND EQUIPMENT:

1 piece of building paper 36″ × 24″ (1 m × 600 mm)
1 framing square
1 pencil
1 chisel

A. Draw a single line plan view of the roof. Do not dimension the drawing (Figure 19.1).

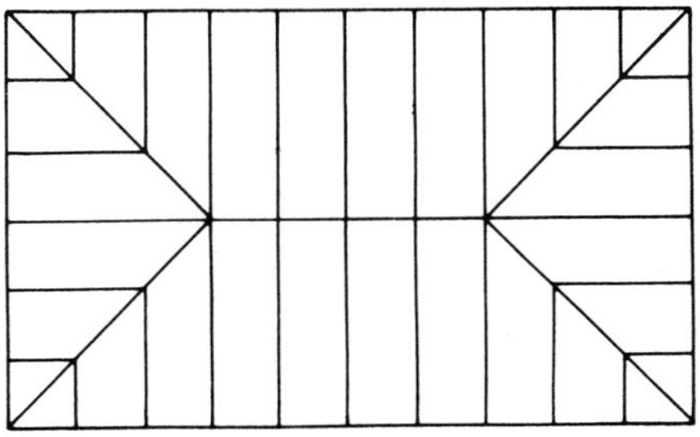

FIGURE 19.1. (Drawings for imperial only)

B. Lay out the pitch board for the CR (see Figure 19.2) using $\frac{1}{12}$ scale for imperial, and $\frac{1}{10}$ scale for metric.

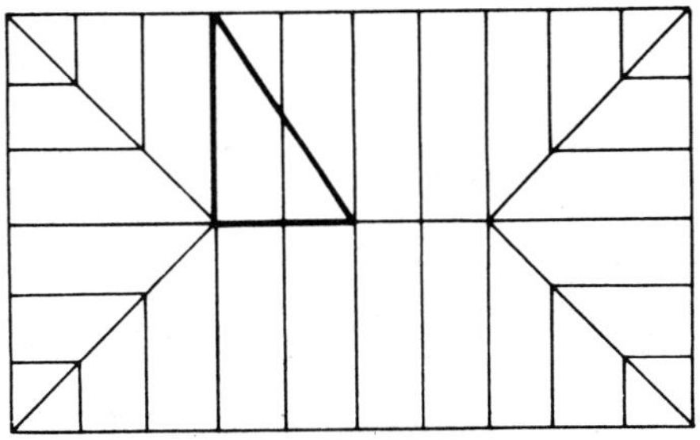

FIGURE 19.2.

Imperial Measurements/Pitch	*Metric Measurements/Slope*
Pitch is $\frac{\text{Rise}}{\text{Span}}$	Slope is $\frac{\text{Rise}}{\text{Run}}$ 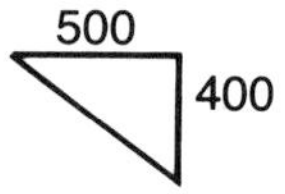
$\frac{1}{3}$ is $\frac{\text{Rise}}{12}$	$0.8 = \frac{\text{Rise}}{2500\text{ mm}}$
4 is Rise (6 is Run)	Rise is 2000 mm
Plumb Cut: 4 and 6, cut on 4 (Graphic Method) or 8 and 12, cut on 8 (Framing square method, with unit run of 12″)	Plumb cut: 2000 and 2500, cut on 2000 or 200 and 250, cut on 200 (Graphic/framing square method)

Scale Measurements

	Imperial		*Metric*	
	1/12	*Full*	*1/10*	*Full*
Run:	6″	6′–0″	250 mm	2500 mm
Rise:	4″	4′–0″	200 mm	2000 mm
LL of CR:	7″ plus	7′–2½″	320.2 mm	3202 mm

Calculation:

```
     14.42
   ×    6
   ------
12 )86.52 ( 7′–2½″
    84
    --
     2
```

C. Lay out the pitch board (slope board) for one hip rafter. See Figure 19.3.

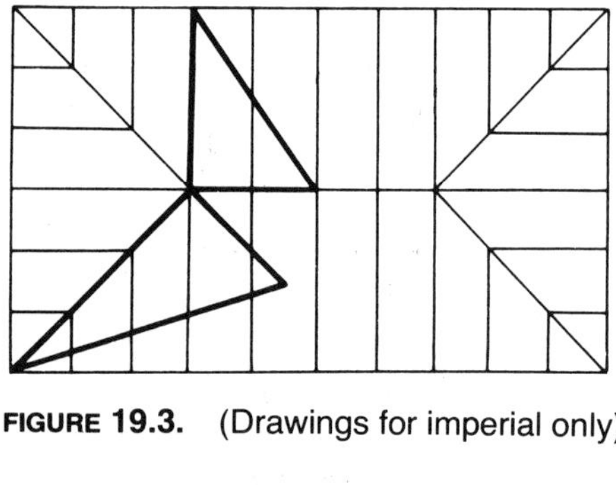

FIGURE 19.3. (Drawings for imperial only)

Imperial Measurements/Pitch	*Metric Measurements/Slope*
Plumb cut: 4 and the diagonal of 6 and 6, cut on 4 (Graphic method) or 8 and 16.97 (17), cut on 8 (Framing square method)	Plumb cut: 200 and the diagonal of 250 and 250 or 200 and 353.6, cut on 200 (Graphic/framing square method)

Scale Measurements

	Imperial		*Metric*	
	1/12	*Full*	*1/10*	*Full*
Run:	$8\frac{6}{12}''$	$8'–5\frac{7}{8}''$	354 (353.6 mm)	3540 (3536 mm)
	Calculation: 16.97 × 6 12)101.82 ($8'–5\frac{7}{8}''$ 96 5			
Rise:	4″	4′–0″	200 mm	2000 mm
LL of HR:	$9\frac{4}{12}''$	$9'–4\frac{1}{2}''$	406+ mm	4062 mm
	Calculation: 18.76 × 6 12)112.56 ($9'–4\frac{1}{2}''$ 108 4			

D. For imperial, lay out the first and second JR pitch boards. For metric, lay out a set of JR slope boards. The plumb cuts for the JRs are the same as for the CR. See Figure 19.4.

Note:

1. *The run of the JR is equal to its distance from the corner.*
2. *The rise of each JR can be measured on the HR pitch board. The measurement of rise A of the first JR is equal to the measurement A on the HR pitch board.*

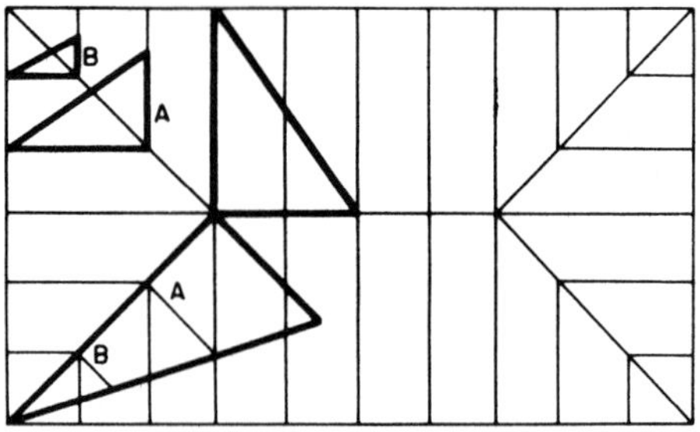

FIGURE 19.4.

Plumb Cuts for JRs

Imperial

1st JR:

8 and 12 cut on 8, or
4 and 6 cut on 4, or
$2\frac{2}{3}$ and 4 (the run) cut on $2\frac{2}{3}$.

2nd JR:

8 and 12 cut on 8, or
4 and 6 cut on 4, or
$1\frac{1}{3}$ and 2 (the run) cut on $1\frac{1}{3}$.

Metric

All	200 and 250 cut on 200
1st	160 and 200 cut on 160
2nd	120 and 150 cut on 120
3rd	80 and 100 cut on 80
4th	40 and 50 cut on 40

Measurements for JRs

Imperial

	1st JR: 1/12 Scale	*1st JR: Full scale*
Rise:	$2\frac{2}{3}''$ or $2\frac{8}{12}''$	2′–8″
Run:	4″	4′–0″
LL:	$4\frac{9}{12}''$	$4'–9\frac{5}{8}''$

Calculation:

$$\begin{array}{r} 14.42 \\ \times \quad 4 \\ \hline 12\,)57.68\ (\ 4'\text{–}9\tfrac{5}{8}'' \\ 48 \\ \hline 9 \end{array}$$

	2nd JR: 1/12 Scale	*2nd JR: Full scale*
Rise:	$1\frac{1}{3}''$ or $\frac{4}{12}''$	1′–4″
Run:	2″	2′–0″
LL:	$2\frac{5}{12}''$	$2'\text{–}4\frac{7}{8}''$

Calculation:

$$\begin{array}{r} 14.42 \\ \times \quad 2 \\ \hline 12\,)28.84\ (\ 2'\text{–}4\tfrac{7}{8}'' \\ 24 \\ \hline 4 \end{array}$$

		Metric		
JR	*Rise*	*Run*	*LL* $\frac{1}{10}$ *scale*	*LL Full scale*
1st	160	200	256	2560
2nd	120	150	192	1920
3rd	80	100	128	1280
4th	40	50	64	640

E. Lay out the pitch boards for the sheathing cut and the JR cheek cut (Figure 19.5).

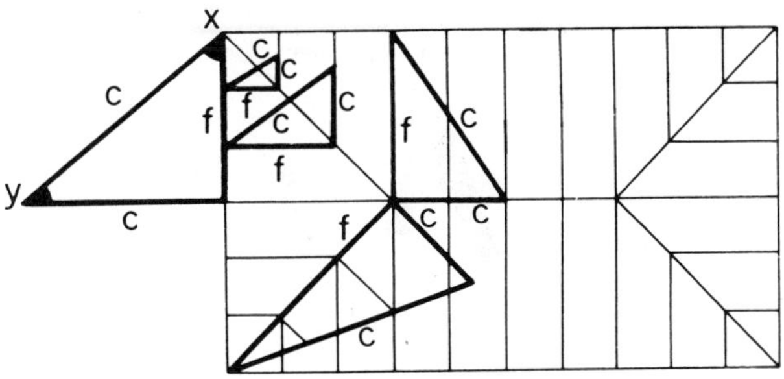

FIGURE 19.5.

	Imperial		Metric	
	1/12 Scale	*Full scale*	*1/10 Scale*	*Full scale*
Base = $\frac{1}{2}$ span =	6″	6′–0″	250 mm	2500 mm
Height = LL of CR =	7″	7′–2$\frac{1}{2}$″	320.2 mm	3202 mm
Hypotenuse = LL of HR =	9$\frac{4}{12}$″	9′–4$\frac{1}{2}$″	406.2 mm	4062 mm

Angle *Y* is similar to the angles to be developed for the JR side cuts. Angle *X* is the sheathing cut face angle.

F. Assemble the pitch boards (Figure 19.6). Cut on C; fold on f.

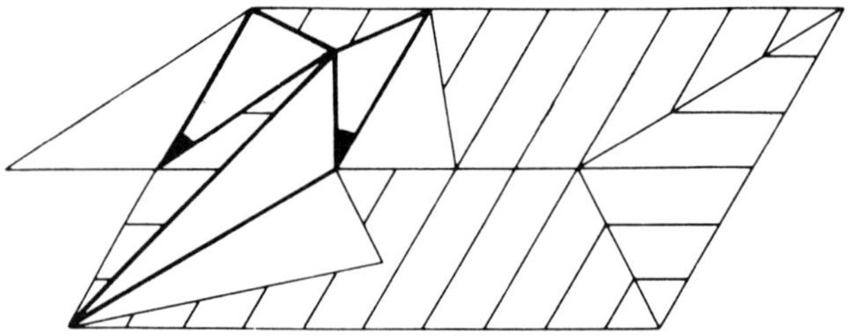

FIGURE 19.6.

By cutting on the "rise" and the "line-length" lines and folding on the "run" lines, the purpose of the pitch-board development is understood.

When the triangles or pitch boards touch one another, the rise edges should be in plumb positions. Since this is one roof, the CR, the HR, and the sheathing-cut pitch boards should rise to the same point.

By turning the layout sheet 180°, the positions of the JR pitch boards can be seen (Figure 19.7).

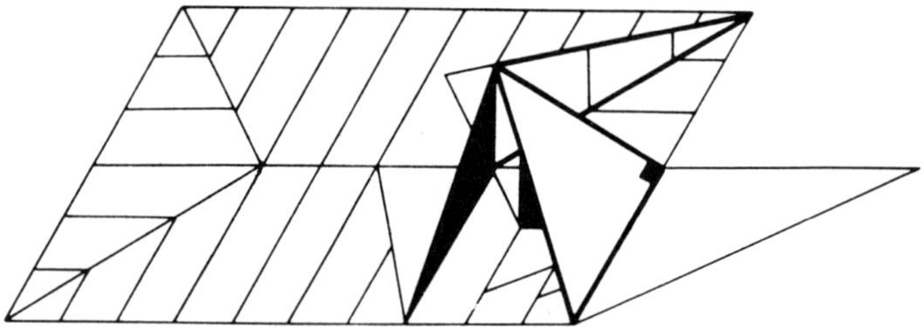

FIGURE 19.7.

G. Lay the framing square on the sheathing pitch board to determine the sheathing cut (Figure 19.8).

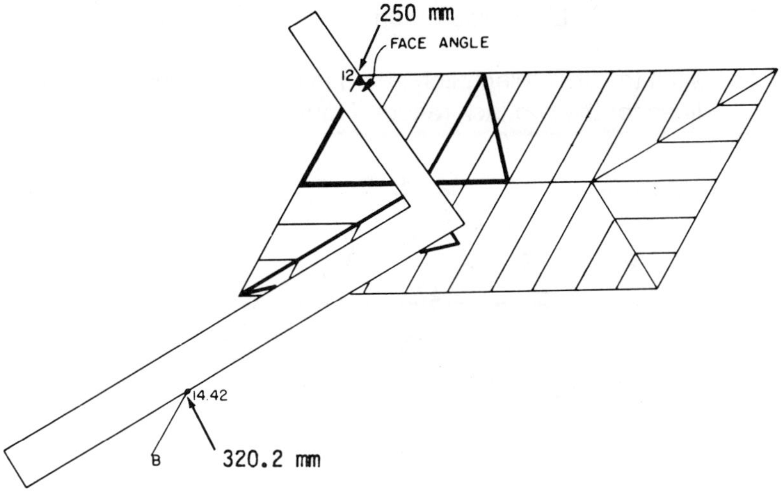

FIGURE 19.8. The sheathing-cut face angle.

Fold the CR, the HR, and the JR pitch boards back to their original positions.

Press the sheathing pitch board down and over the others. Extend the plate line to *B* as shown on the drawing. Place the tongue of the framing square along the sheathing pitch board hypotenuse. The body of the framing square will cross the plate line extension.

For Imperial: Slip the square so that 12 is at the plate line's corner. Read 14.42 where the square touches the plate line extension. The sheathing face cut angle is 12 and 14.42 cut on 12. Observe that 12 and 14.42 are $\frac{1}{3}$ pitch numbers. Keep in mind, too, that sheathing boards run across the roof.

For Metric: Slip the square so that 250 is at the plate lines corner. Read 320.2 when the square touches the plate line extension. The sheathing face cut angle is 250 and 320.2 cut on 250. Observe that 250 and 320.2 are $\frac{4}{5}$ (0.8) slope numbers.

Keep in mind too, that sheathing boards run across the roof.

Rule: *To lay out the sheathing cut face angle, use the run and the LL cut on the run.* The run is the CR unit run measurement. The LL is the CR unit LL measurement.

H. With the sheathing pitch board in its down position, draw the LL lines for the two jack rafters (Figure 19.9).

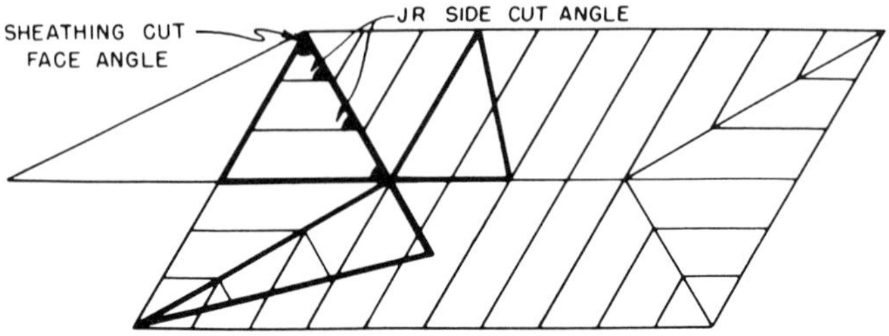

FIGURE 19.9.

Stand the CR, the HR, and the JR pitch boards upright. Place the sheathing pitch board in its roof position. The JRs LL lines on the sheathing pitch board should align with the JRs pitch boards beneath.

Fold the pitch boards to the flat position again. The "sheathing-cut—JR side-cut" pitch board should be on top. The JR side-cut angle is at a different position on the pitch board from that of the sheathing-cut face angle. JRs run up the roof, whereas sheathing runs across.

To determine the side-cut of a JR by means of the framing square, lay the square on the extended-rafter layout-line (Figure 19.10).

Extend the first JR LL line to *C.* Lay the body of the framing square along the pitch board hypotenuse.

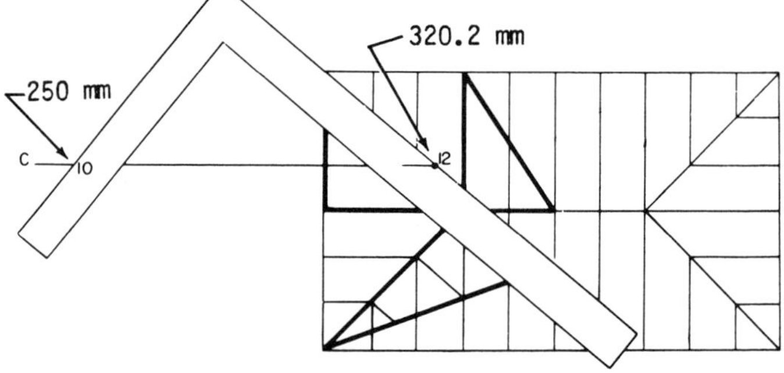

FIGURE 19.10.

For Imperial: Slip the square until the 12 on the body touches the upper end of the second JR LL line. Read 10 on the tongue. According to the rafter table on the framing square, the side-cut for a $\frac{1}{3}$ pitch roof JR is 10 and 12 cut on 12. Only the 10 is stamped on the square. The 12 is the framing square unit and is understood.

Slip the square farther. Instead of reading 12 on the body, read 14.42. The reading on the tongue will be 12. These, too, are $\frac{1}{3}$ pitch numbers, but in this case the cut is on the LL side. For sheathing the cut was on the run side.

For Metric: Slip the square until the 320.2 on the body touches the upper end of the JR LL line. Read 250 on the tongue. By slipping the square(s) compare the following settings: 250 and 320.2; 195 and 250; 200 and 256. Find these settings on the squares in the rafter tables. Observe that, as decimal fractions, they are equivalent.

Remember that on the roof sheathing boards and jack rafters run at right angles to one another.

2. Shop drawing assignment

On a piece of building paper develop the pitch boards, and then determine the cuts for the rafters and the sheathing of an equal-pitch[2] hip roof.

ROOF SPECIFICATIONS:

	Imperial		Metric
Span	12′–0″		5000 mm
Length	20′–0″		7500 mm
Pitch	$\frac{5}{8}$	slope	
Rafters	24″ OC		500 mm
Rafter overhang	Nil		Nil
Scale of drawing	1″ = 1′–0″		1:10

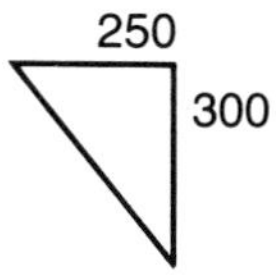

[2] For a metric root use *equal-slope*.

This shop drawing assignment is designed to test your knowledge of roof framing theory and to test your ability to visualize and draw the required pitch boards. (See Figure 19.11.)

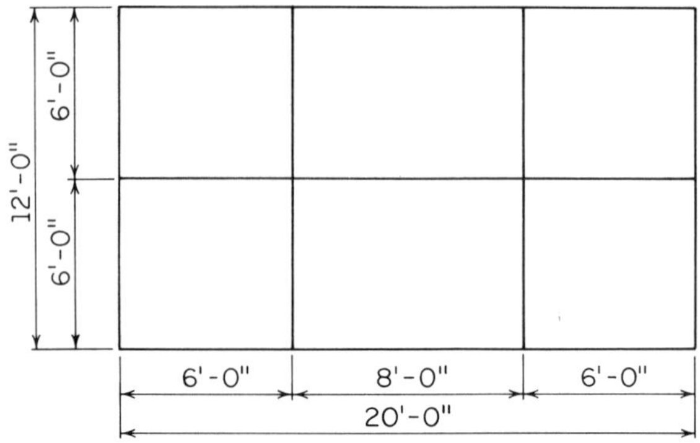

FIGURE 19.11. Assignment 2. Complete the drawing and develop the pitch boards.

It is suggested that the roof builder complete one drawing (see Figure 19.11) on building paper to the scale given. As a guide in relating roofing theory to the shop drawing, a series of partially completed tables of measurements is provided in the following.

A. Prepare the drawing, and cut the pitch boards so that they will fold to their roof positions. Print the name of each measurement on the drawing.

B. Complete the tables of measurements and the assignments following:

The Common Rafter

	Imperial		(refer to specifications on page 198)	Metric	
	1/12 scale	*Full scale*		*1/10 scale*	*Full scale*
Run:	6″	6′–0″	Run:	250 mm	2500 mm
Rise:	$7\frac{1}{2}$″	7′–6″	Rise:	300 mm	3000 mm
Plumb cut:			Plumb cut:		
	(a) _____ (rise) and 6″(run) cut on __________.	(a) _____ and 6′–0″ cut on __________.		(a) _____ (rise) and 250 mm(run) cut on _____.	(a) _____ and 2500 mm cut on _____.
	(b) $\frac{\ }{12}$″ and $\frac{12}{12}$″ (run) cut on __________.	(b) _____ and 12″ run cut on __________.			
Line length:	__________ inches.	_____ feet _____ inches.	Line length:	_____ mm.	_____ mm.

When the unit of run is 12 inches the unit of LL is __________.

When the unit of run is 250 mm the unit of LL is __________.

When the unit of run is 1 m the unit of LL is __________.

The Hip Rafter

	Imperial		(refer to specifications on page 198)	Metric	
	1/12 scale	*Full scale*		*1/10 scale*	*Full scale*
Run:			Run:	353.6 mm	3536 mm
Rise:			Rise:	300 mm	3000 mm
Plumb cut:			Plumb cut:		
	(a) _____ and $8\frac{1}{2}$″ (run) cut on __________.	(a) _____ and 8′–6″ (run) cut on __________.		(a) _____ and 353.6 (run) cut on _____.	(a) _____ and 3536 mm cut on _____.
	(b) $\frac{\ }{12}$″ and $\frac{17}{12}$″ (run) cut on __________.	(b) _____ and 17 (run) cut on __________.			
Line length:	__________ inches.	_____ feet _____ inches.	Line length:	_____ mm.	_____ mm.

When the unit of run is 17 (16.97), the unit of LL is __________.

When the unit of run is 353.5 mm, the unit of LL is __________.

When the unit of run is 1 m, the unit of LL is __________.

The Jack Rafters

Imperial (refer to specifications on page 198) *Metric*

Imperial

There are __________ JRs on one side of one HR.

	Measurements on the 1/12 scale drawing	*Full scale measurements*
Total run of 1st JR		
Total run of 2nd JR		
Total rise of 1st JR		
Total rise of 2nd JR		
LL of 1st JR		
LL of 2nd JR		

When the unit of run for the 1st JR is 12, the unit of LL is __________.

When the unit of run for the 2nd JR is 12, the unit of LL is __________.

When the unit of run for a CR of the roof is 12, the unit of LL is __________.

Metric

There are __________ JRs on one side of one HR.

	1/10 scale	*Full scale*
Total run of:		
1st JR		
2nd JR		
3rd JR		
4th JR		
Total rise:		
1st JR		
2nd JR		
3rd JR		
4th JR		
LL of:		
1st JR		
2nd JR		
3rd JR		
4th JR		
When the unit of run for JRs is 250, the unit of LL is:		
for 1st JR		
for 2nd JR		
for 3rd JR		
for 4th JR		

When the unit of run for a CR of the roof is 250 mm, the unit of LL is __________.

	Measurements on the 1/12 scale drawing	Full scale measurements
Plumb cuts:		
1st JR:	(a) _____ and 4″ cut on __________.	(a) _____ and 4′–0″ cut on __________.
	(b) $\frac{\ \ }{12}''$ and $\frac{12}{12}''$ cut on __________.	(b) _____ and 12″ cut on __________.
2nd JR:	(a) _____ and 2″ cut on __________.	(a) _____ and 2′–0″ cut on __________.
	(b) $\frac{\ \ }{12}''$ and $\frac{12}{12}''$ cut on __________.	(b) _____ and 12″ cut on __________.
CR:	(a) $\frac{\ \ }{12}''$ and $\frac{12}{12}''$ cut on __________.	(b) _____ and 12″ cut on __________.

	1/10 scale	Full scale
Plumb cuts:		
1st JR	_____ and 200 mm cut on __________	_____ and 2000 mm cut on __________
2nd JR	_____ and 150 mm cut on __________	_____ and 1500 mm cut on __________
3rd JR	_____ and 100 mm cut on __________	_____ and 1000 mm cut on __________
4th JR	_____ and 50 mm cut on __________	_____ and 500 mm cut on __________
CR	_____ and 250 mm cut on __________	_____ and 250 mm cut on __________

Imperial

The first JR has a total run of __________ feet.

The first JR is __________ feet from the corner of the building.

The second JR has a total run of __________ feet.

The second JR is __________ feet from the corner of the building.

Draw a line on the HR pitch board which represents the total rise of the first JR; of the second JR.

Metric

The first JR has a total run of _____ mm.

The first JR is _____ mm from the corner of the building.

The second JR has a total run of _____ mm.

The second JR is _____ mm from the corner of the building.

The third JR has a total run of _____ mm.

The third JR is _____ mm from the corner of the building.

The fourth JR has a total run of _____ mm.

The fourth JR is _____ mm from the corner of the building.

Draw a line on the HR slope board which represents the total rise of the first JR; the second JR; the third JR; and the fourth JR.

Sheathing cut and the JR cheek cut

Write the letter *A* beside the sheathing face angle. Write the letter *B* beside the angle on the pitch board (and/or slope board) which is similar to the JR side-cut angle.

Imperial (refer to specifications on page 198) *Metric*

Imperial

The measurements used to lay out the pitch board are:

(1) Base __________

(2) Height __________

(3) Hypotenuse __________

Draw lines on the pitch board which represent the LLs of the first JR and the second JR.

Lay a framing square on the completed drawing. Read it to complete the following statements:

(1) For the sheathing-cut face angle, use 12 and _____ cut on _____.

(2) For the JR side-cut, use $7\frac{1}{2}$ and _____ cut on _____.

(3) For the JR side-cut, use 12 and _____ cut on _____.

Metric

The measurements used to lay out the slope board are:

(1) Base __________

(2) Height __________

(3) Hypotenuse __________

Draw lines on the slope board which represent the LLs of the 1st JR, the 2nd JR, the 3rd JR, and the 4th JR.

Lay a framing square on the completed drawing. Read it to complete the following statements:

(1) For the sheathing-cut face angle, use 250 mm and _____ mm, cut on _____ ; or use 200 mm and _____ mm cut on _____.

(2) For the JR side-cut, use 250 mm and _____ cut on _____ ; or use 200 mm and _____ cut on _____.

Pitch boards and/or slope boards have been developed for CRs, HRs, JRs, and sheathing. Next are those for "odd" JRs, VRs, and VJRs. An understanding with respect to the odd JR further prepares one for the building of an unequal-pitch roof.

20

Shop Drawing: Pitch Boards/Slope Boards and the Framing Square: II

It is useful to show some rafters full thickness when making scaled drawings of roofs for pitch-boards[1] development.

Develop the scale plan view of another equal-pitch hip roof. Show rafters in full-size thickness. Develop three more pitch boards, use **total run** and **total rise** measurements.

ROOF SPECIFICATIONS:

Imperial		*Metric*	
Span	12′–0″	Span	5000 mm
Length	20′–0″	Length	7500 mm
Pitch	$\frac{1}{3}$	Slope	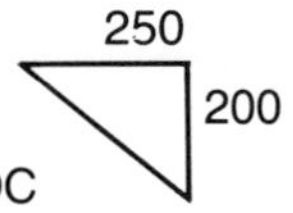
Rafters	24″ OC	Rafters	500 mm OC
Rafter overhang	24″	Rafter overhang	500 mm
Rafter stock	2″ nominal $1\frac{5}{8}$″ actual	Rafter stock	38 mm
Ridge	1″ nominal $\frac{3}{4}$″ actual	Ridge	19 mm

[1]Where metric measurements are used, substitute the term *slope board* for the term *pitch board* and the term *equal slope* for *equal pitch.*

1. On building paper draw a single-line plan view of the roof. Do not dimension the drawing. Scales: Imperial—one twelfth; Metric—one tenth.

   For imperial: show one pair of JRs placed 24″ from the corner.

   For metric: show one pair of JRs placed 500 mm from the corner.

   Only six CRs are required. Note that single lines represent center lines of rafters. See Figure 20.1.

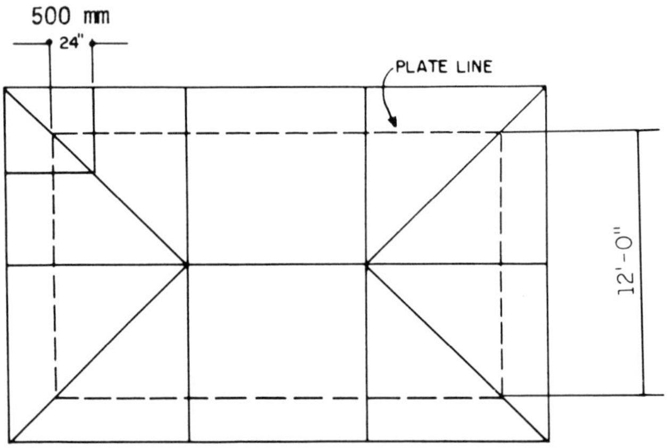

FIGURE 20.1

2. Develop the drawing further by giving the members thickness. Do this for the ridge, three CRs, two HRs, and two JRs. See Figure 20.2.

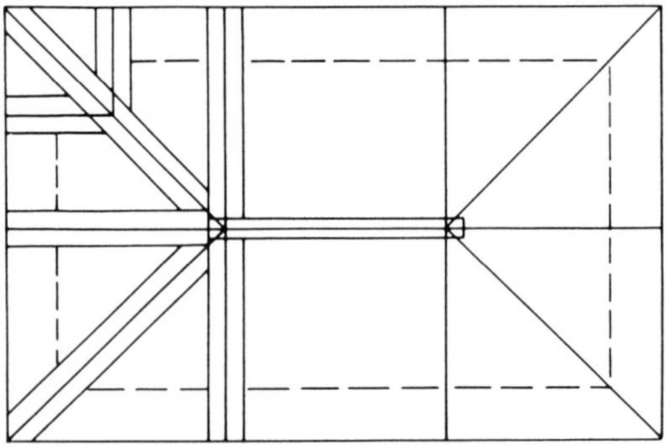

FIGURE 20.2

3. Show rafter measuring lines in black and cutting lines in color. Number the required plumb lines. See Figure 20.3.

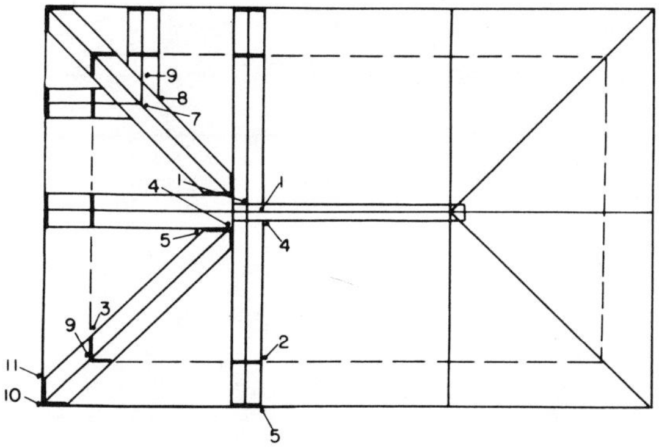

FIGURE 20.3

4. Draw the HR pitch board. Use *total run* and *total rise* measurements. See Figure 20.4.

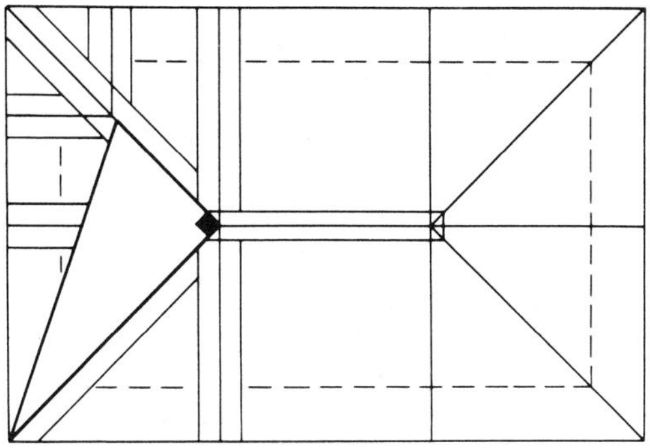

FIGURE 20.4

5. Draw the plan view of a JR placed at an angle other than 90° to the plate line. See Figure 20.5.
6. Lay out the pitch board for this JR. Use *total run* and *total rise* measurements. See Figure 20.6.

Determine the total rise from the HR pitch board. The two perpendiculars to the run lines must be the same length since the JR rises to the HR.

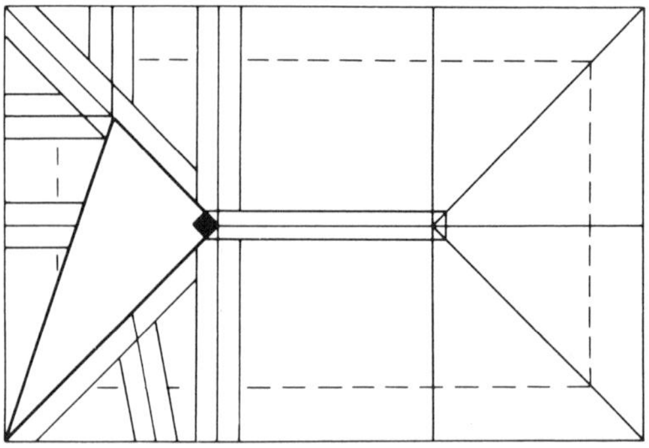

FIGURE 20.5

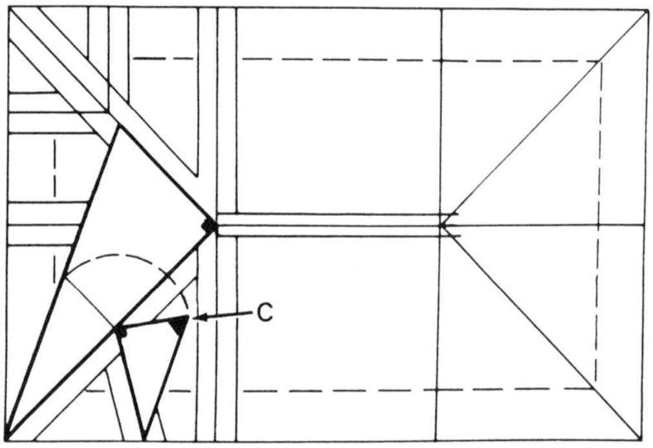

FIGURE 20.6

It is interesting to note that with the pitch board developed and a plan view of the jack rafter drawn full scale in width and $\frac{1}{12}$ scale in length, imperial, ($\frac{1}{10}$ scale in length, metric) enough information is at hand to lay out and cut the actual rafter. Its total run and rise are known. The plumb cut is as at *C* on the pitch board. The shortening and side-cut at the hip rafter are known, as are the cuts at the birdsmouth and tail. The rafter can be accurately placed on the roof since its distance from the corner can be measured.

The birdsmouth may be difficult to visualize and lay out. Since the rafter is not at right angles to the plate line, the birdsmouth will be cut deeper on one side than the other. If this is not done, the rafter will be canted and not in a plumb position. The positions of the two plumb lines of the birdsmouth are shown on the plan view.

The "odd" JR should have one edge backed. That edge, to the center line of the rafter, should have the "rafter above the plate" measurement.

Rafters may be cut for any pitch and roof position if good layout procedures are followed. *When the carpenter can visualize the layout of "odd" jack rafters, he is ready to learn how to cut rafters for the unequal-pitch roof.*

7. With a chisel, cut on the rise and lines of each pitch board. Fold on the run lines. The two pitch boards should meet when in an upright position. It is well to note here that the HR pitch board is larger than the HR pitch board developed in the previous problem, but it has the same shape. The measurements *total rise* and *total run* are usually required when laying out rafters for unequal-pitch roofs. They are not required for equal-pitch roofs, but may be used.

To the roof drawing just completed, add an equal-pitch projection. The specifications for the projection are:

	Imperial		*Metric*	
Span	12′–0″		5000 mm	
Length	6′–0″		2500 mm	250
Pitch	$\frac{1}{3}$	Slope		200
Rafters	24″ OC		500 mm OC	
Rafter overhang	24″		500 mm	

8. Draw the single-line plan view of the extension (Figure 20.7). Include two valley rafters, the ridge, two CRs and one VJR.

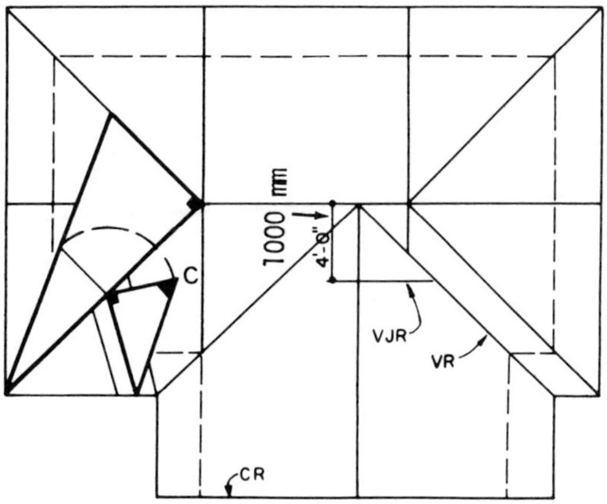

FIGURE 20.7

Draw the pitch boards for the CR, VR, and VJR, using total rise measurements (Figure 20.8).

Cut on the total rise and the LL lines of each new pitch board, and fold on the total run lines. When they are standing, note their relative positions.

A set of pitch boards, inclusive of the one for an "odd" JR, has been completed for an equal pitch roof with overhang. The pitch board development for the odd JR serves as an introduction to the development of pitch boards of an unequal-pitch roof.

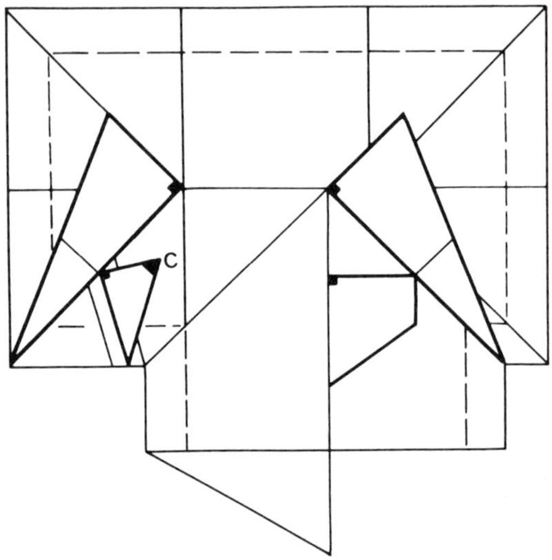

FIGURE 20.8

21

Shop Drawing: Pitch Boards/Slope Boards and the Framing Square: III

Accurate scale-drawings should be used to determine cuts for members of an unequal-pitch roof.

Develop the pitch boards[1] for an unequal-pitch hip roof. Use **total run** and **total rise** measurements.

MAJOR-ROOF SPECIFICATIONS:

	Imperial		*Metric*
Span	12′–0″		5000 mm
Length	20′–0″		8000 mm
Pitch	$\frac{1}{3}$	Slope	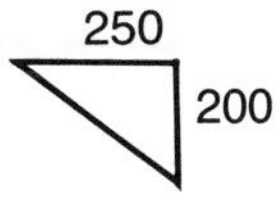
Rafters	24″ OC		500 mm
Rafter overhang	24″		500 mm

[1]Where metric measurements are used, substitute the term *slope board* for the term *pitch board,* and the term *unequal slope* for *unequal pitch.*

MINOR-ROOF SPECIFICATIONS:

	Imperial		*Metric*
Span	8′–0″		3000 mm
Length	6′–0″		1000 mm
Pitch	unknown	Slope	unknown
Rafter overhang	24″		500 mm

This minor roof will have the same total rise as does the major roof.

1. On building paper draw the single-line plan view of the major roof with its unequal-pitch projection. Show the skeleton roof only as in Figure 21.1.
2. Develop three major-roof and three minor-roof pitch boards.

The rise measurements for all the pitch boards must be the same since the two ridge boards are at the same height (Figure 21.2).

With a chisel, cut on the rise and line-length lines. Fold on the run lines so that the pitch boards stand in PLUMB positions. Observe that although they all rise to the same height, they are of different shapes—except perhaps the minor-roof CR and the minor-roof VJR pitch boards.

Since they are of different shapes, it follows that each rafter has its own plumb cut. The plumb cuts are as follows:

	Imperial	*Metric*
Major-roof CR	8 and 12 cut on 8	200 and 250 cut on 200
Major-roof HR	8 and 17 cut on 8	200 and 353.6 cut on 200
Major-roof odd JR	Total run and total rise cut on total rise.	Total run and total rise cut on total rise.
Minor-roof CR	Total run and total rise cut on total rise.	Total run and total rise cut on total rise.
Minor-roof VJR	Same as for minor-roof CR	Same as for minor-roof CR
Minor-roof VR	Total run and total rise cut on total rise.	Total run and total rise cut on total rise.

The total run of the VR may be determined mathematically or by measurement. Its approximate measurement, full scale, would be the drawing measurement multiplied by twelve (imperial) and by ten (metric).

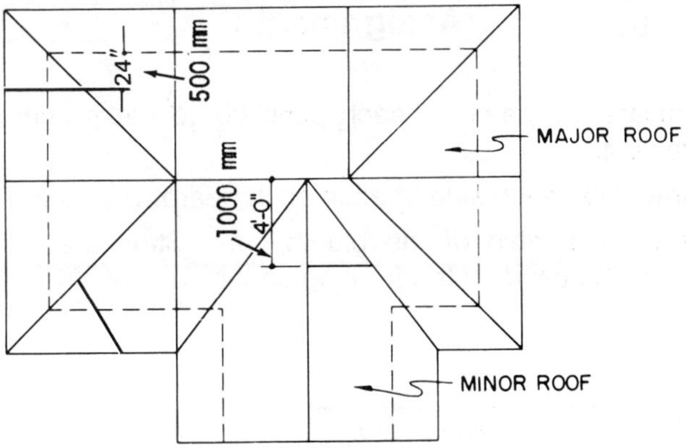

FIGURE 21.1

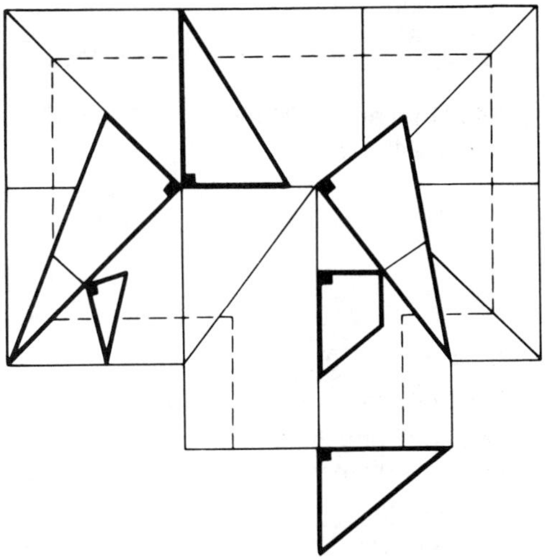

FIGURE 21.2

The actual total rise of the minor-roof rafters may be determined in a number of ways. It is the same as the major-roof CR and HR rise and can be calculated. It can be determined by measuring on the pitch board and multiplying by twelve (imperial) and by ten (metric).

In any case, when scale drawings are to be used to determine cuts and LL of rafters, they must be drawn accurately. The reading of the drawing must also be done carefully.

Assignments

1. Indicate the name of each pitch board measurement on your drawing.
2. Dimension each side of each pitch board. Use actual sizes.
3. Arrange a table of the plumb cuts. State the numbers that should be used on the framing square for each rafter.

Rafter	*Pitch Board Measurements*	*Pitch Board Measurements (reading scale 1″ = 1′0″, for imperial, and 1:10, for metric)*	*Framing-square Cut*
Major-	$5\frac{4}{12}$″ and 8″, cut on $5\frac{4}{12}$″	5′–4″ and 8′–0″, cut on 5′–4″	8 and 12, cut on 8 Note: $\frac{5'-4''}{8'-0''} = \frac{8}{12}$
	240 mm and 300 mm cut on 240 mm	2400 mm and 3000 mm cut on 2400 mm	200 and 250 cut on 200
Major-roof HR:			8 and 17, cut on 8 200 and 353.6 cut on 200
Major-roof JR:			----- and 12, cut on ----- ----- and 250 cut on -----
Minor-roof CR:	$5\frac{4}{12}$″ and 6″, cut on $5\frac{4}{12}$″	5′–4″ and 6′–0″, cut on 5′–4″	----- and 12, cut on -----
	240 mm and 200 mm cut on 240 mm	2400 mm and 2000 mm cut on 2400 mm	----- and 250 cut on -----
Minor-roof VR:	$5\frac{4}{12}$″ and -----, cut on $5\frac{4}{12}$″	5′–4″ and -----, cut on 5′–4″	----- and 12, cut on -----
	240 mm and ----- cut on 240 mm	2400 mm and ----- cut on 2400 mm	----- and 250 cut on -----
Minor-roof JR:			----- and 12, cut on ----- ----- and 250 cut on -----

Note: Imperial measurements are given first; and metric measurements second.

4. Complete a table showing rafter total line lengths of rafters.

Rafter	*Pitch Board Measurement*	*Pitch Board Measurements (reading scale, 1′ = 1′0″, for imperial, and $\frac{1}{10}$ for metric)*	*Calculate Line-Lengths (use rafter tables)*
Major-roof CR:	$9\frac{8}{12}$″ (approx.)	9′–8″ (approx.)	8 × 14.43 = 115.36″ = 9′–7.36″ = 9′–$7\frac{3}{8}$″
	384 mm (approx.)	3840 mm (approx.)	12 × 320.2 = 3842.2 mm
Major-roof HR:			
Major-roof “odd” JR:			
Minor-roof CR:			
Minor-roof HR:			
Minor-roof JR:			

Note: Imperial measurements are given first, and metric measurements second.

When cuts have been determined from accurate scale-drawings, useful settings can be obtained by slipping the square.

In the preceding five chapters some applications of roofing theory have followed a useful learning sequence of rafter layouts, roof framing, and shop drawing, to link pitch boards (and slope boards) with tables of roofing squares.

The next two chapters present a technique for extending the use of pitch boards and slope boards, suggesting less dependency on roofing square tables. Assuming that one has a thorough understanding of roof framing terms, their definitions and relationships, and of how to construct a residential roof, then it seems appropriate to focus on extending such knowledge and experience to more sophisticated methods of layout and cutting. One method is of hopper layouts and construction.

Most applications of this method are for hoppers with mitered joints: (1) to determine face and miter angles of windblocks, purlins, and sheathing, with or without rafter tables, using roof framing terminology;

and (2) to develop data tables for using hand-electric saws and radial-arm saws in the cutting of roof members, following graphic solutions to specific problems.

For the sake of maintaining a unit of learning, hoppers with butt joints are included. Although such hoppers may have fewer applications to roof framing, they have other applications which are often useful in construction: in joining timbers, in bin construction, and in sawhorse-type framing.

22

Shop Drawing: Hopper Construction

Roof framing terms are used to describe the construction of hoppers with miter or butt joints. Some hopper cuts are then related to cuts of roof frame members: cuts made with hand saws, hand-electric saws, or over-head (radial-arm) saws.

Layout Method 1

Lay out a square hopper with mitered joints (Figure 22.1).

HOPPER SPECIFICATIONS (OUTSIDE MEASUREMENTS; USE ONE SET):

	Imperial	*Metric*
Top	11″ × 11″	280 mm × 280 mm
Bottom	7″ × 7″	180 mm × 180 mm
Vertical height	4″	100 mm
Stock size	1″ nominal	19 mm

Tools and Materials for the Shop Drawing

- 1 piece building paper
- 1 black pencil, 2H or harder
- 1 colored pencil
- 2 framing squares
- 1 pair 12″ (300 mm) wing dividers
- 2 bevels

Procedure

1. Draw the exact full-size front and top (plan) views of the hopper. Show only the outside edges of the plan view. Show the section of material in the front view. No dimensioning should be done. See Figure 22.2.
2. Develop the face angle (Figure 22.3).
3. Develop the miter angle on the same drawing (Figure 22.4).
4. "Pick up" the required face angle on the bevel. Transfer this setting to the stock. See Figure 22.5.
5. Pick up the required miter angle on another bevel. Transfer the setting to the edge of the stock as shown in Figure 22.6.
6. Build the hopper. The parts may be cut from the stock and then assembled. See Figure 22.7(a).

Related Information

The inverted hopper has the shape of a miniature hip roof. See Figures 22.7(b) and 22.7(c).

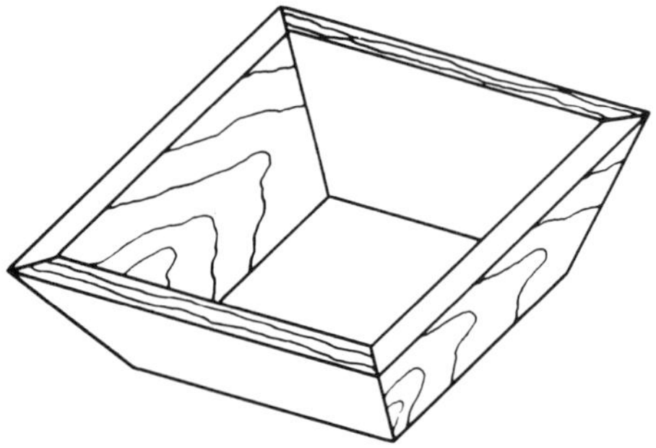

FIGURE 22.1. Square hopper with mitered joints.

Three basic measurements, CR unit run, CR unit rise, and CR unit line length (LL), are common to a hopper and a hip roof of the same pitch.

To relate roofing theory to hopper construction add the measuring lines, as shown below, to the hopper drawing (Figures 22.8 and 22.9).

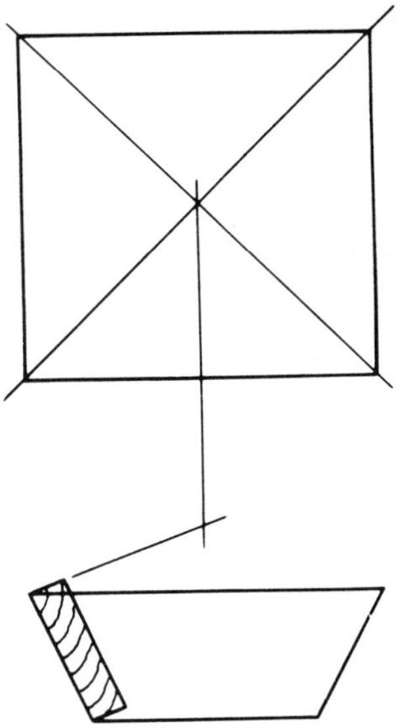

FIGURE 22.2.

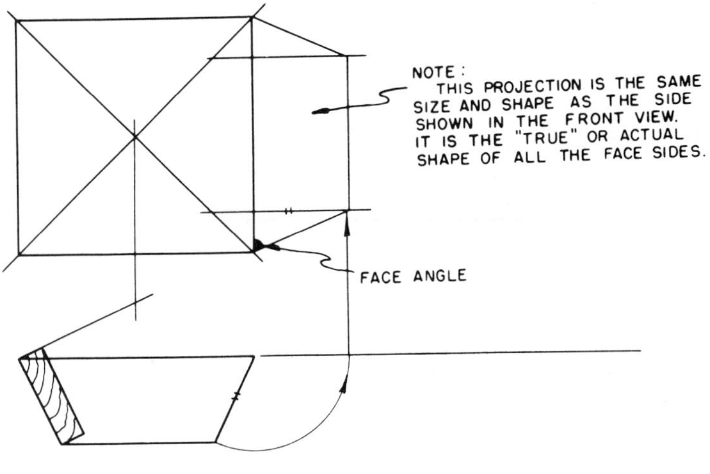

FIGURE 22.3.

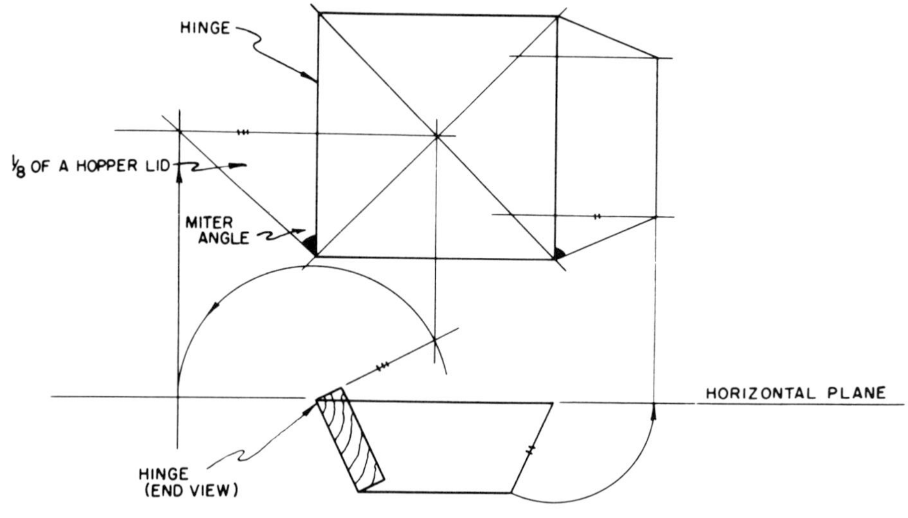

FIGURE 22.4.

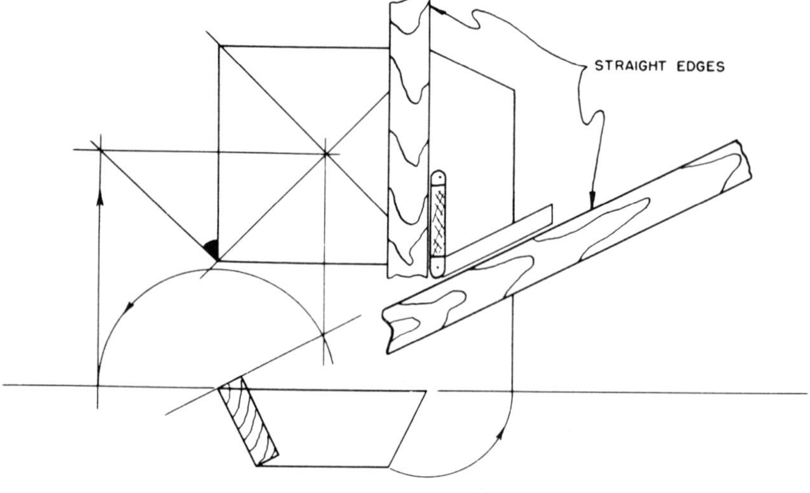

FIGURE 22.5.

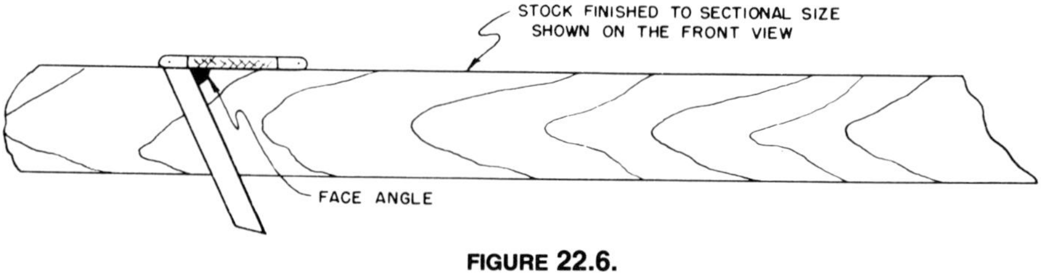

FIGURE 22.6.

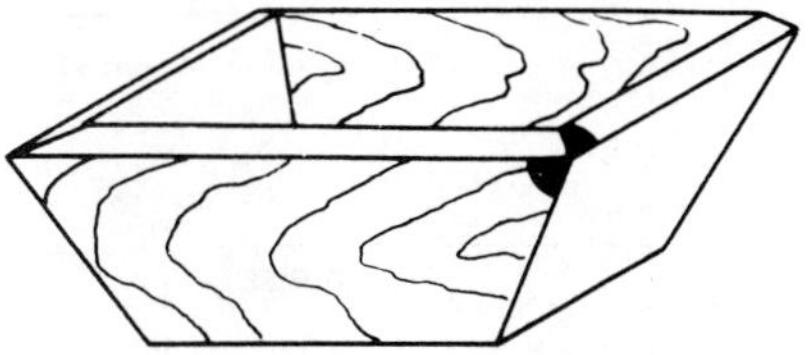

FIGURE 22.7(a). A hopper.

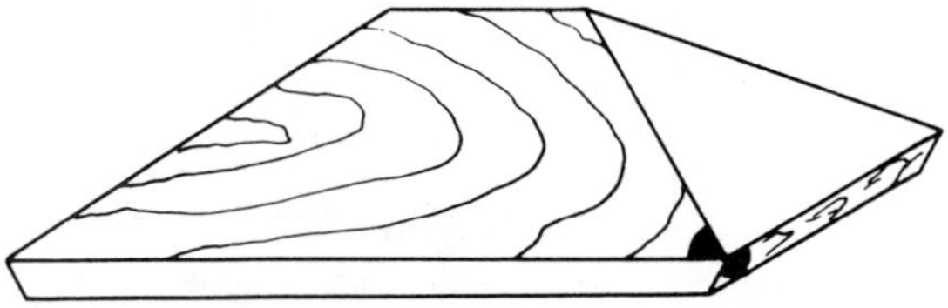

FIGURE 22.7(b). An inverted closed hopper.

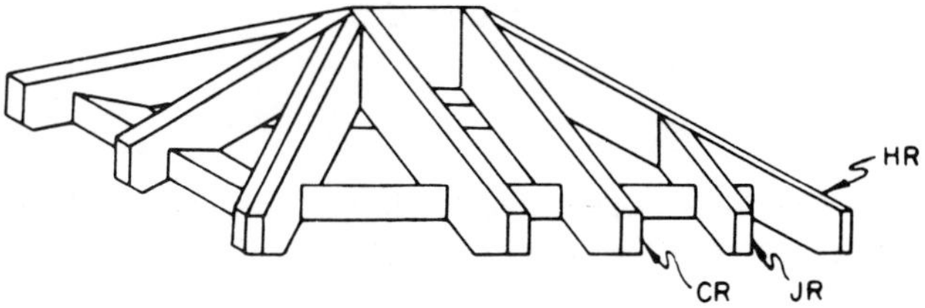

FIGURE 22.7(c). A roof (an incomplete shop model).

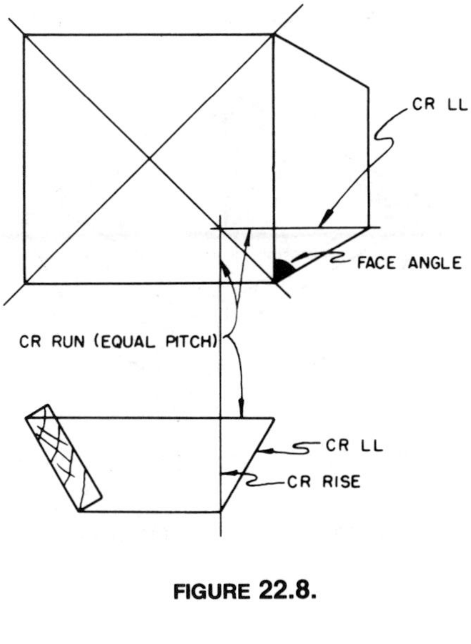

FIGURE 22.8.

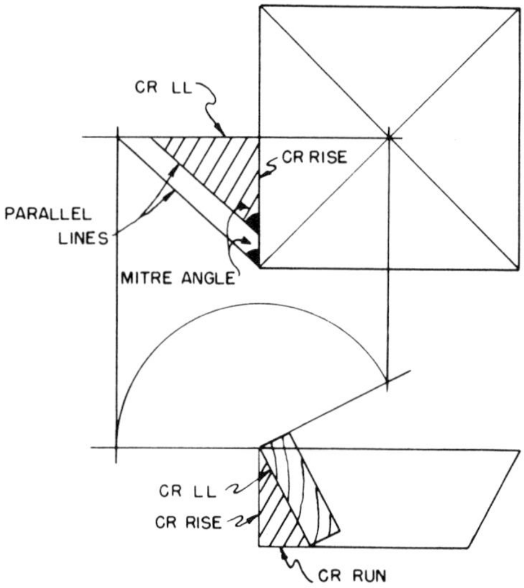

FIGURE 22.9.

RULES

Where the pitch of the hopper is readable on the rafter tables of the framing square, use the CR unit rise and the CR unit LL to lay out the miter angle.

Where the pitch of the hopper is not readable on the rafter tables of the framing square, use the graphic method to lay out the miter angle.

Note: For related information see Figures 13.1 and 13.2 page 119.

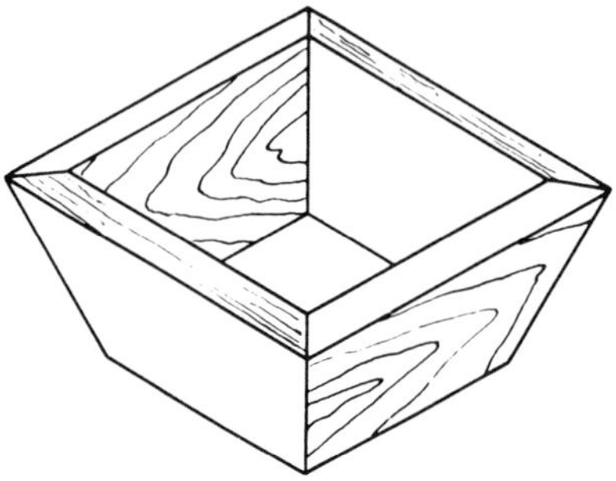

FIGURE 22.10. Equal-pitch square hopper with mitered joints.

Note: Also see Plate VII page 123 for the purlin miter.

Layout Method 2

Lay out an equal-pitch square hopper with mitered joints. (See Figure 22.10.)

HOPPER SPECIFICATIONS (OUTSIDE MEASUREMENTS; USE ONE SET):

	Imperial	*Metric*
Top	11″ × 11″	280 mm × 280 mm
Bottom	7″ × 7″	180 mm × 180 mm
Vertical height	4″	100 mm
Stock size	2″ nominal	38 mm

TOOLS AND MATERIALS FOR THE SHOP DRAWING

1 piece building paper
1 pencil, 2H or harder
1 colored pencil
1 or 2 framing squares
1 pair 12″ (300 mm) wing dividers
2 bevels

PROCEDURE

1. Draw the partial front and top (plan) views of the hopper. See Figure 22.11.
2. Develop the face angle (Figure 22.12).
3. Develop the miter angle (Figure 22.13).
4. "Pick up" the two angles with the bevels. Transfer the settings to the stock.

RELATED INFORMATION

Layout method 2 lends itself to the quick and accurate layout of bevels for hopper cuts. For relating hopper and roof framing theory, it is less useful.

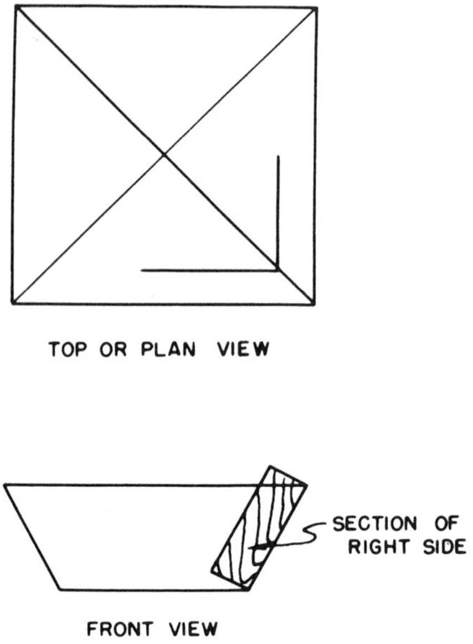

FIGURE 22.11.

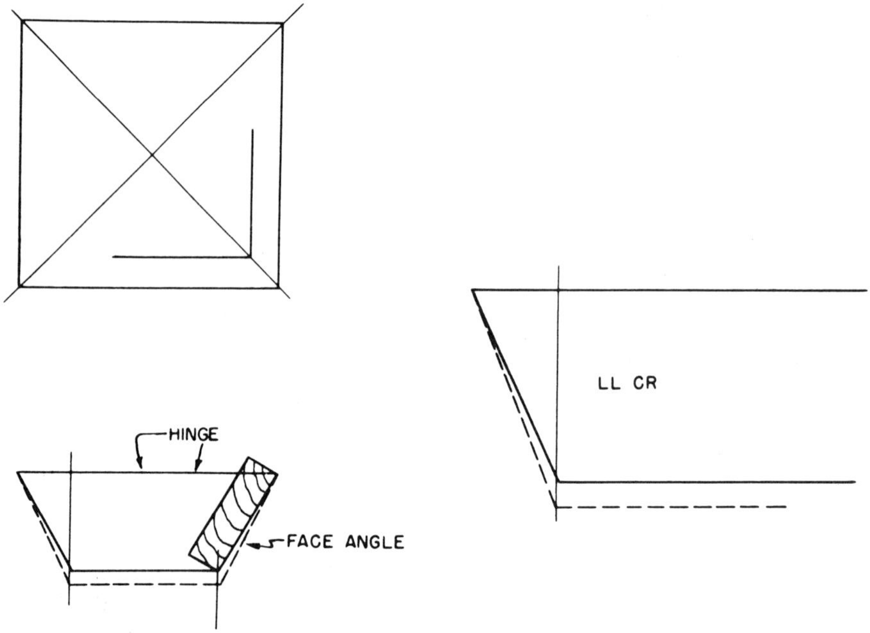

FIGURE 22.12.

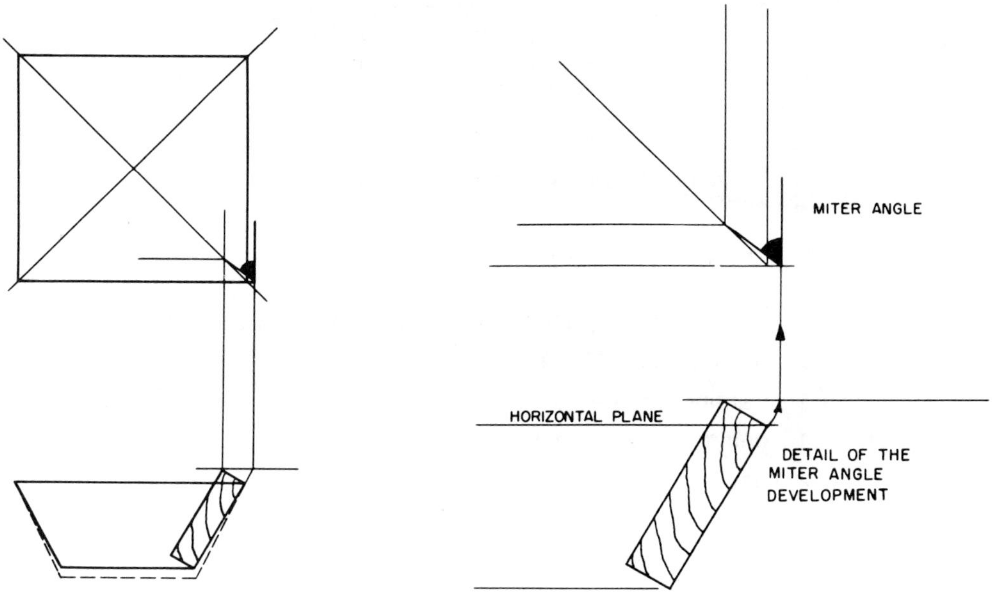

FIGURE 22.13.

Layout Method 3

Lay out an equal-pitch hopper with mitered joints. See Figure 22.14.

HOPPER SPECIFICATIONS (OUTSIDE MEASUREMENT; USE ONE SET):

	Imperial	*Metric*
Top	8″ × 11″ outside measurements	200 mm × 280 mm outside measurements
Pitch[1]	$\frac{5}{8}$″	(slope) 250 / 312
Stock size	2″ × 4″ nominal	38 mm × 89 mm

[1]Pitch and slope measurements may be different.

FIGURE 22.14. Equal-pitch hopper with mitered joints.

TOOLS AND MATERIALS FOR THE SHOP DRAWING

1 piece building paper, composition board, or plywood
1 black pencil 2H or harder
1 framing square
1 pair 12″ (300 mm) wing dividers
2 bevels

PROCEDURE

1. Draw the partial front view of the hopper. From *A* lay out the LL line with a pitch of $\frac{5}{8}$ (of 300 mm in 250 mm) from the horizontal plane, or base, line, as shown in Figure 22.15. Use the framing square in one of the three positions. See Figure 22.16.

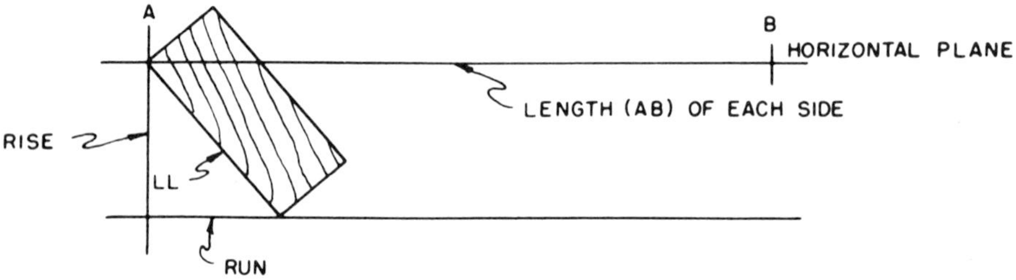

FIGURE 22.15.

2. Read the measurements:

CR LL

CR rise

CR run

3. Lay out the face angle. Set the bevel to the framing square. Use CR, LL, and CR run measurements. Transfer this setting to the prepared stock.

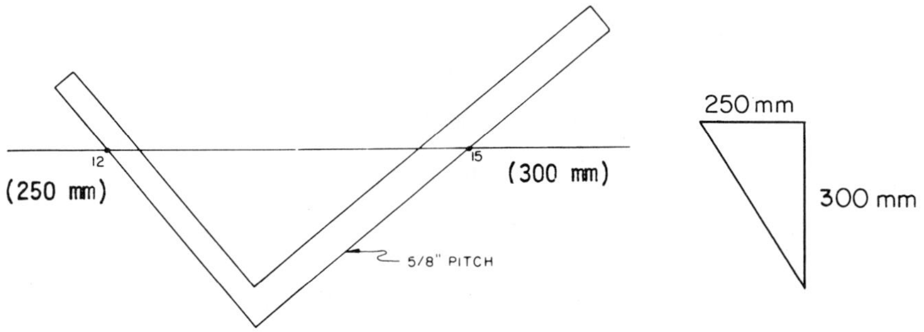

FIGURE 22.16.

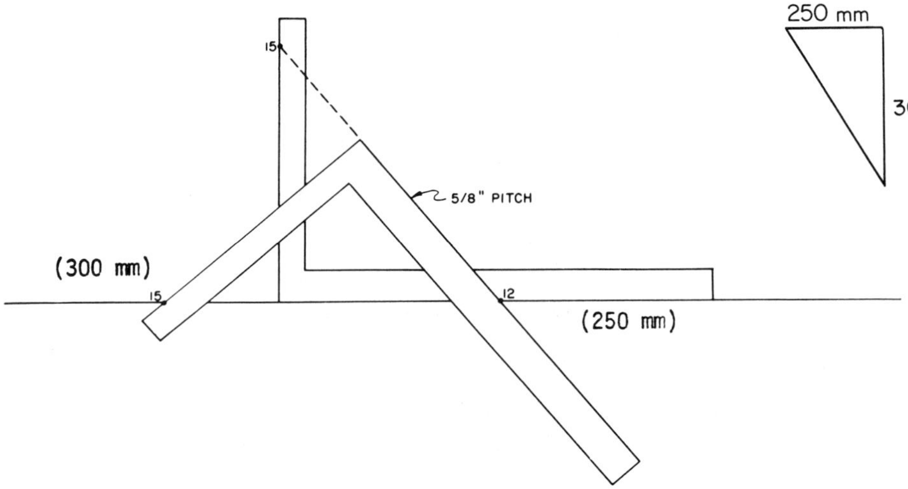

FIGURE 22.16(a).

4. Lay out the miter angle. Set the bevel to the framing square. Use CR, LL, and CR rise measurements. Transfer this setting to the prepared stock.
5. Build the hopper. The parts may be cut from the stock and then assembled.

NOTE

When the pitch of a hip roof is known, the cuts for its sheathing, purlins, and windblocks may be determined directly from the framing square.

For a $\frac{1}{3}$ pitch roof:

the CR unit run is 12

the CR unit rise is 8

the CR unit LL is 14.42.

For the face angles, use 12 and 14.42 cut on 12.
For the miter angles, use 8 and 14.42 cut on 8.

The cuts for a roof of slope 300 in 250, using two different sets of metric data, are developed as follows:

	Data	
	Set 1	*Set 2*
CR unit run	250 mm	1 m
CR unit rise	300 mm	1200 mm
CR unit LL	390.5 mm	1562 mm
	Cuts	
Face angle:	Use 250 and 390.5 cut on 250	1 m and 1562 cut on 1 m
Miter angle:	Use 300 and 390.5 cut on 300	1200 and 1562 cut on 1200

RULES

When the pitch of the hopper is readable on the rafter table of the framing square use: the CR unit rise and the CR unit LL to lay out the miter angle.

When the pitch of the hopper is readable on the rafter table of the framing square use: the CR unit run and the CR unit LL to lay out the face angle.

On a piece of suitable stock lay out the required cut for a windblock which would go against a HR. Rather than making the actual cut with a hand saw, use a hand-electric saw or an overhead (radial-arm) saw.

For a roof of pitch 0.6 (of slope 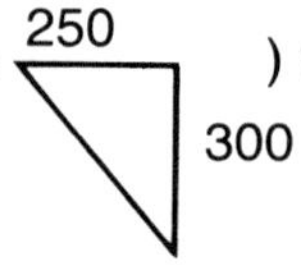

) what are the angular settings of the saw?

For a roof of pitch 0.5 (of slope 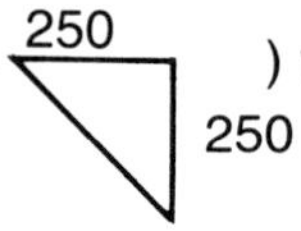

) what would be the settings of the saw?

For common roof pitches or slopes, create a data table for cutting windblocks, purlins, and sheathing, for where they might contact HRs or VRs. Use the outline suggested in Table 22.1 below.

TABLE 22.1

Angular Settings of Hand-Electric and Radial-Arm Saws for Cutting Windblocks, Purlins, and Sheathing

Metric Data				*Hand-Electric Saw*		Radial Arm Saw	
Pitch of Roof	*Slope of Roof*	*Face Angle*	*Miter Angle*	*Tilt Setting (degrees)*	*Face Angle (degrees)*	*Tilt Setting (degrees)*	*Radial-Arm Angle (degrees)*
0.4							
0.5							
0.6	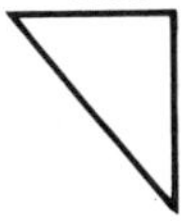	250 and 390.5 cut on 250	300 and 390.5 cut on 300				
0.7							
0.8							
0.9							
1.0							

Questions

1. Which method do you prefer to use when transferring angular measurements of face angles and miter angles to machine settings?

2. Would sets of thin, durable, marked pitch boards or slope boards be useful? Or is a table of measurements more satisfactory?

3. Rather than "pitch of roof" data, would it be better always to give roof slope? Using slope triangles? For example,

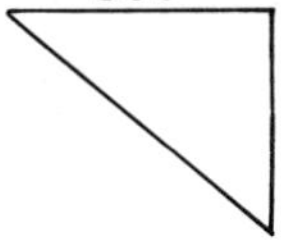

or

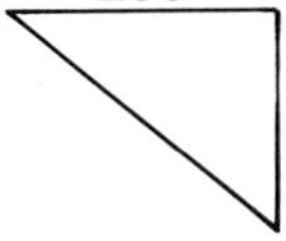

4. For a given roof, are the shapes of pitch boards and slope boards the same? Explain.

Layout Method 4

Lay out an equal-pitch hopper with butt joints. (See Figure 22.17.)

HOPPER SPECIFICATIONS (OUTSIDE MEASUREMENTS; USE ONE SET)

	Imperial	*Metric*
Top	11″ × 11″	280 mm × 280 mm
Bottom	7″ × 7″	180 mm × 180 mm
Vertical height	4″	100 mm
Stock size	2″ nominal	38 mm

TOOLS AND MATERIALS FOR THE SHOP DRAWING

- 1 piece building paper
- 1 black pencil, 2H or harder
- 1 colored pencil
- 2 framing squares
- 1 pair 12″ (300 mm) wing dividers
- 2 bevels

PROCEDURE

1. Draw the exact full-size front and top (plan) views of the hopper. Show only the outside edges of the plan view. Show the section of material in the front view. No dimensioning should be done. See Figure 22.18.
2. Develop the face angle (Figure 22.19).
3. Develop the butt angle (Figure 22.20).
4. Pick up the two angles with the bevels. Transfer the settings to the stock. Observe that two sides are shorter than the other pair of sides. Measure the length of the shorter sides on the front view.

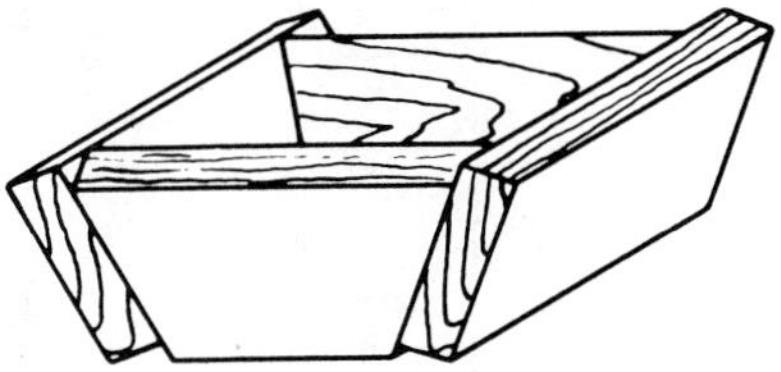

FIGURE 22.17. Equal-pitch hopper with butt joints.

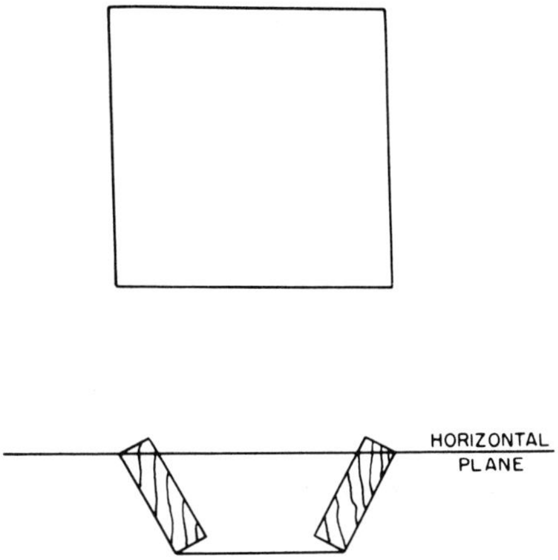

FIGURE 22.18.

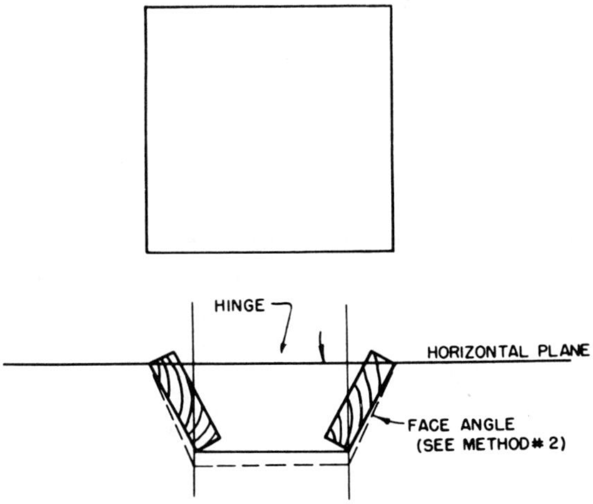

FIGURE 22.19.

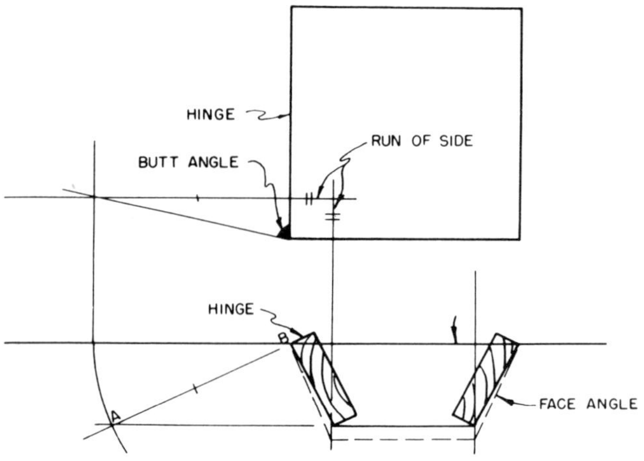

FIGURE 22.20.

RELATED INFORMATION

1. The method of face angle layout has not changed. Use "run and line length cut on run."
2. A rule of thumb method for layout of the butt angle can be determined from the drawing. Use the run of the side as one measurement. Use length *AB* as the other measurement. Only a skeleton drawing, as shown below (Figure 22.21), is required for the layout of both angles.

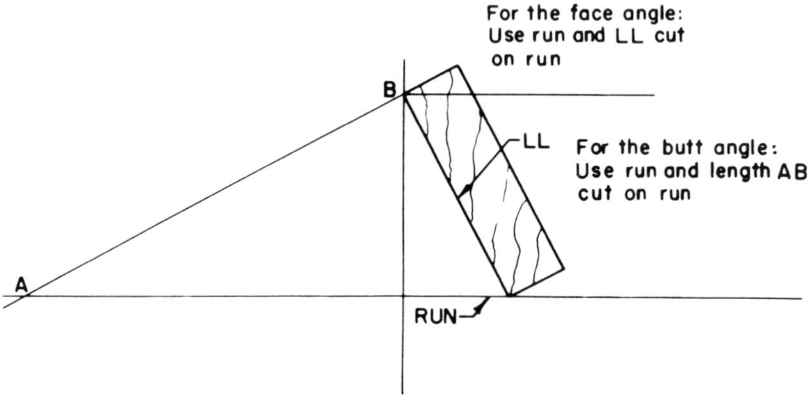

FIGURE 22.21.

Layout Method 5

Lay out an unequal pitch hopper with butt joints. (See Figure 22.22.)

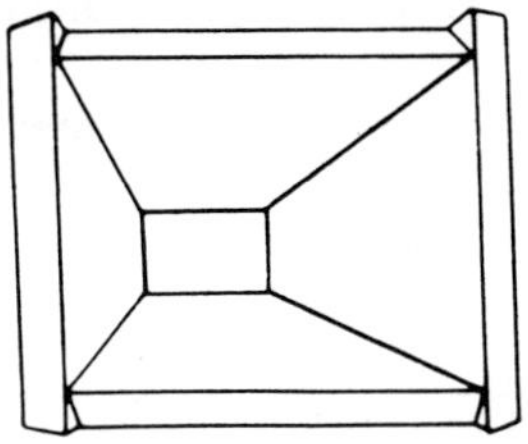

FIGURE 22.22. Unequal pitch hopper with butt joints.

HOPPER SPECIFICATIONS (OUTSIDE MEASUREMENTS; USE ONE SET)

	Imperials	*Metric*
Top	8″ × 11″	200 mm × 280 mm
Bottom	4″ × 4″	100 mm × 100 mm
The bottom is to be centrally located		
Vertical height	$3\frac{1}{2}$″	90 mm
Stock size	1″ nominal	19 mm

TOOLS AND MATERIALS FOR THE SHOP DRAWING

- 1 piece building paper
- 1 black pencil, 2H or harder
- 1 colored pencil
- 2 framing squares
- 1 pair 12″ (300 mm) wing dividers
- 2 bevels

PROCEDURE

1. Draw the exact full-size top (plan) view of the hopper. Show the minimum number of lines required for solving the problem.
2. Draw the front view. Show the sections of material in this view. It should be noted that they are sections of the right and left sides.

3. Draw the right-side view. Show the sections of material in this view (Figure 22.23). These sections are of the front and rear sides.
4. Develop the face angles (2). See Figure 22.24.
5. Develop the butt angles. See Figure 22.25.

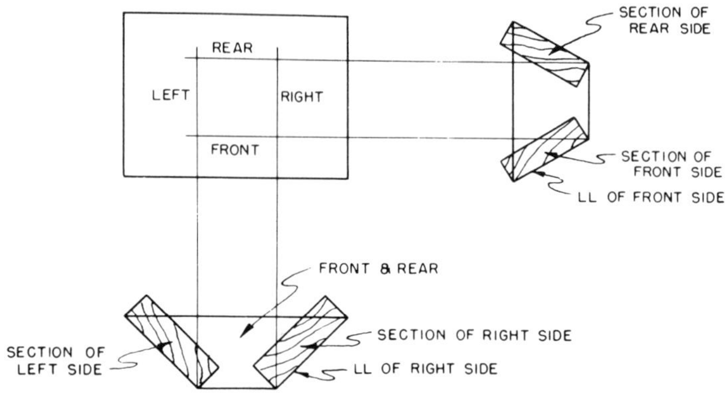

FIGURE 22.23.

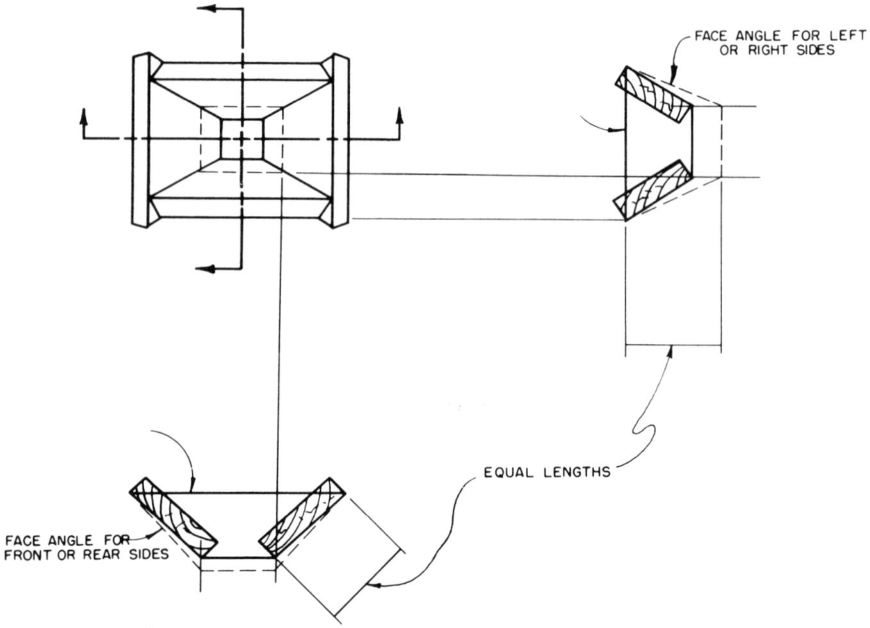

FIGURE 22.24.

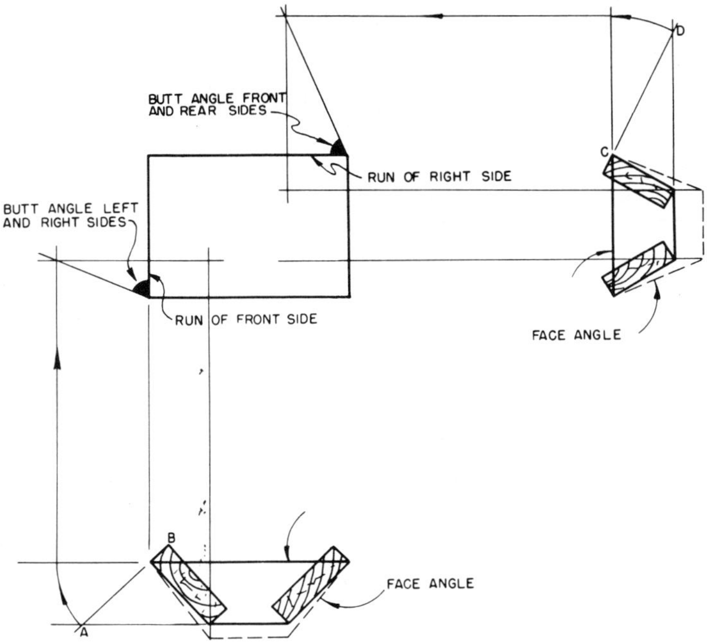

FIGURE 22.25.

6. Machine the stock to the two sizes shown.
7. Pick up the face and butt angles for the front and rear sides.
8. Pick up the face and butt angles for the left and right sides.
9. Cut and assemble the four sides.

RELATED INFORMATION

1. For the front (or rear) side face angle, use the run of the right side and the LL of the front side.
2. For the left (or right) side face angle, use the run of the front side and the LL of the right side.
3. For the left (or right) side butt angle, use the run of the front or rear side and the length *AB*.
4. For the front (or rear) side butt angle, use the run of the left or right side and the length *CD*.

Layout Method 6

Lay out an unequal-pitch hopper with mitered joints. (See Figure 22.26.)

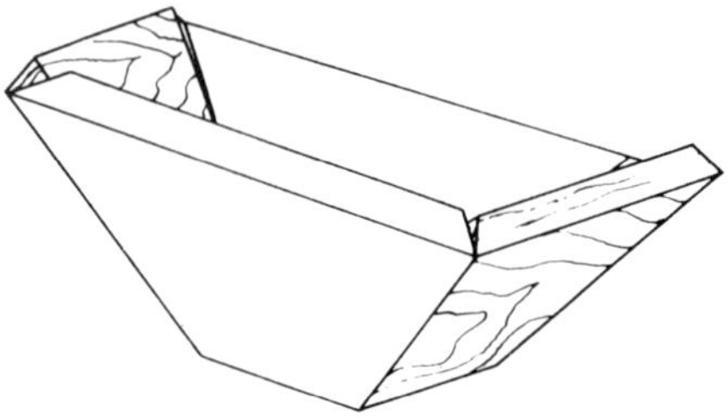

FIGURE **22.26.** Unequal-pitch hopper with mitered joints.

HOPPER SPECIFICATIONS (OUTSIDE MEASUREMENTS; USE ONE SET)

	Imperial	*Metric*
Top	8″ × 11″	200 mm × 280 mm
Bottom	4″ × 4″	100 mm × 100 mm
The bottom is to be centrally located.		
Vertical height	$3\frac{1}{2}$″	90 mm
Stock size	1″ nominal	19 mm

TOOLS AND MATERIALS FOR THE SHOP DRAWING

- 1 piece building paper
- 1 black pencil, 2H or harder
- 1 colored pencil
- 2 framing squares
- 1 pair 12″ (300 mm) wing dividers
- 2 bevels

PROCEDURES

1. Draw the exact full-size top (plan) view of the hopper. Show the minimum number of lines required for solving the problem.
2. Draw the front view. Show the sections of material in this view. Observe that these are sections of the right and left sides.

3. Draw the right-side view. Show the sections of material in this view (Figure 22.27). These are sections of the front and rear sides.
4. Develop the face angles (2). See Figure 22.28.
5. Develop the miter angles (2). See Figure 22.29.

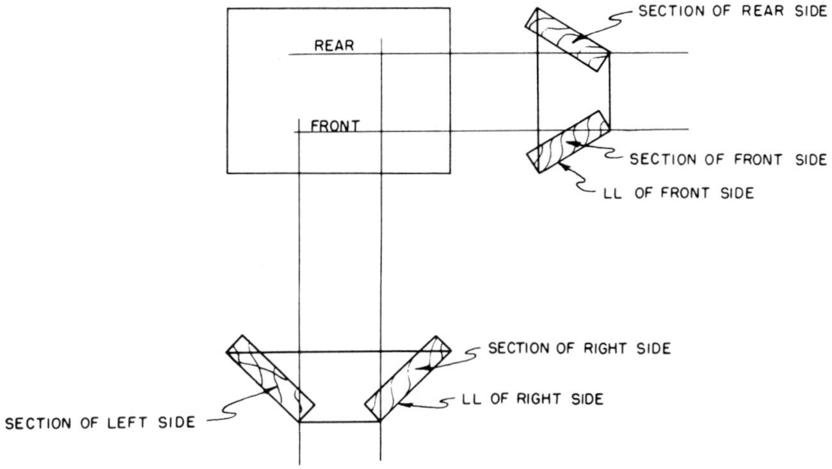

FIGURE 22.27.

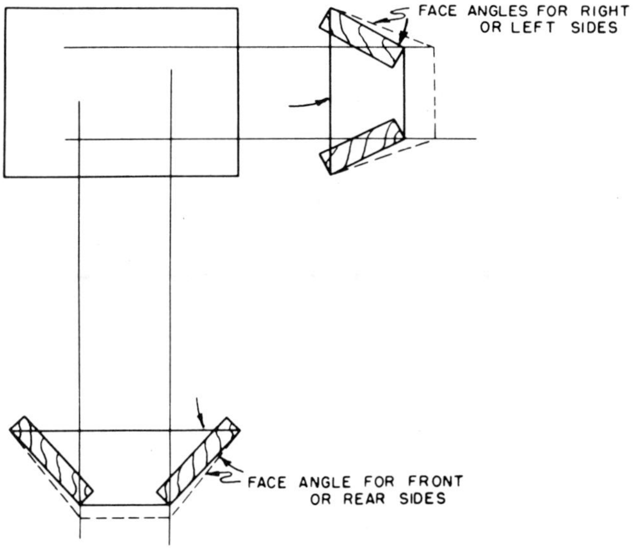

FIGURE 22.28.

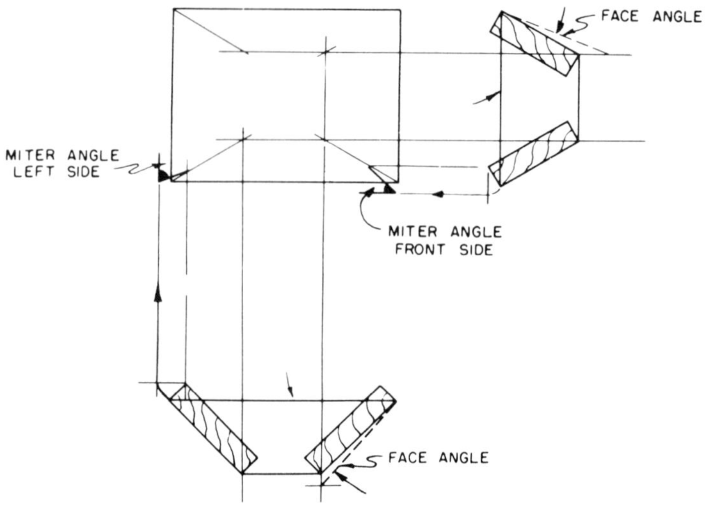

FIGURE 22.29.

6. Machine the stock to the two sizes shown.
7. "Pick up" the face and miter angles for the front and rear sides.
8. "Pick up" the face and miter angles for the left and right sides.
9. Cut and assemble the four sides.

RELATED INFORMATION

1. For the front (and rear) side face angle, use the run of the right side and the LL of the front side.
2. For the left (or right) side face angle, use the run of the front side and the LL of the right side. In other words, use the run of one pitch and the LL of the other pitch.
3. A rule of thumb for the layout of the miter angle is not readily determined from this drawing. However, Layout Method 1 may be applied to this problem as shown below in Figure 22.30.
4. Observe that, when one thickness of stock is used throughout and when the outside faces of the stock meet at the corners as shown in Figure 22.26:
   a. the upper inside edges will not be continuous.
   b. the inside faces of the ends will project farther into the hopper than the inside faces of the sides.

   If it is required to have the inside faces meet:
   a. change the thickness of the stock for one pair of pieces, or
   b. change the layout procedure.

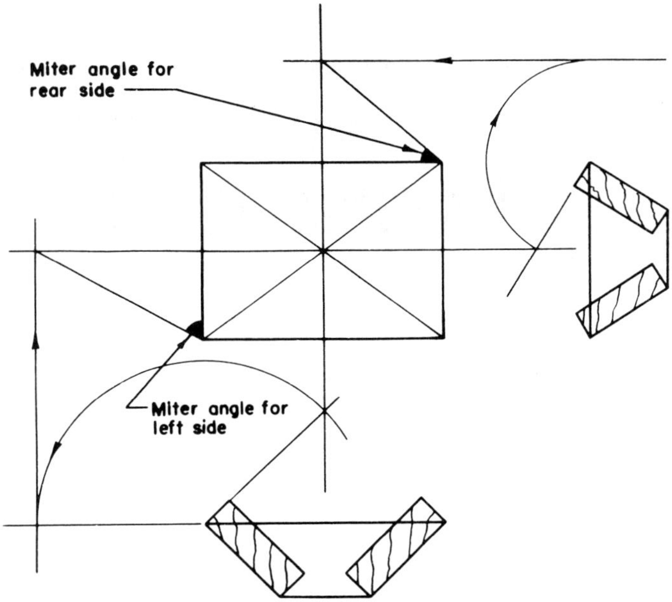

FIGURE 22.30.

Six layout methods have been developed: some with mitered joints and others with butt joints. Those with mitered joints embody applications to residential roof framing and to the cutting of the members. Those with butt joints round out the learning unit.

23

Shop Assignment: Hopper Construction

Two sets of specifications, one imperial and one metric, are given for six hopper construction jobs. Reference is made to the layout methods of the preceding chapter.

Job 1

Lay out and build a square hopper with mitered joints.

HOPPER SPECIFICATIONS (OUTSIDE MEASUREMENTS; USE ONE SET)

	Imperial	*Metric*
Top	7″ × 7″	200 mm × 200 mm
Bottom	4″ × 4″	100 mm × 100 mm
Vertical height	3″	80 mm
Stock size	1″ nominal	19 mm

Use *Layout Method 1* to determine the face and miter angles.

LAYOUT AND BUILDING TIME: ____________________

Questions

1. What is the pitch of the hopper? Answer: --------------------

   What is the slope of the hopper? Answer: --------------------

2. For a roof of the same size and shape:
   (a) What is the run of the CR?

   Answer: --------------------

   (b) What is the rise of the CR?

   Answer: --------------------

   (c) What is the LL of the CR?

   Answer: --------------------

3. What is the mathematical proof that measurements of the rise and the LL are used to determine the miter angle?

4. What are the rules, stated in words, for cutting roof sheathing to fit over the HR?

Job 2

Lay out and build a square hopper with mitered joints.

HOPPER SPECIFICATIONS (OUTSIDE MEASUREMENTS; USE ONE SET)

	Imperial	*Metric*
Top	7″ × 7″	200 mm × 200 mm
Bottom	3″ × 3″	80 mm × 80 mm
Vertical height	3″	80 mm
Stock size	1″ nominal	19 mm

Use *Layout Method 2* to determine the face and miter angles.

LAYOUT AND BUILDING TIME: --------------------

Questions

1. What is the pitch of the hopper? Answer ____________________

   What is the slope? Answer: ____________________

2. For a roof of the same size and shape:
   (a) What is the run of the CR?

   Answer: ____________________

   (b) What is the rise of the CR?

   Answer: ____________________

   (c) What is the LL of the CR?

   Answer: ____________________

3. Does this layout method lend itself to roofing theory?

4. In what way is this method useful in solving other problems?

Job 3

Lay out and build an equal-pitch (equal-slope) hopper with mitered joints.

HOPPER SPECIFICATIONS (OUTSIDE MEASUREMENTS; USE ONE SET)

	Imperial	*Metric*
Top	5″ × 8″	180 mm × 240 mm
Vertical height	3″	90 mm
Pitch	$\frac{5}{8}$″	Slope 250 / 300
Stock size	2″ nominal	19 mm

Use *Layout Method 3* to determine the face and miter angles.

LAYOUT AND BUILDING TIME: ____________________

Questions

1. What is the size of the hopper bottom?

   Answer: ____________________

2. What are the rules, in words, for cutting sheathing boards to fit over the HR?

   Answers:

   (a) To lay out the face angle, use ____________________ and ____________________; cut on ____________________.

   (b) To lay out the miter angle, use ____________________ and ____________________; cut on ____________________.

3. What are the rules, in words, for cutting windblocks to fit between the HR and the adjacent JR?

   Answers:

   (a) To lay out the face angle, use ____________________ and ____________________; cut on ____________________.

   (b) To lay out the miter angle, use ____________________ and ____________________; cut on ____________________.

4. What are the rules, in words, for cutting purlins to fit under and against the HR?

   Answers:

   (a) To lay out the face angle, use ____________________ and ____________________; cut on ____________________.

(b) To lay out the miter angle, use -------------------- and

--------------------; cut on --------------------.

(c) To determine the length --------------------------------

--.

Job 4

Lay out and build an equal-pitch (equal-slope) hopper with butt joints.

HOPPER SPECIFICATIONS (OUTSIDE MEASUREMENTS; USE ONE SET)

	Imperial	*Metric*
Top	8″ × 8″	200 mm × 200 mm
Bottom	$6\frac{1}{4}$″ × $6\frac{1}{4}$″	140 mm × 140 mm
Vertical height	3″	80 mm
Stock size	2″ nominal	38 mm

LAYOUT AND BUILDING TIME: --------------------

Questions

1. What is the pitch of the hopper: Answer: --------------------

   What is the slope? Answer: --------------------

2. What is the rule, in words, for the layout of the face angle with a steel square?

3. What is the rule, in words, for the layout of the butt angle with a steel square?

4. How does this information apply to the layout of a HR?

Job 5

Lay out and build an unequal-pitch (unequal slope) hopper with butt joints.

HOPPER SPECIFICATIONS (OUTSIDE MEASUREMENTS; USE ONE SET)

	Imperial	*Metric*
Top	6″ × 8″	150 mm × 200 mm
Bottom	3″ × 3″	80 mm × 80 mm
The bottom is to be centrally located.		
Vertical height	3″	80 mm
Stock size	1″ nominal	19 mm

LAYOUT AND BUILDING TIME: ____________________

Questions

What are the cuts for the brace of a sawhorse of splay 6″, spread 8″, and vertical height 24″?

Answer ____________________

or

What are the cuts for the brace of a sawhorse of splay 150 mm, spread 200 mm, and vertical height 600 mm?

Answer ____________________

Job 6

Lay out and build an unequal-pitch (unequal slope) hopper with mitered joints.

HOPPER SPECIFICATIONS (OUTSIDE MEASUREMENTS; USE ONE SET)

	Imperial	*Metric*
Top	6″ × 8″	150 mm × 200 mm
Bottom	3″ × 3″	80 mm × 80 mm
The bottom is to be centrally located.		
Vertical height	3″	80 mm
Stock size	1″ nominal	19 mm

LAYOUT AND BUILDING TIME: ____________________

Questions

1. Suppose that the 3″ × 3″ bottom had been moved 1″ to the right and $1\frac{1}{2}$″ back, or suppose that the 80 mm × 80 mm bottom had been moved 25 mm to the right and 35 mm back. How many different face angles would be required?

   Answer ____________________

2. What particular difficulty has the carpenter over the sheet-metal worker in the building of a hopper?

Hopper construction has been presented to help one to understand residential roofing theory.

24

Roof Trusses

Chapter 24 relates rafter layout to construction of residential roof trusses. Attention is given to: reasons for using trusses; preparation on-the-job for placing them, types, and common truss systems. Several truss shapes are described.

Technical data selected for this chapter and for Chapter 25 are supplied by the Canadian Wood Council, and TECO Products and Testing Corporation (Washington and Oregon) and published with permission.

1. Some reasons for using roof trusses

a. Trusses may be assembled on the subfloor of the building and stacked, ready for erection.

b. If prefabricated trusses are to be used they may be stored on the job or elsewhere until the wall plates are in place.

c. It is easier to assemble a roof made of trusses than of rafters. Time and effort is saved by doing as much work as possible on the subfloor, rather than above the top plate.

d. A building may be closed-in more quickly, providing a weather controlled area sooner—for electricians, plumbers and sheet-metal workers.

e. Truss systems, properly designed and fabricated, may provide increased flexibility with respect to the design of a building: longer spans, fewer load bearing walls, and a variety of shapes of interior ceilings and exterior roofs.

f. Use of trusses saves construction time, permitting quicker turnover of housing units and more profit for the builder.

2. Preparations on the job for placing residential roof trusses

After the trusses are assembled:

a. the walls are framed and sheathed on the subfloor, then raised and fastened in place.

b. interior walls are assembled, raised, and fastened to the outside walls, and to the framing beneath.

c. diagonal braces are nailed to the inside framing to give rigidity to the structure in case of gusty winds.

3. Completing the roof assembly

After the walls are in place:

a. trusses are raised to the top plate and secured at the specified on-center spacings.

b. with appropriate nails and/or fasteners, roof sheathing is applied.

c. the roof covering is applied, according to the specifications of the manufacturer.

4. Roof construction with trusses

Plates XIII, XIV, XV, XVI, and XVII provide general information about roof trusses: wood frame construction; truss shapes; types of roofs; trussed rafter systems; rafter and joist framing; roof framing; and roof to wall

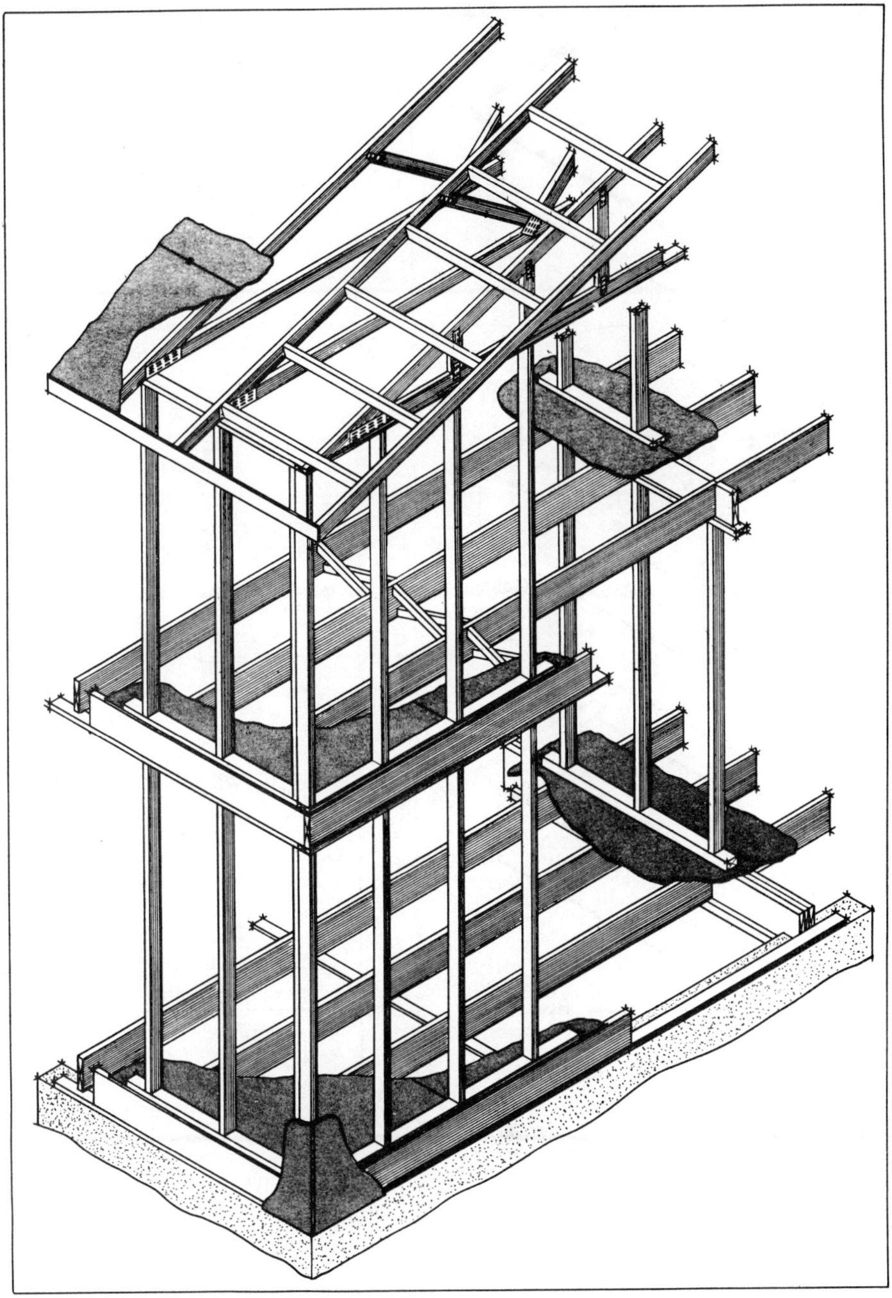

PLATE XIII*. Wood-Frame Construction

**Framing Techniques,* CWC Datafile WB-1, p. 2. Published by the Canadian Wood Council, 170 Laurier Avenue West, Ottawa, Canada K1P 5V5.

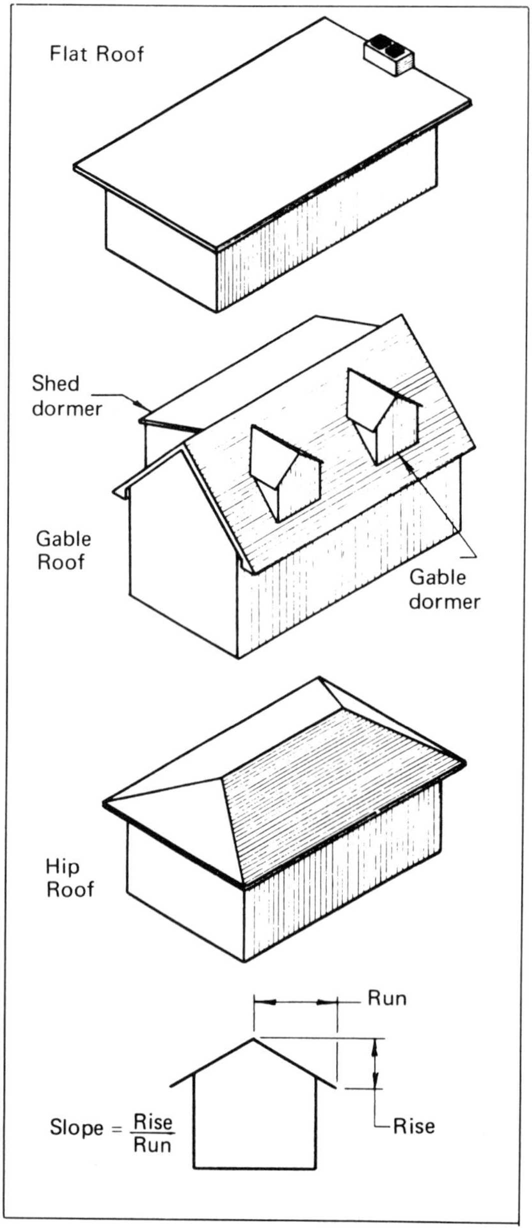

Types of Roofs[1]

Type of Truss	Typical Uses
Common Trusses	
King-Post	Short span requirements for all types of construction.
Queen Post	Short span requirements with heavy top chord loading.
Fink	Commonly used in all types of construction up to 40 ft.
Howe	Commonly used where lower chord support is required.
Fan	For heavy top chord loading common in agricultural buildings.
Special King-Post (or Modified Queen Post)	For applications where light bottom loads exist up to 60 ft.
Belgian (or double Fink)	For long span applications up to 60 ft.
4 Panel Scissor	Vaulted ceiling common in houses and churches.
2-2 Mono	For short span, single slope applications.
3-2 Mono	For heavy top chord load applications.
4-3 Mono	Long span applications.
Pratt	Flat truss used for all types of construction in roof and floor systems.
Warren	Flat truss used for all types of construction in roof and floor systems.
Gable End	For gable roof ends; no special design required.
Special Trusses	**Typical Uses**
	Used where dual slope profile is required.
	Hip or mansard truss type commonly used in residential construction.
	For single slope hip or mansard roof systems.
	Attic frame where living space is required in the roof.
	Combination trusses for dual slope roofs and ceilings.

Truss Shapes[2]

PLATE XIV

[1] *Framing Techniques,* CWC Datafile WB-1, p. 18.

[2] *Light Frame Trusses,* CWC Datafile WC-3, p. 3.

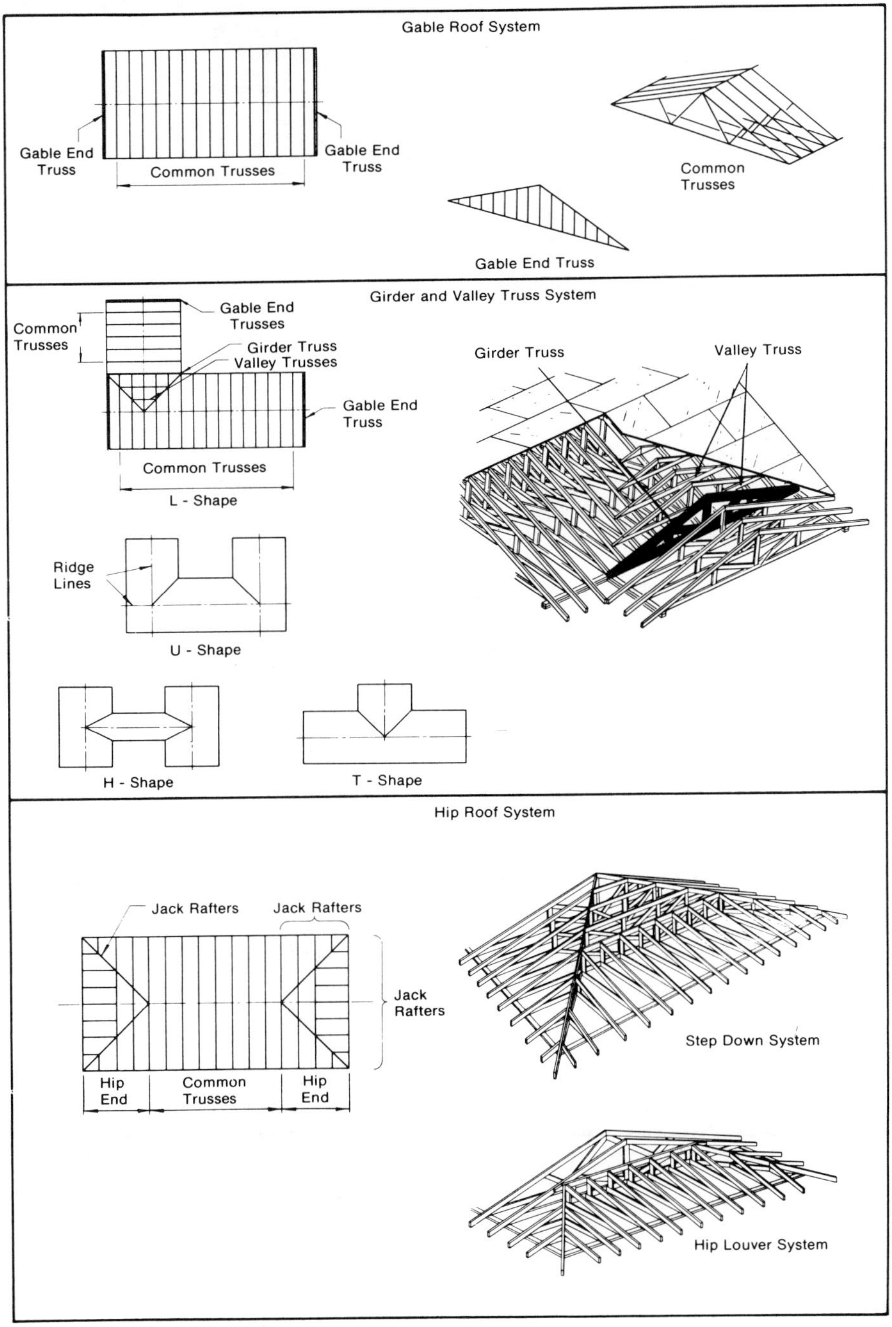

PLATE XV. Trussed Rafter Systems*

*_Light Frame Trusses,_ CWC Datafile WC-3, p. 3.

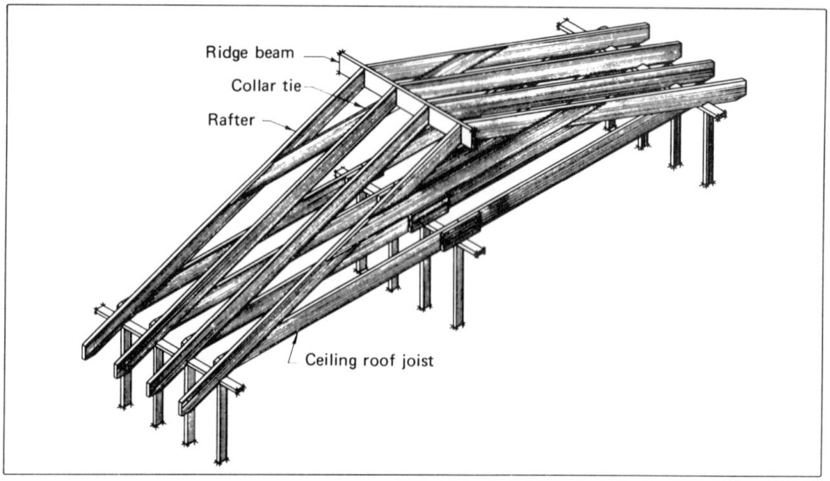

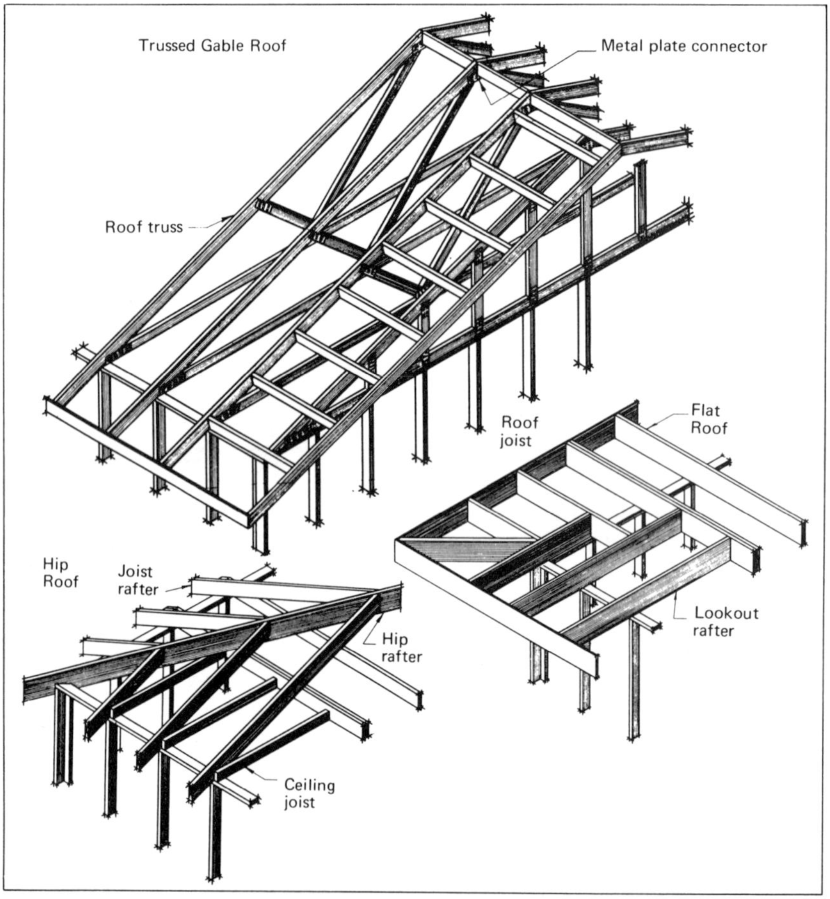

PLATE XVI. Roof Framing*

**Framing Techniques*, CWC Datafile WB-1, pp. 19 and 20.

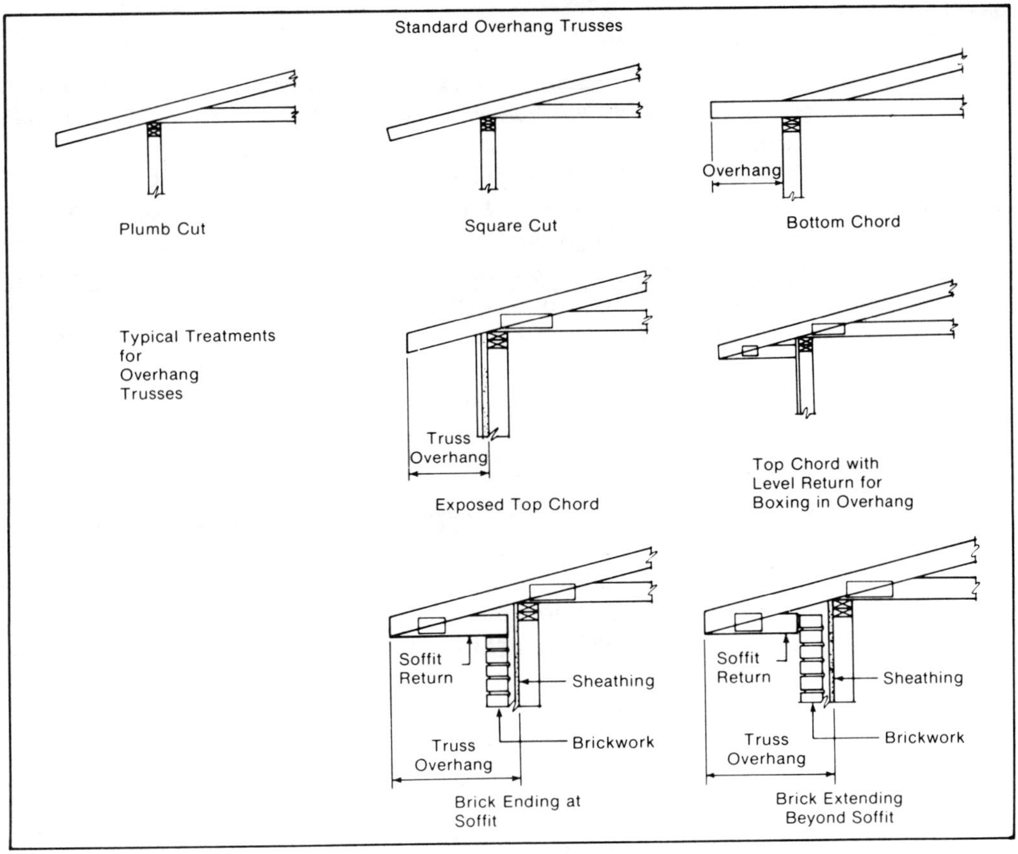

PLATE XVII. Roof To Wall Framing*

Light Frame Trusses, CWC Datafile WC-3, p. 8.

framing. Plate XV, Trussed Rafter Systems, may be of particular value in the general context of the book, when applying rafter layout skills to truss construction.

Plate XV describes three systems graphically: gable roof, girder and valley, and hip roof. Selection of a system for a residential building is an important decision. The ordering of prefabricated trusses or the on-site building of trusses involves other important decisions with respect to design, size, slopes of trusses, and appearance of the finished roof. Aside from the mathematical (engineering) considerations of truss design which are not offered here, the builder needs to become knowledgeable in terms of how different trusses are constructed and best used.

5. Freehand sketches of typical construction details

The following six tasks are designed to provide direction with respect to the reading of the data sheets of this chapter. They help with the learning about truss designs, without involvement with truss load and bar stress calculations.

a. Use single line sketches to indicate the shapes of typical gable, common, girder, and valley trusses.
b. Sketch the plan view of a hip roof assembly using common trusses, HRs, JRs, and CRs.
c. Sketch the plan view of a roof assembly using the following trusses: gable end, common, girder, and valley.
d. Sketch the plan view of a hip roof assembly with a step down system. Describe the step-down system by means of elevation sketches.
e. Sketch the plan view of a gable roof with gable end trusses. Then sketch an addition of different span. Show how valley trusses would be used.
f. Draw to scale a hip louver system. Using a sketch, describe the hip assembly. On full-size rafter stock, lay out the hip beam. Be specific about which roof-to-wall framing detail is to be used.

The presentation in this chapter was about roof truss shapes and roof assemblies. Additional readings might be done about truss designs, fasteners, storm resistant construction, and energy conservation, in terms of provisions for insulation.

Knowledge of rafter layout, applied when building trusses, is used as well in post/beam/plank construction as outlined in Chapter 25.

25

Post/Beam/Plank Construction

"The post and beam system is essentially a framework made up of decking, beams and posts supported on a foundation. The floor and roof decks transfer loads to the beams which, in turn, carry them to posts and on down to the foundation." (p. 2 CWC Datafile WB-2)

Six plates which follow are included to help describe post/beam/plank construction.

1. Plate XVIII compares post and beam framing with conventional framing.
2. Plate XIX illustrates typical post and beam framing.
3. Plate XX is a drawing of post and beam framing combined with conventional framing in a two-story house.
4. Plate XXI provides six post and beam framing details: spaced beam; post to footing connection; first floor framing; solid beam bearing over post; and bearing of solid roof beam on the exterior wall post.
5. Plate XXII describes additional construction details: support for nonbearing partitions; arrangements of beams to absorb horizontal thrust; and methods of supporting the roof beam on the ridge to absorb horizontal thrust.
6. Plate XXIII shows applications of connectors in plank and beam construction.

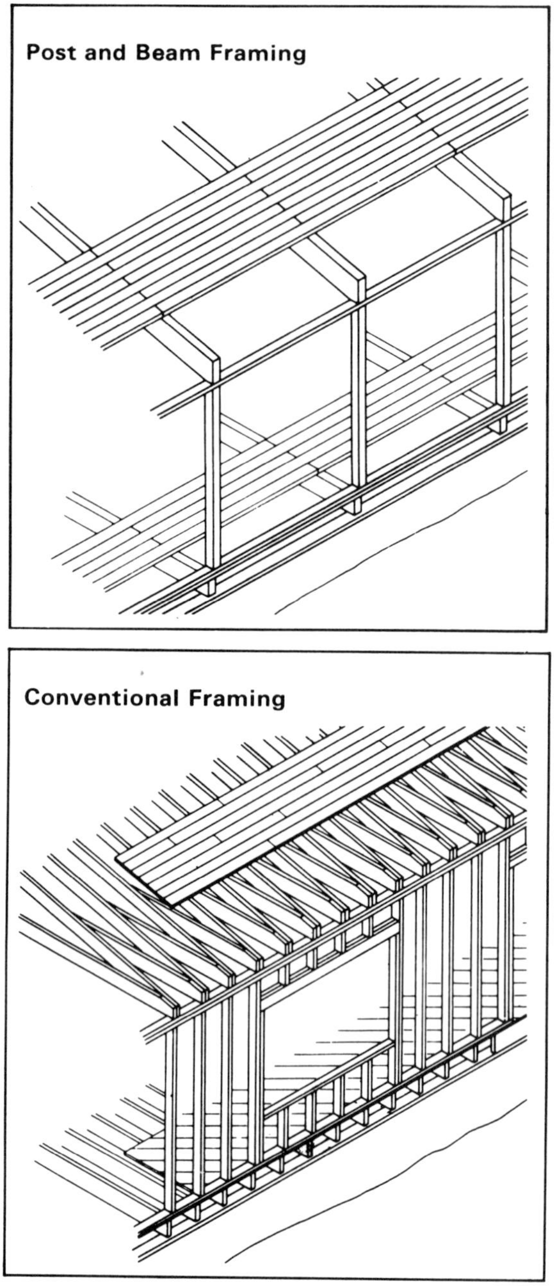

PLATE XVIII. Post/Beam and Conventional Framing*

**Post and Beam,* CWC Datafile WB-2, p. 3.

Typical Post and Beam Framing

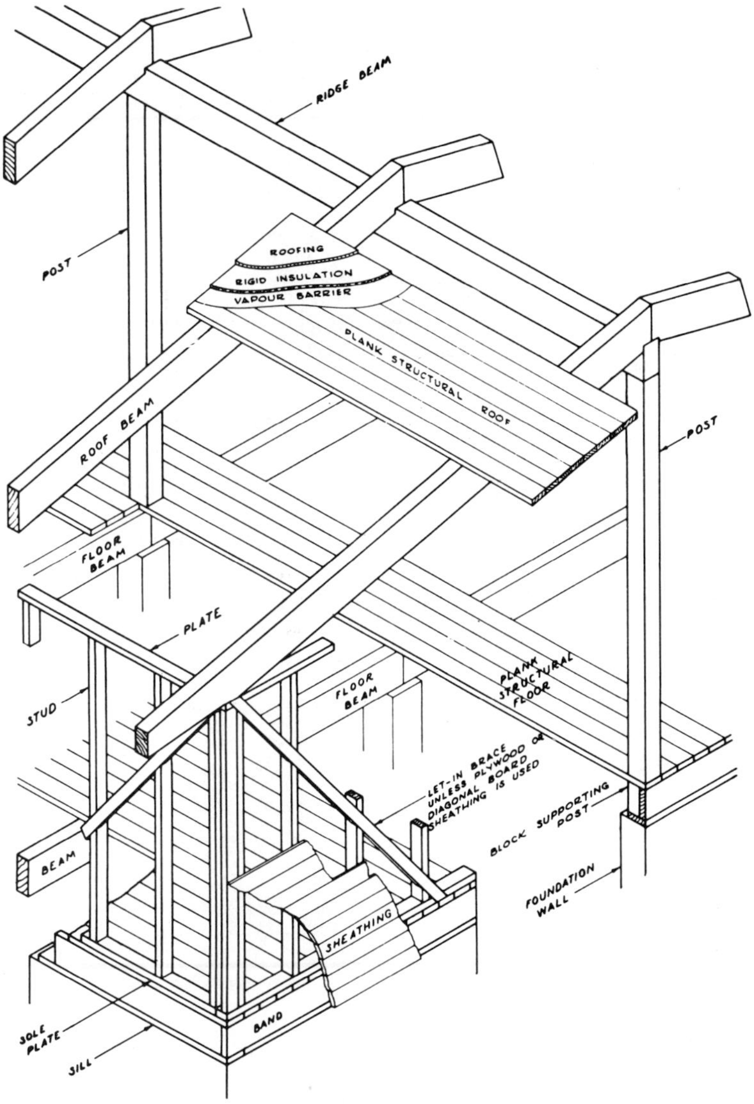

PLATE XIX. Typical Post and Beam Framing*

Post and Beam, CWC Datafile WB-2, p. 4.

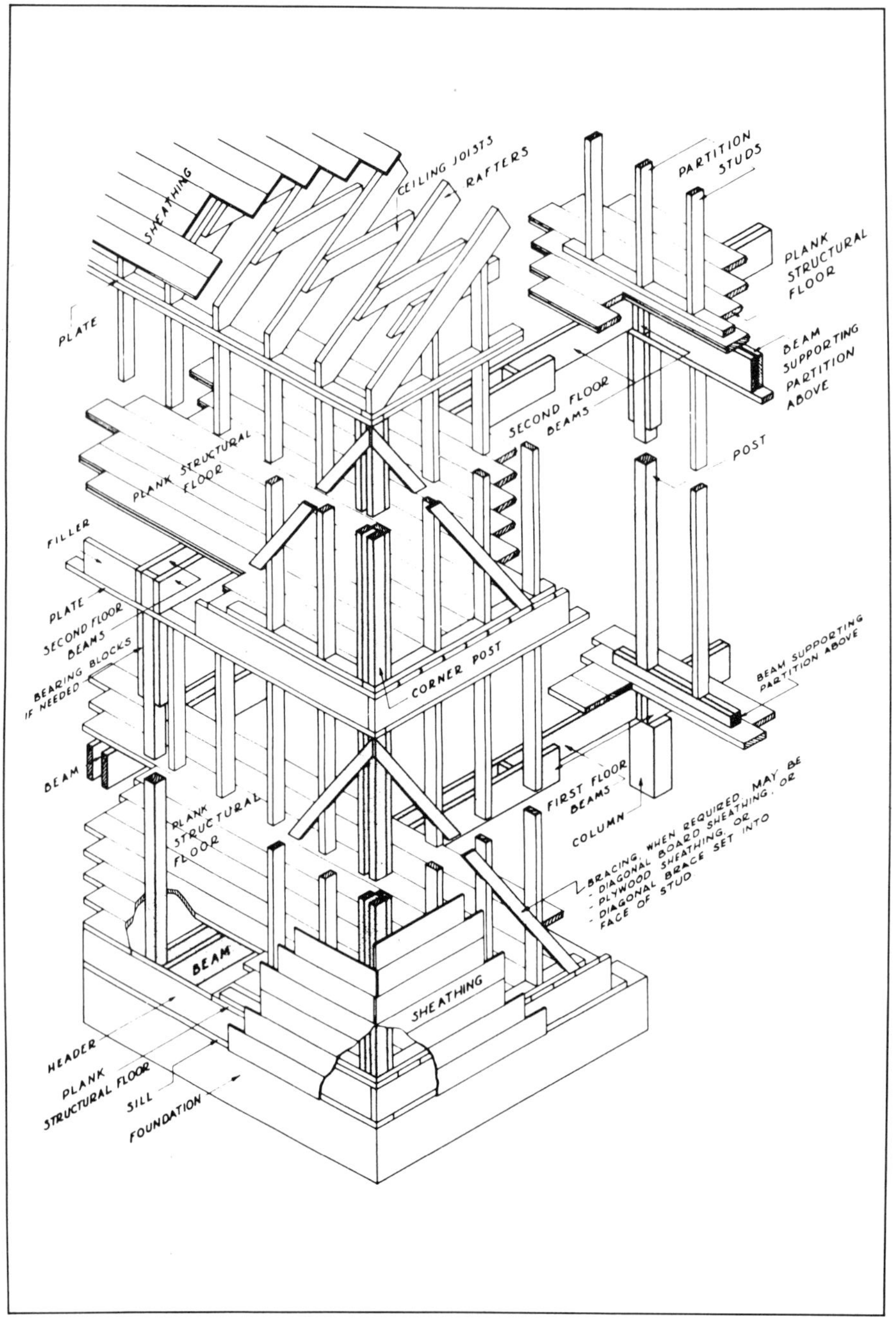

PLATE XX. Post and Beam Framing Combined with Conventional Framing in Two-Story House*

Post and Beam, CWC Datafile WB-2, p. 5.

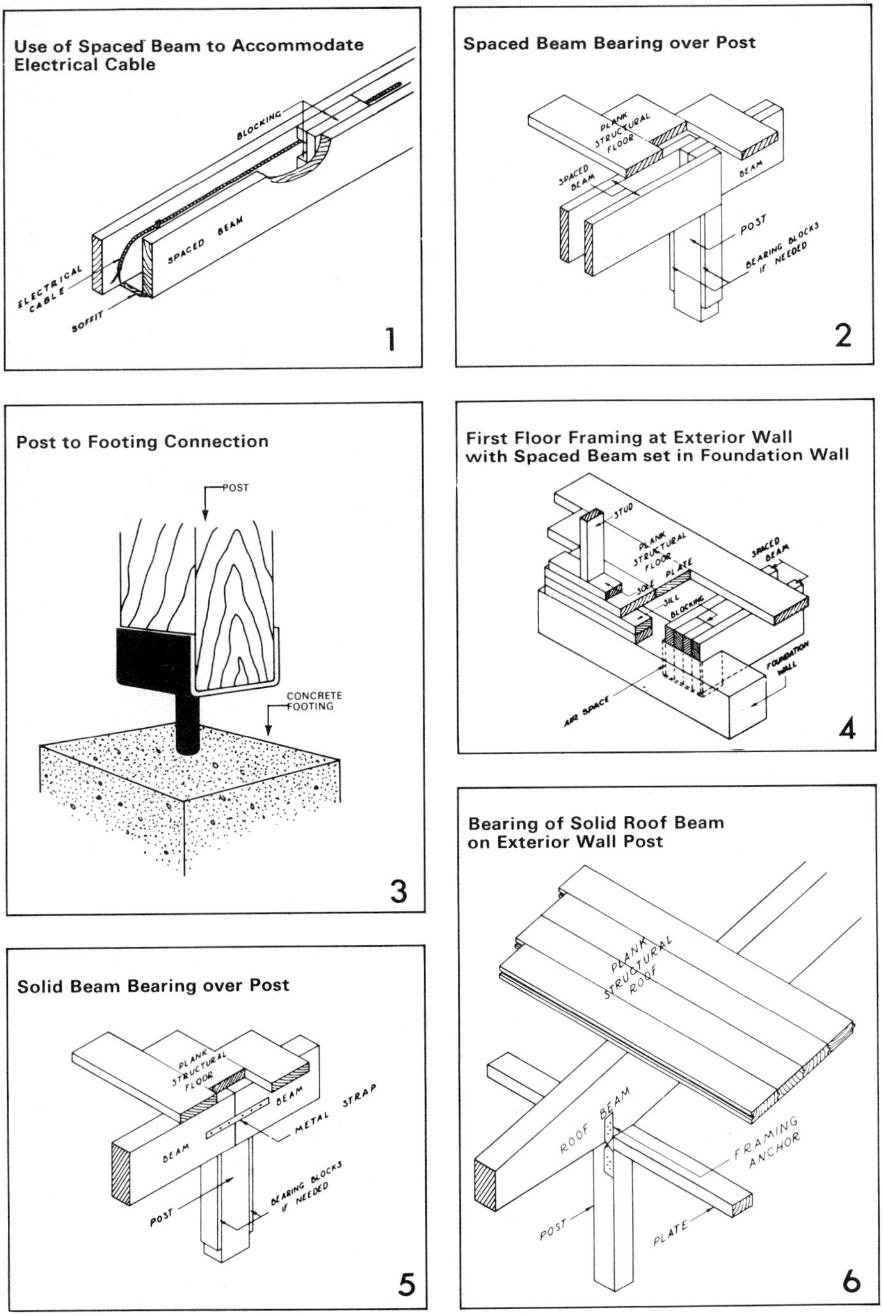

PLATE XXI. Post and Beam Framing Details*

Post and Beam, CWC Datafile WB-2, p. 6.

Support for Non-bearing Partition Parallel to Plank

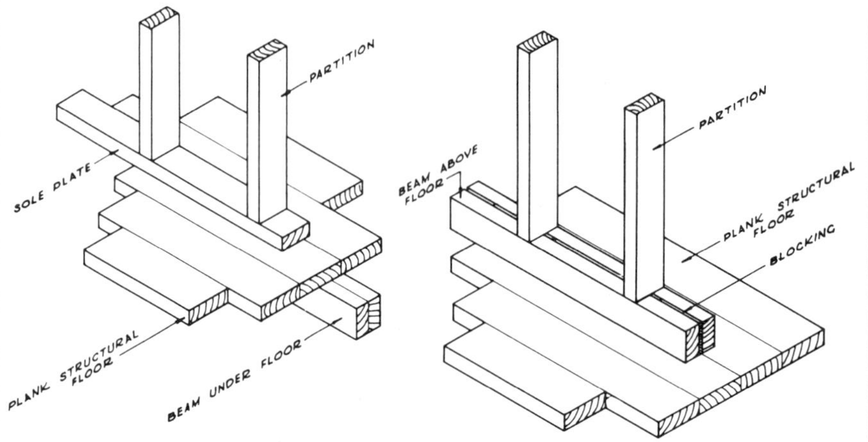

PARTITION SUPPORTED BY BEAM UNDER THE FLOOR

PARTITION SUPPORTED BY BEAM ABOVE THE FLOOR

1

Arrangement of Beams to Absorb Horizontal Thrust

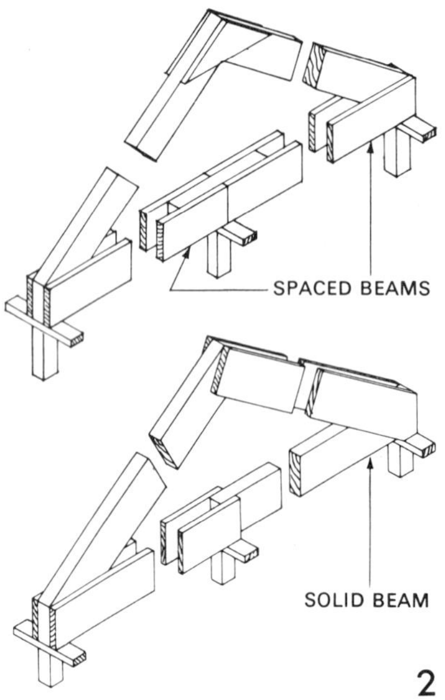

2

Methods of Supporting Roof Beam on Ridge to Absorb Horizontal Thrust

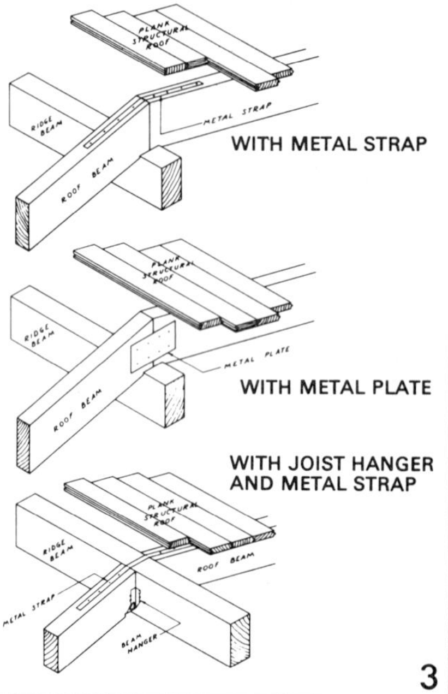

3

PLATE XXII. Plank and Beam Construction Details*

**Post and Beam,* CWC Datafile WB-2, p. 10.

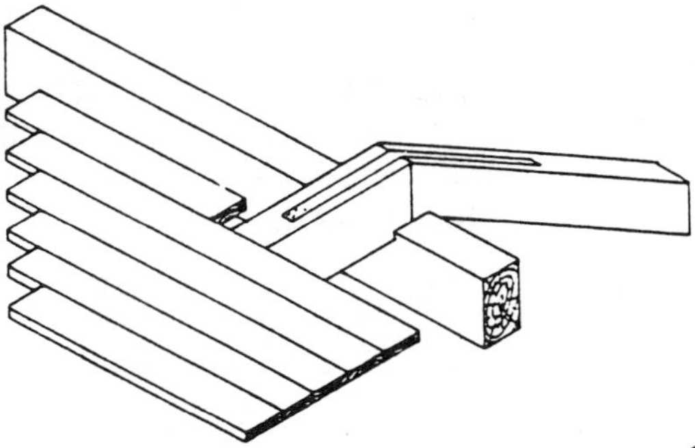

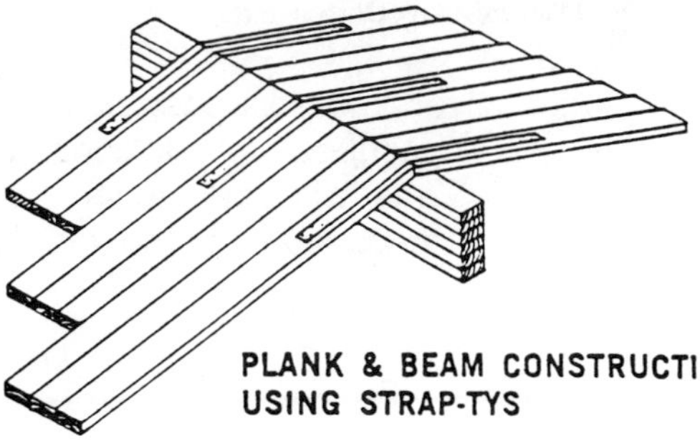

PLANK & BEAM CONSTRUCTION:
USING STRAP-TYS

PLANK & BEAM CONSTRUCTION:
USING STRAP-TYS

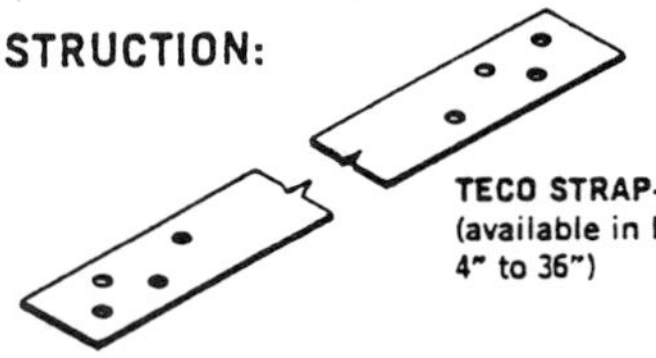

TECO STRAP-TY
(available in lengths
4" to 36")

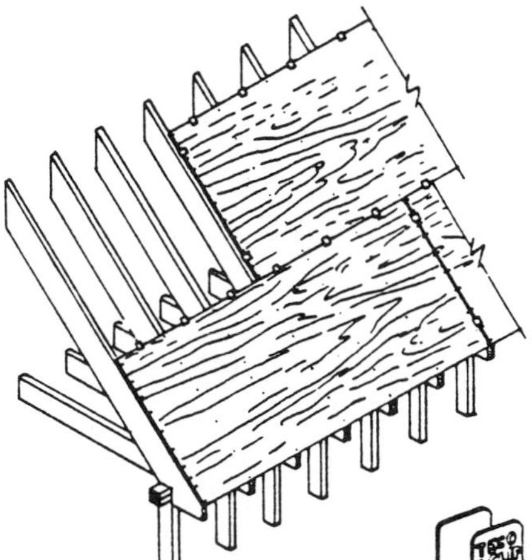

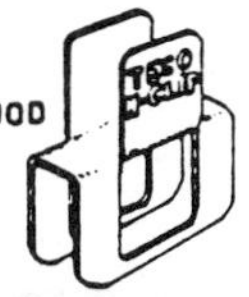

TECO H-CLIP PLYWOOD
SUPPORT OR CLIP
(available for ⅜",
$^{7}/_{16}$", ½", ⅝", ¾"
thickness plywood)

PLYWOOD SHEATHING EDGE SUPPORT:
USING TECO H-CLIPS TO SAVE
WOOD BLOCKING
(also provides automatic end spacing for expansion &
contraction of plywood)

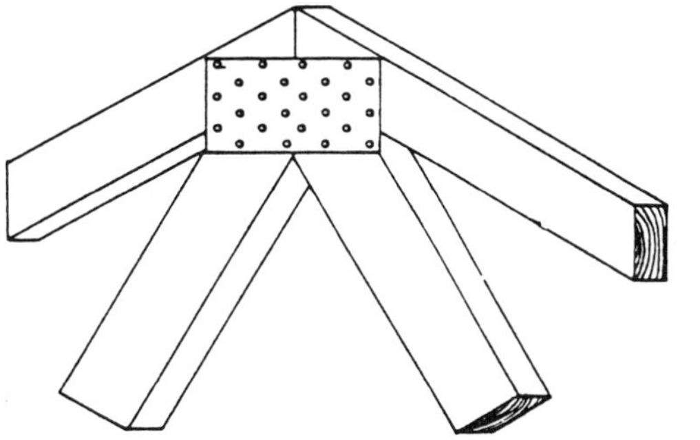

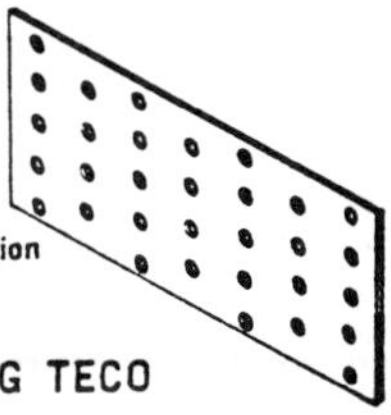

TECO NAIL-ON TRUSS PLATE
(available in 25 sizes;
consult TECO catalog for
specifics)

also extensively used for splice
plates in all phases of construction

TRUSS CONSTRUCTION USING TECO
NAIL-ON TRUSS PLATES

PLATE XXIII. Plank and Beam Construction with Connectors*

---

*Information shown in Plate XXIII has been excerpted from a booklet published by TECO Product and Testing Corporation, 5530 Wisconsin Avenue, Washington D.C. 20015 and 86 305 College View Road, Eugene, Oregon 97405.

In this chapter roof framing principles have been associated with post/beam/plank construction. In the previous chapter the association was with roof truss construction. Other similar associations may be made, as with timber construction and metal truss fabrication.

Glossary of Housing Terms

The following material has been excerpted from *Estimating in Building Construction* by Frank R. Dagostino, 1973, published by Reston Publishing Co., Inc., a Prentice-Hall Company.

Addenda. Statements or drawings that modify the basic contract documents after the latter have been issued to the bidders, but prior to the taking of bids.

Air-dried lumber. Lumber that has been piled in yards or sheds for any length of time. For the United States as a whole, the minimum moisture content of thoroughly air-dried lumber is 12 to 15 percent and the average is somewhat higher. In the South, air-dried lumber may be no lower than 19 percent.

Airway. A space between roof insulation and roof boards for movement of air.

Alligatoring. Coarse checking pattern characterized by a slipping of the new paint coating over the old coating to the extent that the old coating can be seen through the fissures.

Alternates. Proposals required of bidders reflecting amounts to be added to or subtracted from the basic proposal in the event that specific changes in the work are ordered.

Anchor bolts. Bolts used to anchor structural members to concrete or the foundation.

Approved equal or. The term used to indicate that material or product finally supplied or installed must be equal to that specified and as approved by the architect (or engineer).

Apron. The flat member of the inside trim of a window placed against the wall immediately beneath the stool.

Areaway. An open subsurface space adjacent to a building used to admit light or air or as a means of access to a basement.

As-built drawings. Drawings made during the progress of construction, or subsequent thereto, illustrating how various elements of the project were actually installed.

Asphalt. Most native asphalt is a residue from evaporated petroleum. It is insoluble in water but soluble in gasoline and melts when heated. Used widely in building for waterproofing roof coverings of many types, exterior wall coverings, flooring tile, and the like.

Astragal. A closure between the two leafs of a double-swing or double-slide door to close the joint. This can also be a piece of molding.

Attic ventilators. In houses, screened openings provided to ventilate an attic space. They are located in the soffit area as inlet ventilators and in the gable end or along the ridge as outlet ventilators. They can also consist of power-driven fans used as an exhaust system. (See also **Louver.**)

Axial. Anything situated around, in the direction of, or along an axis.

Backband. A simple molding sometimes used around the outer edge of plain rectangular casing as a decorative feature.

Backfill. The replacement of excavated earth into a trench around and against a basement foundation.

Balusters. Usually small vertical members in a railing used between a top rail and the stair treads or a bottom rail.

Balustrade. A railing made up of balusters, top rail, and sometimes bottom rail, used on the edge of stairs, balconies, and porches.

Bank or pit run gravel. Natural aggregates as dug from a pit.

Barge board. A decorative board covering the projecting rafter (fly rafter) of the gable end. At the cornice, this member is a fascia board.

Base or baseboard. A board placed against the wall around a room next to the floor to finish properly between floor and plaster.

Base molding. Molding used to trim the upper edge of interior baseboard.

Baseplate. A plate attached to the base of a column which rests on a concrete or masonry footing.

Base shoe. Molding used next to the floor on interior baseboard. Sometimes called a carpet strip.

Batten. Narrow strips of wood used to cover joints or as decorative vertical members over plywood or wide boards.

Batter board. One of a pair of horizontal boards nailed to posts set at the corners of an excavation, used to indicate the desired level, also as a fastening for stretched strings, to indicate outlines of foundation walls.

Bay. The space between column center lines or primary supporting members, lengthwise in a building. Usually the crosswise dimension is considered the span or width module, and the lengthwise dimension is considered the bay spacing.

Bay window. Any window space projecting outward from the walls of a building, either square or polygonal in plan.

Beam. A structural member that is normally subjected to bending loads and is usually a horizontal member carrying vertical loads. (An exception to this is a purlin.)
There are three types of beams:

1. *Continuous beam.* A beam that has more than two points of support.
2. *Cantilevered beam.* A beam that is supported at only one end and is restrained against excessive rotation.

3. *Simple beam.* A beam that is freely supported at both ends, theoretically with no restraint.

Beam and column. A primary structural system consisting of a series of beams and columns; usually arranged as a continuous beam supported on several columns with or without continuity, that is, subjected to both bending and axial forces.

Beam bearing plate. Steel plate with attached anchors that is set in top of a masonry wall so that a purlin or a beam can rest on it.

Bearing partition. A partition that supports any vertical load in addition to its own weight.

Bearing wall. A wall that supports any vertical load in addition to its own weight.

Bed molding. A molding in an angle, as between the overhanging cornice, or eaves, of a building and the side walls.

Bench mark. A fixed point used for construction purposes as a reference point in determining the various levels of floor, grade, etc.

Bid. Proposal prepared by prospective contractor specifying the charges to be made for doing the work in accordance to the contract documents.

Bid bond. A surety bond guaranteeing that a bidder will sign a contract, if offered, in accordance with his proposal.

Bid security. A bid bond, certified check, or other forfeitable security guaranteeing that a bidder will sign a contract, if offered, in accordance with his proposal.

Bill of materials. A list of items or components used for fabrication, shipping, receiving, and accounting purposes.

Bird screen. Wire mesh used to prevent birds from entering the building through ventilators or louvers.

Blind-nailing. Nailing in such a way that the nailheads are not visible on the face of the work—usually at the tongue of matched boards.

Blind stop. A rectangular molding, usually $\frac{3}{4}$ by $1\frac{3}{8}$ inches or more in width, used in the assembly of a window frame. Serves as a stop for storm and screen or combination windows and to resist infiltration.

Blue stain. A bluish or grayish discoloration of the sapwood caused by the growth of certain moldlike fungi on the surface and in the interior of a piece, made possible by the same conditions that favor the growth of other fungi.

Bodied linseed oil. Linseed oil that has been thickened in viscosity by suitable processing with heat or chemicals. Bodied oils are obtainable in a great range of viscosity from a little greater than that of raw oil to just short of a jellied condition.

Boiled linseed oil. Linseed oil in which enough lead, manganese, or cobalt salts have been incorporated to make the oil harden more rapidly when spread in thin coatings.

Bolster. A short horizontal timber or steel beam on top of a column to support and decrease the span of beams or girders.

Bond. Masonry units interlocked in the face of a wall by overlapping.

Bonded roof. A roof that carries a printed or written warranty, usually with respect to weather-tightness, including repair and/or replacement on a prorated cost basis for a stipulated number of years.

Bonus and penalty clause. A provision in the proposal form for payment of a bonus for each day the project is completed prior to the time stated, and for a charge against the contractor for each day the project remains uncompleted after the time stipulated.

Boston ridge. A method of applying asphalt or wood shingles at the ridge or at the hips of a roof as a finish.

Brace. An inclined piece of framing lumber applied to wall or floor to stiffen the structure. Often used on walls as temporary bracing until framing has been completed.

Brace rods. Rods used in roofs and walls to transfer wind loads and/or seismic forces to the foundation (often used to plumb building but not designed to replace erection cables when required).

Brick veneer. A facing of brick laid against and fastened to sheathing of a frame wall or tile wall construction.

Bridging. Small wood or metal members that are inserted in a diagonal position between the floor joists at midspan to act both as tension and compression members for the purpose of bracing the joists and spreading the action of loads.

Buck. Often used in reference to rough frame opening members. Door bucks used in reference to metal door frame.

Building paper. Used in many areas of building construction, it may form a dust trap between a hardwood floor and its subfloor.

Built-up roof. A roofing composed of three to five layers of asphalt felt laminated with coal tar, pitch, or asphalt. The top is finished with crushed slag or gravel. Generally used on flat or low-pitched roofs.

Butt joint. The junction where the ends of two timbers or other members meet in a square-cut joint.

Camber. A permanent curvature designed into a structural member in a direction opposite to the deflection anticipated when loads are applied.

Canopy. Any overhanging or projecting structure with the extreme end unsupported. It may also be supported at the outer end.

Cantilever. A projecting beam supported and restrained only at one end.

Cant strip. A triangular-shaped piece of lumber used at the junction of a flat deck and a wall to prevent cracking of the roofing which is applied over it.

Cap. The upper member of a column, pilaster, door cornice, molding, and the like.

Cap plate. A horizontal plate located at the top of a column.

Casement frames and sash. Frames of wood or metal enclosing part or all of the sash, which may be opened by means of hinges affixed to the vertical edges.

Cash allowances. Sums that the contractor is required to include in his bid and contract sum for specific purposes.

Casing. Molding of various widths and thicknesses used to trim door and window openings at the jambs.

Caulk. To seal and make weathertight the joints, seams, or voids by filling with a waterproofing compound or material.

Cavity wall. Two separate leaves of masonry units forming a wall with a space of 2″ minimum between them and secured together with metal ties or bonding units. The leaves are called wythes or withes and may be of different-sized units.

Cement. Keene's. A white finish plaster that produces an extremely durable wall. Because of its density, it excels for use in bathrooms and kitchens and is also used extensively for the finish coat in auditoriums, public buildings, and other places where walls may be subjected to unusually hard wear or abuse.

Certificate of occupancy. Statement issued by the governing authority granting permission to occupy a project for a specific use.

Certificate of payment. Statement by an architect informing the owner of the amount due a contractor on account of work accomplished and/or materials suitably stored.

Change order. A work order, usually prepared by the architect and signed by the owner or his agent, authorizing a change in the scope of the work and a change in the cost of the project.

Channel. A steel member whose formation is similar to that of a "C" section without return lips; may be used singularly or back-to-back.

Checking. Fissures that appear with age in many exterior paint coatings, at first superficial, but which in time may penetrate entirely through the coating.

Checkrails. Meeting rails sufficiently thicker than a window to fill the opening between the top and bottom sash made by the parting stop in the frame of double-hung windows. They are usually beveled.

Clip. A plate or angle used to fasten two or more members together.

Clip angle. An angle used for fastening various members together.

Collar beam. Nominal 1- or 2-inch thick members connecting opposite roof rafters. They serve to stiffen the roof structure.

Collateral loads. A load, in addition to normal live, wind or dead loads, intended to cover loads that are either unknown or uncertain (sprinklers, lighting, etc.).

Column. In architecture: A perpendicular supporting member, circular or rectangular in section, usually consisting of a base, shaft, and capital. In engineering: A vertical structural compression member which supports loads acting in the direction of its longitudinal axis.

Combination doors or windows. Combination door or windows used over regular openings. They provide winter insulation and summer protection and often have self-storing or removable glass and screen inserts. This eliminates the need for handling a different unit each season.

Concrete plain. Concrete either without reinforcement, or reinforced only for shrinkage or temperature changes.

Condensation. In a building: Beads or drops of water (and frequently frost in extremely cold weather) that accumulate on the inside of the exterior covering of a building when warm, moisture-laden air from the interior reaches a point where the temperature no longer permits the air to sustain the moisture it holds. Use of louvers or attic ventilators will reduce moisture condensation in attics. A vapor barrier under the gypsum lath or dry wall on exposed walls will reduce condensation in them.

Conduit, electrical. A pipe, usually metal, in which wire is installed.

Construction dry-wall. A type of construction in which the interior wall finish is applied in a dry condition, generally in the form of sheet materials or wood paneling, as contrasted to plaster.

Construction, frame. A type of construction in which the structural parts are wood or depend upon a wood frame for support. In codes, if masonry veneer is applied to the exterior walls, the classification of this type of construction is usually unchanged.

Contract documents. Working Drawings, Specifications, General Conditions, Supplementary General Conditions, the Owner-Contractor Agreement and all Addenda (if issued).

Corbel out. To build out one or more courses of brick or stone from the face of a wall, to form a support for timbers.

Corner bead. A strip of formed sheet metal, sometimes combined with a strip of metal lath, placed on corners before plastering to reinforce them. Also, a strip of wood finish three-quarters-round or angular placed over a plastered corner for protection.

Corner boards. Used as trim for the external corners of a house or other frame structure against which the ends of the siding are finished.

Corner braces. Diagonal braces at the corners of frame structure to stiffen and strengthen the wall.

Crawl space. Space between the lowest member of a floor and the ground beneath, affording two or three feet of clearance.

Curb. A raised edge on a concrete floor slab.

Curtain wall. Perimeter walls that carry only their own weight and wind load.

Dampproofing. Material used to render a surface impervious to the passage of damp by an emulsion or with special mortar for masonry units near ground level.

Datum. Any level surface to which elevations are referred (see **Bench mark**).

Dead load. The weight of the structure itself, such as floor, roof, framing and covering members, plus any permanent loads.

Deflection. The displacement of a loaded structural member or system in any direction, measured from its no-load position, after loads have been applied.

Demurrage. Fee charged for detention of vessel or vehicle over agreed time.

Design loads. Those loads specified by building codes, state or city agencies, or owner's or architect's specifications to be used in the design of the structural frame of a building. They are suited to local conditions and building use.

Door guide. An angle or channel guide used to stabilize and keep plumb a sliding or rolling door during its operation.

Downspout. A hollow section such as a pipe used to carry water from the roof or gutter of a building to the ground or sewer connection.

Drain. Any pipe, channel, or trench for which waste water or other liquids are carried off; i.e., to a sewer pipe.

Drip. (a) A member of a cornice or other horizontal exterior-finish course that has a projection beyond the other parts for throwing off water. (b) A groove in the underside of a sill or drip cap to cause water to drip off on the outer edge instead of drawing back and running down the face of the building.

Drip cap. A molding placed on the exterior top side of a door or window frame to cause water to drip beyond the outside of the frame.

Dry-wall. Interior covering material, such as gypsum board or plywood, which is applied in large sheets or panels.

Ducts. In a house, usually round or rectangular metal pipes for distributing warm air from the heating plant to rooms, or air from a conditioning device or for use as cold air returns. Ducts are also made of asbestos and composition materials.

Eave. The line along the sidewall formed by the intersection of the inside faces of the roof and wall panels; the projecting lower edges of a roof, overhanging the walls of a building.

Erection. The assembly of components to form the completed portion of a job.

Equal or. (See **Approved equal.**)

Expansion joint. A connection used to allow for temperature-induced expansion and contraction of material.

Fabrication. The manufacturing process performed in the plant to convert raw material into finished metal building components. The main operations are cold forming, cutting, punching, welding, cleaning, and painting.

Facia or fascia. A flat board, band, or face, used sometimes by itself but usually in combination with moldings, often located at the outer face of the cornice.

Field. The jobsite or building site.

Field fabrication. Fabrication performed by the erection crew or others in the field.

Field welding. Welding performed at the jobsite, usually with gasoline-powered machines.

Filler (wood). A heavily pigmented preparation used for filling and leveling off the pores in open-pored woods.

Fire-resistive. In the absence of a specific ruling by the authority having jurisdiction, applies to materials for construction not combustible in the temperatures of ordinary fires and that will withstand such fires without serious impairment of their usefulness for at least 1 hour.

Fire-retardant chemical. A chemical or preparation of chemicals used to reduce flammability or to retard spread of flame.

Fire stop. A solid, tight closure of a concealed space, placed to prevent the spread of fire and smoke through such a space. In a frame wall, this will usually consist of 2 by 4 cross blocking between studs.

Fishplate. A wood or plywood piece used to fasten the ends of two members together at a butt joint with nails or bolts. Sometimes used at the junction of opposite rafters near the ridge line.

Flagstone (flagging or flags). Flat stones, from 1 to 4 inches thick, used for rustic walks, steps, floors, and the like.

Flashing. Sheet metal or other material used in roof and wall construction to protect a building from water seepage.

Flat paint. An interior paint that contains a high proportion of pigment and dries to a flat or lusterless finish.

Flue. The space or passage in a chimney through which smoke, gas, or fumes ascend. Each passage is called a flue, which together with any others and the surrounding masonry make up the chimney.

Flue lining. Fire clay or terra-cotta pipe, round or square, usually made in all ordinary flue sizes and in 2-foot lengths, used for the inner lining of chimneys with the brick or masonry work around the outside. Flue lining in chimneys runs from about a foot below the flue connection to the top of the chimney.

Fly rafters. End rafters of the gable overhang supported by roof sheathing and lookouts.

Footing. A masonry section, usually concrete, in a rectangular form wider than the bottom of the foundation wall or pier it supports.

Foundation. The supporting portion of a structure below the first-floor construction, or below grade, including the footings.

Framing, balloon. A system of framing a building in which all vertical structural elements of the bearing walls and partitions consist of single pieces extending from the top of the foundation sill plate to the roofplate and to which all floor joists are fastened.

Framing, platform. A system of framing a building in which floor joists of each story rest on the top plates of the story below or on the foundation sill for the first story, and the bearing walls and partitions rest on the subfloor of each story.

Frieze. In house construction, a horizontal member connecting the top of the siding with the soffit of the cornice.

Frostline. The depth of frost penetration in soil. This depth varies in different parts of the country. Footings should be placed below this depth to prevent movement.

Fungi, wood. Microscopic plants that live in damp wood and cause mold, stain, and decay.

Fungicide. A chemical that is poisonous to fungi.

Furring. Strips of wood or metal applied to a wall or other surface to even it and normally to serve as a fastening base for finish material.

Gable. In house construction, the portion of the roof above the eave line of a double-sloped roof.

Gable end. An end wall having a gable.

Girder. A large or principal beam of wood or steel used to support concentrated loads at isolated points along its length.

Girths or girts. Wood members tightly fitted between wood walls as stiffeners and as fire traps.

Gloss (paint or enamel). A paint or enamel that contains a relatively low proportion of pigment and dries to a sheen or luster.

Grain. The direction, size, arrangement, appearance, or quality of the fibers in wood.

Grain, edge (vertical). Edge-grain lumber has been sawed parallel to the pith of the log and approximately at right angles to the growth rings; i.e., the rings form an angle of 45° or more with the surface of the piece.

Grain, flat. Flat-grain lumber has been sawed parallel to the pith of the log and approximately tangent to the growth rings, i.e., the rings form an angle of less than 45° with the surface of the piece.

Grain, quartersawn. Another term for edge grain.

Grounds. Guides used around openings and at the floorline to strike off plaster. They can consist of narrow strips of wood or of wide subjambs at interior doorways. They provide a level plaster line for installation of casing and other trim.

Grout. Mortar made of such consistency (by adding water) that it will just flow into the joints and cavities of the masonry work and fill them solid.

Gusset. A flat wood, plywood, or similar type member used to provide a connection at intersection of wood members. Most commonly used at joints of wood trusses. They are fastened by nails, screws, bolts, or adhesives.

Gutter or eave trough. A shallow channel or conduit of metal or wood set below and along the eaves of a house to catch and carry off rainwater from the roof.

Gypsum plaster. Gypsum formulated to be used with the addition of sand and water for base-coat plaster.

Head. The top of a door, window, or frame.

Header. (a) A beam placed perpendicular to joists and to which joists are nailed in framing for chimney, stairway, or other opening. (b) A wood lintel.

Hearth. The inner or outer floor of a fireplace, usually made of brick, tile, or stone.

Heartwood. The wood extending from the pith to the sapwood, the cells of which no longer participate in the life processes of the tree.

Hip. The external angle formed by the meeting of two sloping sides of a roof.

Hip roof. A roof that rises by inclined planes from all four sides of a building.

Humidifier. A device designed to increase the humidity within a room or a house by means of the discharge of water vapor. They may consist of individual room-size units or larger units attached to the heating plant to condition the entire house.

I-beam. A steel beam with a cross section resembling the letter **I.** It is used for long spans as basement beams or over wide wall openings, such as a double garage door, when wall and roof loads are imposed on the opening.

Impact load. The assumed load resulting from the motion of machinery, elevators, cranes, vehicles, and other similar moving equipment.

INR (Impact Noise Rating). A single figure rating which provides an estimate of the impact sound-insulating performance of a floor-ceiling assembly.

Instructions to bidders. A document stating the procedure to be followed by bidders.

Insulation. Any material used in building construction for the protection from heat or cold.

Insulation board, rigid. A structural building board made of coarse wood or cane fiber in $\frac{1}{2}$- and $\frac{25}{32}$-inch thicknesses. It can be obtained in various size sheets, in various densities, and with several treatments.

Insulation, thermal. Any material high in resistance to heat transmission that, when placed in the walls, ceiling, or floors of a structure, will reduce the rate of heat flow.

Interior finish. Material used to cover the interior framed areas, or materials of walls and ceilings.

Invitation to bid. An invitation to a selected list of contractors furnishing information on the submission of bids on a subject.

Jack rafter. A rafter that spans the distance from the wallplate to a hip, or from a valley to a ridge.

Jamb. The side and head lining of a doorway, window, or other opening.

Joint. The space between the adjacent surfaces of two members or components joined and held together by nails, glue, mortar, or other means.

Joint cement. A powder that is usually mixed with water and used for joint treatment in gypsum-wallboard finish. Often called "spackle."

Joist. One of a series of parallel beams, usually 2 inches in thickness, used to support floor and ceiling loads, and supported in turn by larger beams, girders, or bearing walls.

Kiln-dried lumber. Lumber that has been kiln-dried, often to a moisture content of 6 to 12 percent. Common varieties of softwood lumber, such as framing lumber, are dried to a somewhat higher moisture content.

Kip. A unit of weight, force, or load equal to 1,000 pounds.

Knot. In lumber, the portion of a branch or limb of a tree that appears on the edge or face of the piece.

Landing. A platform between flights of stairs or at the termination of a flight of stairs.

Lath. A building material of wood, metal, gypsum, or insulating board that is fastened to the frame of a building to act as a plaster base.

Lattice. A framework of crossed wood or metal strips.

Lavatory. A bathroom-type sink.

Leader. (See **Downspout.**)

Ledger strip. A strip of lumber nailed along the bottom of the side of a girder on which joists rest.

Liens. Legal claims against an owner for amounts due those engaged in or supplying materials for the construction of the building.

Light. Space in a window sash for a single pane of glass. Also, a pane of glass.

Lintel. A horizontal structural member that supports the load over an opening such as a door or window.

Liquidated damages. An agreed-to sum chargeable against the contractor as reimbursement for damages suffered by the owner because of the contractor's failure to fulfill his contractual obligations.

Live load. The load exerted on a member or a structure due to all imposed loads except dead, wind, and seismic loads. Examples include snow, people, movable equipment, etc. This type of load is movable and does not necessarily exist on a given member or structure.

Loads. Anything that causes an external force to be exerted on a structural member. Examples of different types are:

1. *Dead load.* In a building, the weight of all permanent constructions, such as floor, roof, framing, and covering members.
2. *Impact load.* The assumed load resulting from the motion of machinery, elevators, craneways, vehicles, and other similar kinetic forces.

3. *Roof live load.* All loads exerted on a roof (except dead, wind, and lateral loads) and applied to the horizontal projection of the building.
4. *Seismic load.* The assumed lateral load due to the action of earthquakes and acting in any horizontal direction on the structural frame.
5. *Wind load.* The load caused by wind blowing from any horizontal direction.

Lookout. A short wood bracket or cantilever to support an overhang portion of a roof or the like, usually concealed from view.

Louver. An opening provided with one or more slated, fixed, or movable fins to allow flow of air, but to exclude rain and sun or to provide privacy.

Lumber. Lumber is the product of the sawmill and planing mill not further manufactured other than by sawing, resawing, and passing lengthwise through a standard planing machine, crosscutting to length, and matching.

Lumber, boards. Yard lumber less than 2 inches thick and 2 or more inches wide.

Lumber, dimension. Yard lumber from 2 inches to, but not including, 5 inches thick and 2 or more inches wide. Includes joists, rafters, studs, planks, and small timbers.

Lumber, dressed size. The dimension of lumber after shrinking from green dimension and after machining to size or pattern.

Lumber, matched. Lumber that is dressed and shaped on one edge in a grooved pattern and on the other in a tongued pattern.

Lumber, shiplap. Lumber that is edge-dressed to make a close rabbeted or lapped joint.

Lumber, timbers. Yard lumber 5 or more inches in least dimension. Includes beams, stringers, posts, caps, sills, girders, and purlins.

Lumber, yard. Lumber of those grades, sizes, and patterns which are generally intended for ordinary construction, such as framework and rough coverage of houses.

Mantel. The shelf above a fireplace. Also used in referring to the decorative trim around a fireplace opening.

Masonry. Stone, brick, concrete, hollow-tile, concrete-block, gypsum-block, or other similar building units or materials or a combination of the same, bonded together with mortar to form a wall, pier, buttress, or similar mass.

Mastic. A pasty material used as a cement (as for setting tile) or a protective coating (as for thermal insulation or waterproofing).

Metal lath. Sheets of metal that are slit and drawn out to form openings. Used as a plaster base for walls and ceilings and as reinforcing over other forms of plaster base.

Millwork. Generally all building materials made of finished wood and manufactured in millwork plants and planing mills are included under the term "millwork." It includes such items as inside and outside doors, window and door frames, blinds, porchwork, mantels, panelwork, stairways, moldings, and interior trim. It normally does not include flooring, ceiling, or siding.

Mullion. The large vertical piece between windows. (It holds the window in place along the edge with which it makes contact.)

Nonbearing partition. A partition which supports no weight except its own.

Parapet. That portion of the vertical wall of a building that extends above the roof line at the intersection of the wall and roof.

Partition. A material or combination of materials used to divide a space into smaller spaces.

Performance bond. A bond that guarantees to the owner, within specified limits, that the contractor will perform the work in accordance with the contract documents.

Pier. A structure of masonry (concrete) used to support the bases of columns and bents. It carries the vertical load to a footing at the desired load-bearing soil.

Pilaster. A flat rectangular column attached to or built into a wall masonry or pier; structurally, a pier, but treated architecturally as a column with a capital, shaft, and base. It is used to provide strength for roof loads or support for the wall against lateral forces.

Precast concrete. Concrete that is poured and cast in some position other than the one it will finally occupy; cast either on the jobsite and then put into place or away from the site to be transported to the site and erected.

Prestressed concrete. Concrete in which the reinforcing cables, wires, or rods are tensioned before there is load on the member.

Progress payments. Payments made during progress of the work, on account, for work completed and/or suitably stored.

Progress schedule. A diagram showing proposed and actual times of starting and completion of the various operations in the project.

Punch list. A list prepared by the architect or engineer of the contractor's uncompleted or work to be corrected.

Purlin. Secondary horizontal structural members located on the roof extending between rafters, used as (light) beams for supporting the roof covering.

Rafter. A primary roof support beam usually in an inclined position, running from the tops of the structural columns at the eave to the ridge or highest portion of the roof. It is used to support the purlins.

Recess. A notch or cut-out, usually referring to the blockout formed at the outside edge of a foundation, and providing support and serving as a closure at the bottom edge of wall panels.

Reinforcing steel. The steel placed in concrete to carry the tension, compressions, and shear stresses.

Retainage. A sum withheld from each payment to the contractor in accordance with the terms of the owner-contractor agreement.

Rolling doors. Doors that are supported on wheels that run on a track.

Roof overhang. A roof extension beyond the end or side walls of a building.

Roof pitch. The angle or degree of slope of a roof from the eave to the ridge. The pitch can be found by dividing the height, or rise, by the span; for example, if the height is eight feet and the span is sixteen feet, the pitch is $\frac{8}{16}$ or $\frac{1}{2}$ and the angle of pitch is 45°. (See **Roof slope.**)

Roof slope. The angle that a roof surface makes with the horizontal. Usually expressed as a certain rise in 12 inches of run.

Sandwich panel. An integrated structural covering and insulating component consisting of a core material with inner and outer metal or wood skins.

Schedule of values. A statement furnished to the architect by the contractor reflecting the amounts to be allotted for the principal divisions of the work. It is to serve as a guide for reviewing the contractor's periodic application for payment.

Sealant. Any material that is used to close up cracks or joints.

Separate contract. A contract between the owner and a contractor other than the general contractor for the construction of a portion of a project.

Sheathing. Rough boarding (usually plywood) on outside of a wall or roof over which is placed siding or shingles.

Shim. A piece of steel used to level or square beams or column baseplates.

Shipping list. A list that enumerates by part, number, or description each piece of material to be shipped.

Shop drawings. Drawings that illustrate how specific portions of the work shall be fabricated and/or installed.

Sill. The lowest member beneath an opening such as a window or door; also, the horizontal framing members at floor level, such as sill girts or sill angles; the member at the bottom of a door or window opening.

Sill, lug. A sill that projects into the masonry at each end of the sill. It must be installed as the building is being erected.

Sill, slip. A sill that is the same width as the opening—it will slip into place.

Skylight. An opening in a roof or ceiling for admitting daylight; also, the reinforced plastic panel or window fitted into such an opening.

Sleeper. Usually, a wood member embedded in concrete, as in a floor, that serves to support and to fasten subfloor or flooring.

Snow load. In locations subject to snow loads, as indicated by the average snow depth in the reports of the United States Weather Bureau, the design loads shall be modified accordingly.

Soffit. The underside of any subordinate member of a building, such as the undersurface of a roof overhang or canopy.

Soil boring. A boring made on the site in the general location of the proposed building to determine soil type, depth of the various types of soils, and water table level.

Soil cover (ground cover). A light covering of plastic film, roll roofing, or similar material used over the soil in crawl spaces of buildings to minimize moisture permeation of the area.

Soil pressure. The allowable soil pressure is the load per unit area a structure can safely exert on the substructure (soil) without exceeding reasonable values of footing settlements.

Soil stack. General term for the vertical main of a system of soil, waste, or vent piping.

Solid bridging. A solid member placed between adjacent floor joists near the center of the span to prevent joists from twisting.

Spall. A chip or fragment of concrete that has chipped, weathered, or otherwise broken from the main mass of concrete.

Span. The distance between structural supports such as walls, columns, piers, beams, girders, and trusses.

Spandrel beam. A beam from column to column carrying an exterior wall and/or the outermost edge of an upper floor.

Specifications. A statement of particulars of a given job as to size of building, quality and performance of men and materials to be used, and the terms of the contract. A set of specifications generally indicates the design loads and design criteria.

Splash block. A small masonry block laid with the top close to the ground surface to receive roof drainage from downspouts and to carry it away from the building.

Square. A unit measure of 100 square feet usually applied to roofing material. Sidewall coverings are sometimes packed to cover 100 square feet and are sold on that basis.

Stain, shingle. A form of oil paint, very thin in consistency, intended for coloring wood with rough surfaces, such as shingles, without forming a coating of significant thickness or gloss.

Stair carriage. Supporting member for stair treads. Usually a 2-inch plank notched to receive the treads; sometimes called a "rough horse."

Stair landing. (see **Landing.**)

STC (Sound Transmission Class). A measure of sound stopping of ordinary noise.

Stile. An upright framing member in a panel door.

Stock. A unit that is standard to its manufacturer. It is not custom-made.

Stool. A shelf across an inside bottom of a window.

Storm sash or storm window. An extra window usually placed on the outside of an existing one as additional protection against cold weather.

Story. That part of a building between any floor and the floor or roof next above.

Strip flooring. Wood flooring consisting of narrow, matched strips.

String, stringer. A timber or other support for cross members in floors or ceilings. In stairs, the support on which the stair treads rest; also stringboard.

Stucco. Most commonly refers to an outside plaster made with Portland cement as its base.

Stud. A vertical wall member to which exterior or interior covering or collateral material may be attached. Load-bearing studs are those which carry a portion of the loads from the floor, roof, or ceiling above as well as the collateral material on one or both sides. Non-load-bearing studs are used to support only the attached collateral materials and carry no load from the floor, roof, or ceiling above.

Subcontractor. A separate contractor for a portion of the work (hired by the general contractor).

Subfloor. Boards or plywood laid on joists over which a finish floor is to be laid.

Substantial completion. For a project or specified area of a project, the date when the construction is sufficiently completed in accordance with the contract documents, as modified by any change orders agreed to by the parties, so that the owner can occupy the project or specified area of the project for the use for which it was intended.

Supplementary general conditions. One of the contract documents, prepared by the architect, that may modify provisions of the General Conditions of the contract.

Suspended ceiling. A ceiling system supported by hanging it from the overhead structural framing.

Tail beam. A relatively short beam or joist supported in a wall on one end and by a header at the other.

Temperature reinforcing. Lightweight deformed steel rods or wire mesh placed in concrete to resist possible cracks from expansion or contraction due to temperature changes.

Termites. Insects that superficially resemble ants in size, general appearance, and habit of living in colonies; hence, they are frequently called "white ants." Subterranean termites establish themselves in buildings not by being carried in with lumber, but by entering from ground nests after the building has been constructed. If unmolested, they eat out the woodwork, leaving a shell of sound wood to conceal their activities, and damage may proceed so far as to cause collapse of parts of a structure before discovery. There are about 56 species of termites known in the United States; but the two major ones, classified by the manner in which they attack wood, are ground-inhabiting or subterranean termites (the most common) and dry-wood termites, which are found almost exclusively along the extreme southern border and the Gulf of Mexico in the United States.

Termite shield. A shield, usually of noncorrodible metal, placed in or on a foundation wall or other mass of masonry or around pipes to prevent passage of termites.

Terneplate. Sheet iron or steel coated with an alloy of lead and tin.

Threshold. A strip of wood or metal with beveled edges used over the finish floor and the sill of exterior doors.

Time of completion. The number of days (calendar or working) or the actual date by which completion of the work is required.

Toenailing. To drive a nail at a slant with the initial surface in order to permit it to penetrate into a second member.

Tread. The horizontal board in a stairway on which the foot is placed.

Trim. The finish materials in a building, such as moldings, applied around openings (window trim, door trim) or at the floor and ceiling of rooms (baseboard, cornice, and other moldings).

Trimmer. A beam or joist to which a header is nailed in framing for a chimney, stairway, or other opening.

Truss. A frame or jointed structure designed to act as a beam of long span, while each member is usually subjected to longitudinal stress only, either tension or compression.

Turpentine. A volatile oil used as a thinner in paints and as a solvent in varnishes. Chemically, it is a mixture of terpenes.

Undercoat. A coating applied prior to the finishing or top coats of a paint job. It may be the first of two or the second of three coats. In some usage of the word it may become synonymous with priming coat.

Under layment. A material placed under finish coverings, such as flooring, or shingles, to provide a smooth, even surface for applying the finish.

Valley. The internal angle formed by the junction of two sloping sides of a roof.

Vapor barrier. Material used to retard the movement of water vapor into walls and prevent condensation in them. Usually considered as having a perm value of less than 1.0. Applied separately over the warm side of exposed walls or as a part of batt or blanket insulation.

Varnish. A thickened preparation of drying oil or drying oil and resin suitable for spreading on surfaces to form continuous, transparent coatings, or for mixing with pigments to make enamels.

Vehicle. The liquid portion of a finishing material; it consists of the binder (nonvolatile) and volatile thinners.

Veneer. A thin covering of valuable material over a less expensive body; for example, brick on a wood frame building.

Vent. A pipe or duct which allows flow of air as an inlet or outlet.

Vermiculite. A mineral closely related to mica, with the faculty of expanding on heating to form lightweight material with insulation quality. Used as bulk insulation and also as aggregate in insulating and acoustical plaster and in insulating concrete floors.

Volatile thinner. A liquid that evaporates readily and is used to thin or reduce the consistency of finishes without altering the relative volumes of pigments.

Wainscot. Protective or decorative covering applied or built into the lower portion of a wall.

Wall, bearing. In cases where a floor, roof, or ceiling rests on a wall, the wall is designed to carry the load exerted. These types of walls are also referred to as load-bearing walls.

Wall covering. The exterior wall skin consisting of panels or sheets and including their attachment, trim, facia and weather sealants.

Wall, nonbearing. Wall not relied upon to support a structural system.

Wane. Bark, or lack of wood from any cause, on edge or corner of a piece of wood.

Water closet. More commonly known as a toilet.

Water-repellant preservative. A liquid designed to penetrate into wood and impart water repellency and a moderate preservative protection. It is used for millwork, such as sash and frames, and is usually applied by dipping.

Weatherstrip. Narrow or jamb-width sections of thin metal or other material to prevent infiltration of air and moisture around windows and doors. Compression weather stripping prevents air infiltration, provides tension, and acts as a counter balance.

Wood rays. Strips of cells extending radially within a tree and varying in height from a few cells in some species to 4 inches or more in oak.

Working drawing. The actual plans (drawings and illustrations) from which the building will be built. They show how the building is to be built and are included in the contract documents.

Index

H

I

J

L

M

T

V